Games and Puzzles for Elementary and Middle School Mathematics

Readings from the *Arithmetic Teacher*

edited by

SEATON E. SMITH, JR.
CARL A. BACKMAN

Faculty of Elementary Education
The University of West Florida
Pensacola, Florida

National Council of Teachers of Mathematics

Seventh printing 1990

Library of Congress Cataloging in Publication Data:

Main entry under title:

Games and puzzles for elementary and middle school mathematics.

1. Mathematics—Study and teaching (Elementary)—Addresses, essays, lectures.
2. Mathematics—Study and teaching (Secondary)—Addresses, essays, lectures.
3. Mathematical recreations—Addresses, essays, lectures.
I. Smith, Seaton E. II. Backman, Carl A. III. The Arithmetic Teacher
QA135.5.G26 510'.7'8 75-16349
ISBN 0-87353-054-3

Printed in the United States of America

Contents

3. Numeration

4. Integers

8. Reasoning and Logic 199

9. Multipurpose Games and Puzzles 221

Introduction

During the past decade, the use of games and puzzles in the teaching of mathematics has increased dramatically. This current interest in games and puzzles is evidenced by the growing number of presentations on the subject at NCTM regional and national meetings in recent years. It is also highly visible in the new products offered by commercial producers of books and materials related to the teaching of mathematics.

Although most teachers are well aware of the importance of good planning in all instruction, it seems worthy to note that games and puzzles should be used primarily within the framework of a well-planned sequence of instruction. When selecting a game or puzzle for classroom use, the teacher should make his or her selection because the game or puzzle can serve a particular purpose related to the mathematical content to be studied.

It is also important to note that mathematical games and puzzles per se are not the ultimate solution for all our attempts to individualize instruction, to provide enrichment, or to provide experiences in a mathematics laboratory. It is true, however, that games and puzzles may be used effectively as one of many components in each of these areas.

One of the general purposes for using mathematical games and puzzles is to stimulate interest and develop favorable attitudes toward mathematics. In addition, most games and puzzles can be keyed to one or more of the following instructional purposes:

1. To develop concepts
2. To provide drill and reinforcement experiences
3. To develop perceptual abilities
4. To provide opportunities for logical thinking or problem solving

A most important consideration in the effective use of games and puzzles in mathematics education is preparing the students to play the games properly. Instruction in the use of games and puzzles should be given the same careful attention that is given to any other aspects of a lesson. Children should be conscious of the fact that a particular game or puzzle has been selected because it will help them learn something relative to mathematics and because it is hoped that they will have fun while they are learning. They should also realize that if the game loses its mathematical focus and becomes simply a recreational activity, they can expect it either to be brought back into focus or terminated.

Through the years since its conception in 1954, the *Arithmetic Teacher* has served as a forum for sharing ideas about teaching mathematics in the elementary and middle schools. Many excellent articles on games and puzzles have been published during this time. For this book of readings more than one hundred of these articles have been selected and organized on the basis of the following major strands of elementary and middle school mathematics: whole numbers, numeration, integers, rational numbers, number theory and patterns, geometry and measurement, reasoning, and logic. As you read these articles, try and envision multiple adaptations and improvements for the many games and puzzles so that they may be used more effectively for your own purposes.

We hope that you will find this publication a meaningful and useful resource and that you will be stimulated to share some of your ideas for mathematical games and puzzles with your fellow teachers in the future, perhaps through the *Arithmetic Teacher*.

1

Using Games and Puzzles in Mathematics Instruction

Mathematical games and puzzles can serve many functions in the elementary school classroom. In the opening article, Donovan Johnson provides an excellent discussion of some of these functions as well as a set of guidelines for using mathematical games "at the right time, for the right purpose, in the right way." Additional support for the idea of using mathematical games, puzzles, and riddles is offered by Dohler, who illustrates her ideas with sample activities.

Golden suggests that child-made games can foster enthusiasm in the classroom and contribute to learning in a number of ways. She supports her point of view with a discussion of many interesting games that were created by first- and second-grade students. Kerr provides a refreshing discussion about the involvement of a group of students with a mathematical game that resulted in some excellent analytical thinking and ultimately in a much improved version of the original game.

Bradfield proposes to enliven the mathematics classroom by providing more comedy, mystery, and drama. He then classifies and analyzes a number of enrichment problems accordingly and suggests that a "teacher is limited only by his own ingenuity and creativity."

In the concluding article of this chapter, Rea and French provide some evidence from a small-scale research study to support the idea that mental computation, number puzzles, and enrichment activities can produce an increase in mathematics achievement as measured by standardized tests. They are also careful to note some of the limitations of their study, and they describe and illustrate some of the mental computation and enrichment activities that were employed with sixth-grade students.

Commercial Games for the Arithmetic Class

DONOVAN A. JOHNSON
University of Minnesota, Minneapolis

AMUSEMENT AND PLEASURE ought to be combined with instruction in order to make the subject more interesting. There should be games of various kinds such as a game played with different kinds of coins mixed together. There should also be problems connected with boxing and wrestling matches. These things make a pupil useful to himself and more wide awake."—Plato

Even as Plato proclaimed, *some* learning of arithmetic can be accomplished through games. Games are usually considered a recreational activity. They are a means for releasing boredom or tension. Games break down barriers between strangers and quickly establish a friendly group spirit. At the same time games may be a way to learn to follow rules, to be cooperative, to be observant, or to practice diligently. Games also give participants a chance to develop restraint, to contribute as a leader and follower, to accept responsibility for individual as well as group action. However, learning arithmetic is not all a game. Other types of individual and independent group activities are obviously essential. But the variety and activity of a game may break up the monotony so frequently present in practice or drill lessons.

Unquestionably the key to learning arithmetic is through meaningful experiences and practical applications. However, the skills of computation need to be nurtured by a variety of systematic practice and drill. Thus the total approach to learning arithmetic uses meaning, practice, and application. The meaning of numbers, the understanding of a process, the mathematical structure involved precedes the practice. Practice then is the part of the learning process which builds accuracy, efficiency, and retention. Purposeful practice in the right amount at the right time will help build the arithmetical competence that business, science, industry, and education are demanding. Arithmetic games are ideally suited for a systematic program of practice. However, games can serve a variety of functions in your classroom, such as:

- Build desirable attitudes towards arithmetic
- Provide for individual differences
- Provide appropriate "homework" for parent-child activity
- Make practice periods pleasant and successful

Right now there are a great variety of arithmetic games available from commercial publishers. Most of these games are available at school supply stores or toy stores. They are usually attractive and durable with complete playing instructions included. Of course you can make up a variety of games with flash cards or number cards. However, the convenience of commercial cards makes them highly desirable. Try them sometime. You will be pleasantly surprised with the vigor pupils will exhibit in learning combinations in order to win a game instead of working problems to win the teachers favor.

Using Arithmetic Games in the Classroom

The success of an arithmetic game, like any classroom material or technique, is highly dependent on how it is used. If an arithmetic game is to serve a real function it should be used at the right time, for the right purpose, in the right way.

1. *Select a game according to the needs of the class.* The basic criteria is that the game make a unique contribution to learn-

ing of arithmetic that cannot be attained as well or better by any other material or technique. The material of the game should be closely related to that of the regular classwork. Whatever game is selected should involve important skills and concepts. Major emphasis should be on the learning of concepts or skills rather than on the pleasure of playing the game itself.

2. *Use the game at the proper time.* If games are to make the maximum contribution to the learning of arithmetic they should be used at the time when the ideas or skills are being taught or reviewed. However, many teachers prefer to use games after completing a topic, on the day before a vacation, or during the pupils free time at noon, in study periods or homerooms. Others use them during days of heavy absence due to storms, concerts, or excursions. Some teachers use them as rewards for work well done while others use them for remedial work. Usually games should be played a relatively short time so that pupils do not lose interest.

3. *Arrange the game situation so that ALL pupils will be participating in every play.* Even though only one person is working on a certain problem at a given time every team member must be responsible for its solution too. Games must also avoid embarrassing the person who cannot solve a problem. Keep comments positive, commend good work rather than making unfavorable comparisons. Whenever possible pupils should compete with their peers, and should be working on material according to their ability.

4. *Plan and organize the game carefully so that the informality and excitement of the setting does not defeat its purpose.* Teach the playing of games in a planned, organized way as you would present other activities. Have all the materials at hand so that the game can proceed in an orderly fashion. When a new, complex game is being played, start with a few essential rules and then add other rules as they are needed during the game. Use a few practice plays to help get started. Then expect good work, as neat and accurate as regular classwork. Pupils may referee the competition as well as play the game. Before beginning the game the participants should be instructed on the purpose of the game, the rules of the game, and the way to participate. Often the pupils can establish ground rules, so that everyone (including the teacher) may enjoy the activities. "Coaching" or "kibitizing" should never be allowed. The loss of points for the breaking of rules will usually be sufficient to maintain appropriate behavior. Avoid the choosing of team members by pupil captains so that low ability pupils will not be embarrassed by being last choice.

5. *Emphasize the responsibility of learning something from the game.* Follow-up activities such as discussions, readings, or tests will emphasize this responsibility. As the teacher, evaluate the results by asking yourself how successful the game was in promoting desired learning.

EDITORIAL COMMENT.—Although this article originally appeared in 1958, the message may be even more to the point today, when mathematical games are riding a crest of popularity. Donovan Johnson has noted that much value can be found in good mathematical games. He has also noted that games alone will not suffice to teach mathematics. Much depends on *how* a game is used.

The role of games, puzzles, and riddles in elementary mathematics

DORA DOHLER *Owatonna Public Schools, Owatonna, Minnesota*

What can we do to make math more inviting for our students? How can we lead them into the interesting experiences which are so much a part of mathematics?

I have found that puzzles, riddles, and games can not only arouse interest, but they can also help further mathematics knowledge and understanding.

Careful analysis of so-called "trick riddles" can show fallacies in thinking and bring out the necessity of careful study before drawing conclusions. This analysis also can help concept formation by pointing out the "whys." A thorough understanding of mathematics is necessary if a child is to tell why a certain incorrect solution was

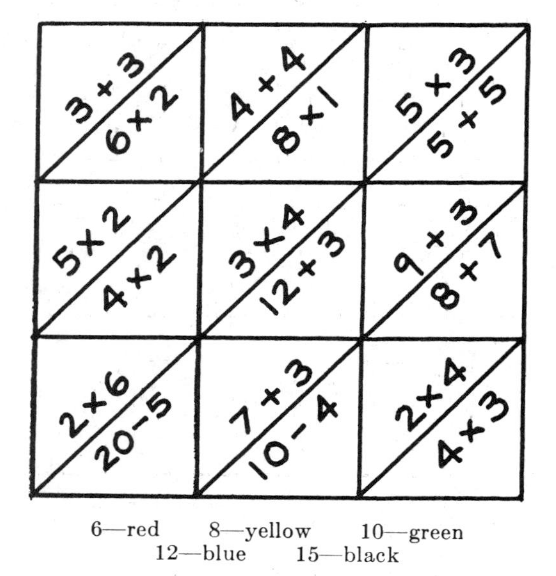

6—red 8—yellow 10—green
12—blue 15—black

Figure 1

obtained or why a certain problem always "works out."

Much meaningful practice can be presented in an exciting way through the use of games. Dot-to-dot puzzles can be constructed where the sums or products (depending upon the grade level) are the numerals to be connected. This is good practice with the mathematics facts.

Another game for practice with the various operations is a squared sheet where each triangle is to be colored to correspond to the given number representing the simplest name for the expression in the triangle (see Fig. 1).

The children consider work with a number line as a real "fun" experience. We can imagine we are grasshoppers and can jump a given number of units. This offers practice in skip-counting, addition, subtraction, or multiplication. The grasshopper may jump thirteen units beginning at 7. Where will he be after 4 jumps of thirteen units each? The children may arrive at the solution in many different ways, which they should be encouraged to share with the rest of the class.

Two points, *A* and *B*, may be plotted on a number line. The children find a point half way between the two points, or they find another point equidistant from the second point. They could also plot two other points which would be the same distance apart as the first two points. In this way children naturally work into a study of fractional numbers.

Addition-subtraction and multiplication-division "wheels" provide practice with the various operations as well as showing the inverse relationship between multiplication-division and addition-subtraction. The sum (or product) of the numbers recorded in the two inner circles equals the numbers recorded in the outer circle (see Fig. 2).

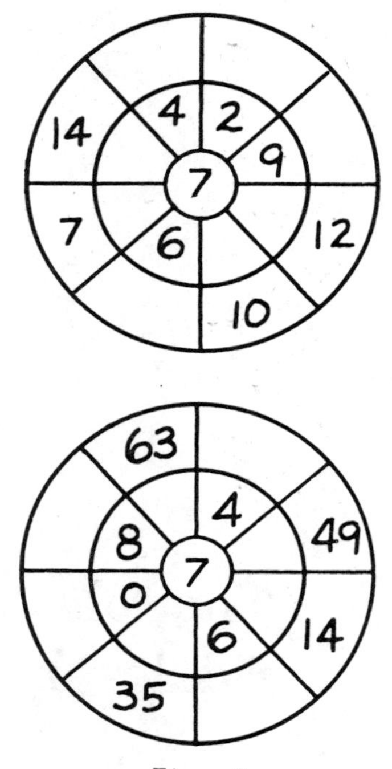

Figure 2

Puzzles can offer an early introduction to geometry. For example, the children can find the number of triangles in objects such as these:

or the number of squares in an object:

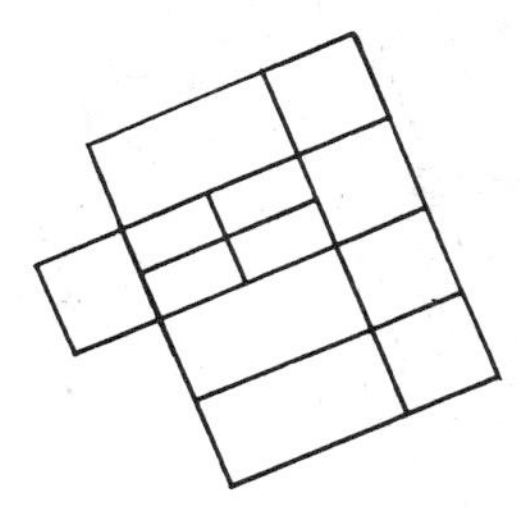

A good puzzle for that spare moment is the oral "string problem" where we can attempt to solve a complex equation involving many varied operations such as in the following illustration: Take 7; multiply by 3; add 4; divide by 5; multiply by 9; add 3; divide by 6; multiply by 4; add 4; divide by 9. The answer is 4. Depending upon the ability of the class, we can speed up or slow down the rate of presenting the operations. Success depends upon knowing the mathematics facts, an ability to solve equations mentally, and rapid thinking.

Children enjoy reconstruction problems if a careful developmental sequence is presented. One may begin using examples such as the following.

$$2+\square=6$$
$$\square+3=4$$
$$10+\square=(7+3)+4$$
$$20+\square=17+4$$
$$4\times6=\square$$
$$\square\times4=12$$
$$(\square\times7)+6=20$$
$$24=(\square\times6)+(2\times6)$$
$$70\times8=\square\times10$$
$$\tfrac{1}{2}+\square=1\tfrac{1}{4}$$
$$\square-\tfrac{3}{4}=5\tfrac{1}{4}$$
$$\$1,294+\square=\$4,927-\$1,090$$
$$\square\times\tfrac{1}{2}=\tfrac{5}{8}$$
$$\square\text{ pt.}+\square\text{ qt.}=1\text{ gal.}$$
$$1,692,046+\square=7,298,102$$
$$4\tfrac{1}{2}\times\square=27$$

Work with inequalities and equalities gives experience with the operations. Examples: Supply the sign =, < (less than), or > (greater than).

$$1,469,201 \bigcirc 469,201$$
$$1\tfrac{1}{2}+\tfrac{1}{2} \bigcirc \tfrac{1}{2}+\tfrac{3}{4}+\tfrac{3}{4}+\tfrac{1}{4}$$

(The $\bigcirc$ serves as a placeholder for the equals sign or for the signs of inequality.)

$$(4+5)\times2 \bigcirc 18$$
$$(6+3)+9 \bigcirc 6\times3$$
$$4\times9 \bigcirc (5\times6)+(4\times6)$$
$$2,462,910 \bigcirc 4,262,910$$

Equations with operation symbols to be supplied by the pupils prove quite challenging to many children besides supplying a means of practicing computation. Examples are:

$$6 \stackrel{\triangle}{\times} 4 = 8 \stackrel{\triangle}{+} 6 \stackrel{\triangle}{+} 10$$

(The $\triangle$ serves as a placeholder for the operational sign.)

$$9 \triangle 3 = (4 \triangle 3) \triangle (5 \triangle 3)$$

(Distributive Property of Multiplication)

Number sequences are a good method of developing skill and accuracy of computation. The students find the pattern and supply the missing terms. This can be easily adapted to any grade level or concept being developed. Examples are:

a. 1, 3, 5, ______, ______, 11, ______, ______, ______

b. 5, 10, 15, ______, ______, ______, ______, 40, ______, 50

c. $7\frac{1}{2}$, 9, $10\frac{1}{2}$, ______, ______, ______, ______

d. ______, ______, ______, $30\frac{1}{2}$, 25, $19\frac{1}{2}$, ______, ______

e. 2:00, ______, ______, 1:00, 12:40, ______, ______, ______

f. 5¢, 35¢, ______, ______, ______, $1.55

g. 1, 3, 6, 10, ______, ______, ______, ______, ______, 55

These and other puzzles and riddles developed by both the teacher and members of the class supply a little extra "spark" to begin or end a class period, an enjoyable way to practice skills, an effective enrichment or remedial lesson, and an interest-catcher for our mathematics program. They offer the dessert for a mathematics meal that can be enjoyed by all.

Fostering enthusiasm through child-created games

SARAH R. GOLDEN

Sarah Golden is a classroom teacher of a combination first- and second-grade class at Halecrest School in Chula Vista City School District, Chula Vista, California. Mathematics is a favorite subject for her and for her class.

How can the young child develop enthusiasm for mathematics, maintain needed skills, gain mathematical insight, and work at his individual level in a relaxed classroom climate of success?

One of the possible ways developed as a result of a recent visit by Donald Cohen, Madison Project Resident Coordinator in New York City, to my combination first-second class. He introduced an original Guess the Rules board game involving movement of imaginary traffic. Following this experience, one boy tried to make up his own Guess the Rules game while the other children played with commercial and teacher-made games. I noted that in playing some of the games, variations of the original rules were used spontaneously by small groups to increase total scores possible. For example, the chalkboard scores for Ring Toss ran into the thousands, although the indicated score for each toss was only 5 or 15. "Oh," said Jeff, sitting on the floor with one foot extended, "When he rings my foot, it counts for one hundred. When he rings Robert's foot, that's five hundred." Excitement ran high.

The next day we discussed the possibility of each child creating his own board game. The following criteria were established:

1. The game must be fun to play.
2. The game must have rules.
3. The game should enable children to play together and become friends.
4. The children must learn from the game.

Mathematics, science, reading, and spelling were suggested by the children as the content areas. The mathematics games might include, they said, go ahead and go back, counting, addition, subtraction, and multiplication. Dismissal time arrived all too quickly.

Traci (Grade 2)
Object: Get to finish line.
Rules: Use spinner or die to indicate number of moves. Correct response, stay on that box. Incorrect response, back to previous box.[1]

FINISH
1219 + 0 = □
970 − 1 = □
468 + 1 = □
□ + 1 = 10
10 = □ + △
□ = 8 − 10
□ + □ = 12
□ = 8 + 12
15 = △ + □
□ = 2 − 15
16 = □ + □
□ = 3 − 8
19 = △ + □
□ = 10 − 20
18 = □ + □
□ = 15 − 30
□ = 10 − 30
30 = □ + □
100 = □ + □
START

FIG. 1. Number Game

1. It was surprising that Traci deliberately planned to have negative numbers as responses in her equations.

The following morning one girl submitted a sketch of her game which used dice with three golf tees as pawns. Her explanations were quickly grasped by the class. As she proceeded to make a more durable copy of her game, children used the following vocabulary to describe the situation: Traci's game, Traci's pattern, Traci's copy, Traci's original, Traci's guide, Traci's directions, Traci's instructions.

Eager to start on his own, each child then developed a game on a sheet of chipboard about 12 by 22 inches. Dice, spinners, or number cubes were used to indicate direction and number of moves. In some cases, answers on cards were to be matched with the corresponding problem in the game. The games showed a wide variation in difficulty, ranging from simple counting to the use of negative numbers. Danny asked, "What if you landed on 3 and your card said, 'Go back 6'? Where would you be? Off of the board?" He went on, "I guess that would be a negative number, and you'd have to get 'Go ahead 6' to get back to where you were on 3." Danny thought about this and incorporated negative numbers in his Bang Bang Chitty Chitty game by using a starting line with negative numbers behind it leading to the Repair Shop. The twenty-seven games included such names as Try and Guess; The Happy Game; Go Ahead, Go Back; Treasure Island; The Whacky Racer; Streets and Numbers; The Counting Game; and Bang Bang Chitty Chitty.

In making games that involved rounded shapes, the children encountered the problem of dividing the rounded portions into congruent regions. They then learned how to use simple protractors to construct such regions. Several lessons on curves followed, including simple, complex, open, and closed curves.

Upon completion of the games the children were given repeated opportunities to explain the object and rules of their games and to play them with their classmates. Scores were tallied and comparisons such as "I beat him by 35 points" were made.

Deborah's Streets and Numbers game led to a discussion of odd and even numbers. We took a walk in the neighborhood to find out how the houses were numbered.

For two players. Robert (Grade 1)
Object: Get Snoopy from the yard to his dog house.
Rules: Use spinner to show number of boxes to advance. Correct answer allows player to remain there until next spin. Wrong answer makes player go back to his preceding position.[2]

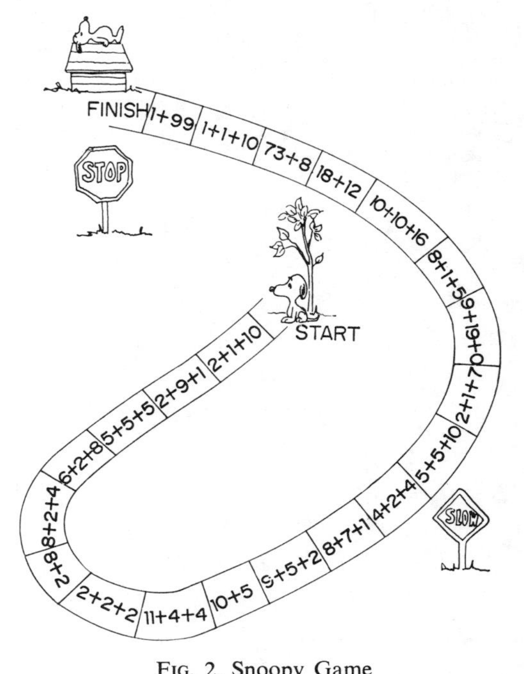

FIG. 2. Snoopy Game

2. Robert, a first grader, was interested in using three addends.

Each child made a map of his street and the houses on it and numbered the houses. We invited a member of the Chula Vista City Engineering Staff to come to school

For one or two players. Kathy (Grade 1)
Object: Put the greatest number of cards where they belong in the counting ladder.
Rules: Players take turns picking top card from a shuffled pack. Cards show numerals 6, 12, 14, etc. Player puts the card on his side of the ladder. If incorrect, player loses turn. Each player counts his correct responses at end of game to determine winner.

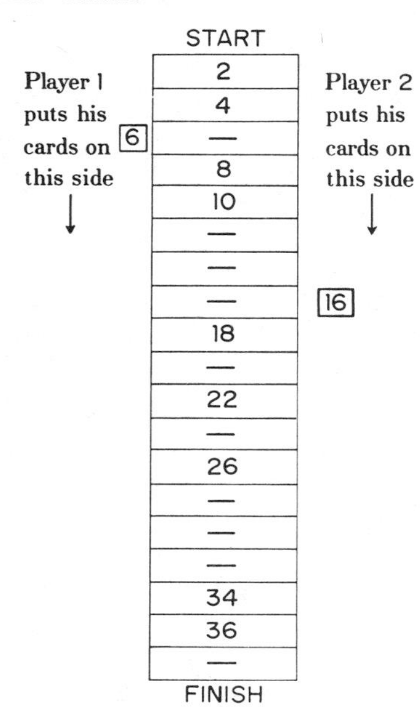

Fig. 3. Counting Game

to explain how numbers were assigned to new homes in the city. Deborah revised her game to reflect her new learnings in this area—odd numbers on one side of the street, even numbers on the opposite side.

The Bang Bang Chitty Chitty board game was devised by four boys as an outcome of their interest in mathematics, model cars, races, and the story of *Chitty Chitty Bang Bang*. Speed limits were discussed. The boys consulted *Sports Encyclopedia,* almanacs, and other books for information on speed records and reasonableness of posted speed signs. Of course, the intricate speedway had to be made wide enough to accommodate the toy racing cars used. Much experience in measuring resulted.

Summary

Child-made games foster enthusiasm and contribute to learning in several ways. Children learn through a pleasurable medium of their own creation. The varying degrees of complexity in the games are commensurate with a child's mathematical concepts and interests. In devising the games, the children must try out their ideas and pursue them to a reasonable outcome. Language development is advanced as the children are highly motivated to explain the games clearly to others. There is an

Leslie (Grade 1)
Object: Get to home.
Rules: Use die to indicate number of boxes to advance. Then follow directions in box.

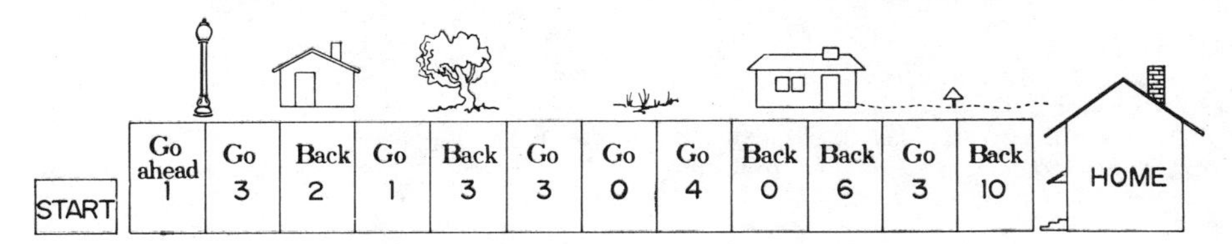

FIG. 4. Happy Game

exchange of information between players, especially when more than one correct response is possible. The teacher gains information concerning individuals and their levels of thinking. For example, it was surprising to note some children's use of difficult combinations and negative numbers and their free use of commutative and as-

For two players. Danny (Grade 2), Chris (Grade 1), Robert (Grade 1), Bobby (Grade 2)
Object: Get to finish line.
Rules: From a stack of cards, each player takes a card, which tells him what to do.[3]

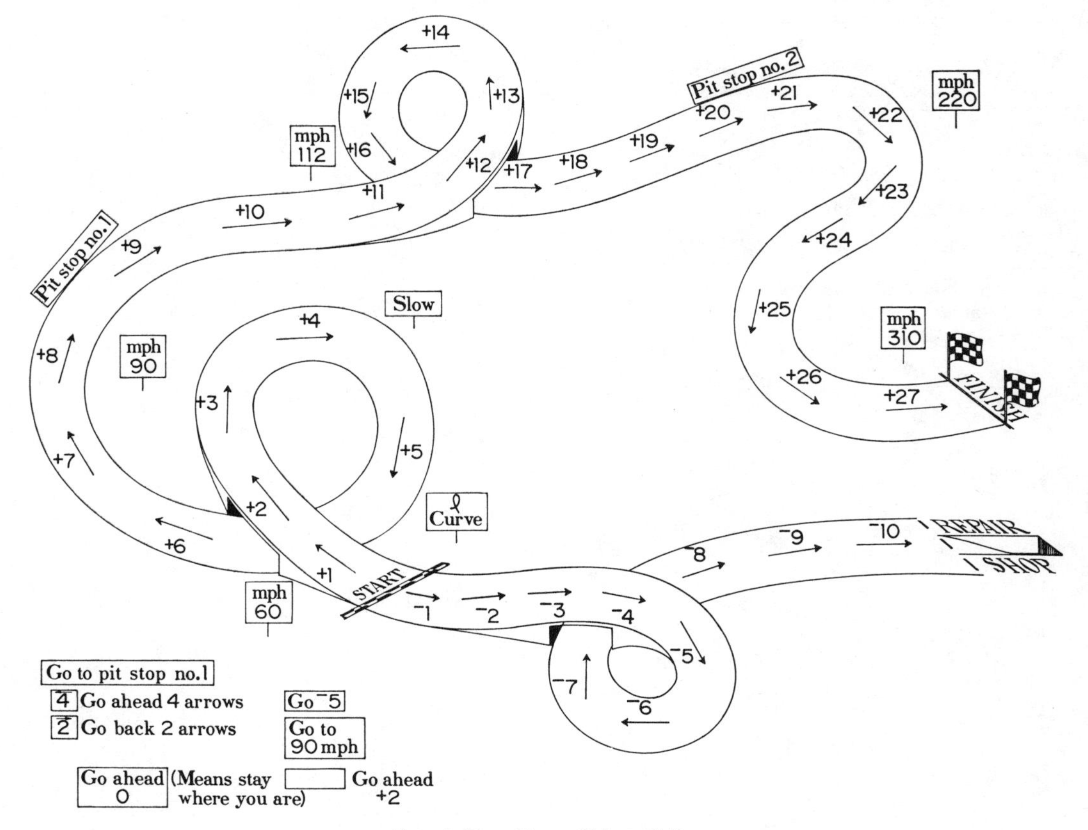

FIG. 5. Bang Bang Chitty Chitty

3. Numbers back of starting line were negative numbers and led back to the repair shop with its grease rack, etc.

sociative properties of addition in computing total scores. The games provide teachable moments for the teacher to introduce pertinent material in greater depth.

Bobby (Grade 2)

Object: Get to treasure at end.

Rules: Use spinner to indicate number of moves. If answer is correct, stay until next spin. If answer is not correct, go back to preceding place.[4]

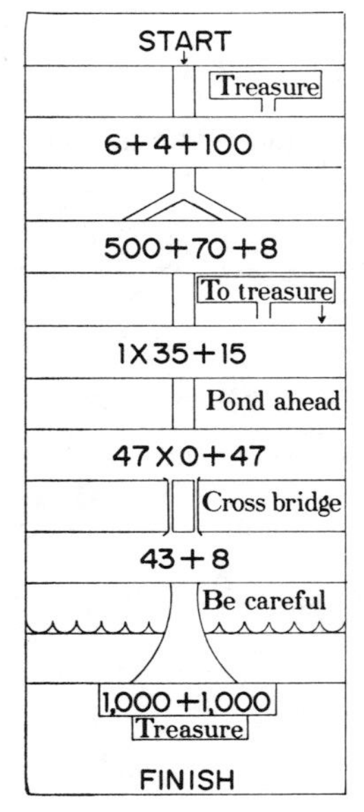

Fig. 6. Treasure Island Game

4. Each child has a marker of a different color or of a different kind.

Chris (Grade 1)

This game is played similarly to the Snoopy Game (fig. 2). This first grader included multiplication and subtraction in his game.

Object: Get to finish line.

Rules: Use number cube to indicate number of boxes for move. Correct answer allows player to remain there until next spin. Wrong answer makes player go back to his preceding position. If player lands on "Go back," he must go back the number of places indicated in the box.

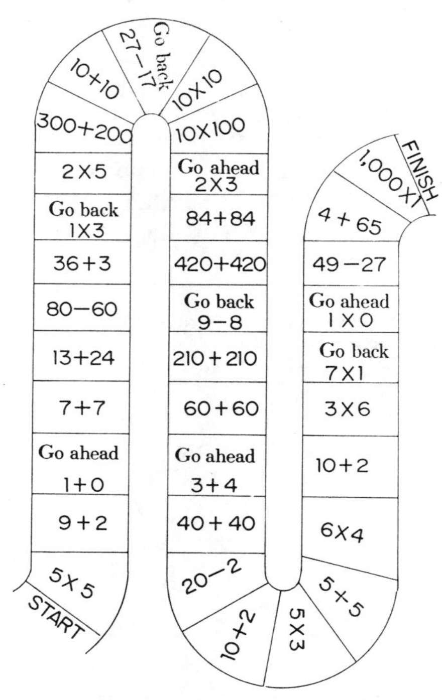

Fig. 7. Try and Guess Game

Mathematics games in the classroom

DONALD R. KERR, JR.

The Indiana University Mathematics Education Development Center is currently at work on a program that combines mathematics content, mathematics methods, and public school experience for prospective elementary school teachers. Donald Kerr is assistant director of the Center.

Games are fun, and it is important to have fun. Mathematical games in school are good because, in playing them, children have fun associated with a topic that is not always considered to be enjoyable. It is neither possible nor desirable to organize the bulk of mathematics instruction around games, but it is sometimes possible to develop games in such a way that they complement the regular mathematics instruction and thereby justify more classroom time. This article chronicles one classroom experience of the author in which a simple game was introduced to a fifth-grade class on a Friday afternoon.

Our first game-playing involved a version of the well-known game Multo. The children were given copies of three-by-three arrays of squares (see fig. 1) and were told to put some number between 1 and 20 in each of the nine squares. They were also asked to tear up bits of paper to use to cover the squares. Every child in the class knew how to play bingo, so no other instructions were needed except to say that numbers would be called by giving a multiplied pair—two times four, three times five, and so on. The author called off the numbers for the first game, and the winner thereafter served as caller.

Class impatience with repeated numbers and a few disputes over winners soon dictated that the caller keep track of the numbers he had called. Disenchantment with the limitation to the numbers from

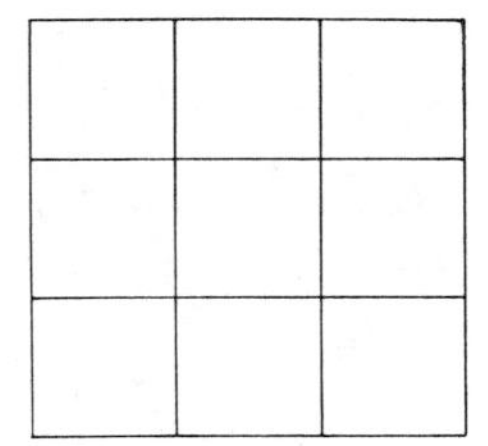

Fig. 1

1 to 20 gave rise to using other intervals, to using longer intervals, and to discussing the effect of longer intervals on the pace of the game.

The inevitable cry of "He wins all the time," led to a discussion of which numbers were the best to choose. A child kept a tally on the frequency of called numbers, and this led to a discussion of why some numbers seemed to be called more frequently than others. At this point, caller-player competition started to develop. Some callers tried to call numbers that were low on the frequency chart; some players clearly understood the greater likelihood of composites over primes.

The pupils were quite content to continue to play the same game with minor modifications during last period on Friday afternoons, but the author decided that it was time (if you'll pardon the pun) to add a new dimension to the game. So three-dimensional Multo was introduced. The children were given copies of three three-by-three arrays of squares (see fig. 2) and were instructed to fill them in with numbers between 1 and 30.

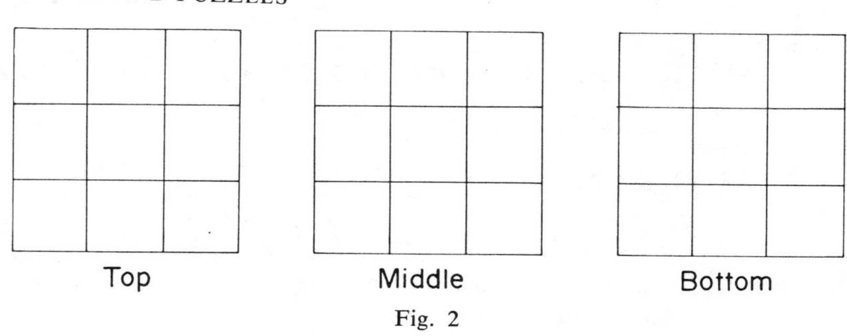

Fig. 2

Since none of the pupils was familiar with the three-dimensional versions of bingo or tic-tac-toe, some explanation was needed.

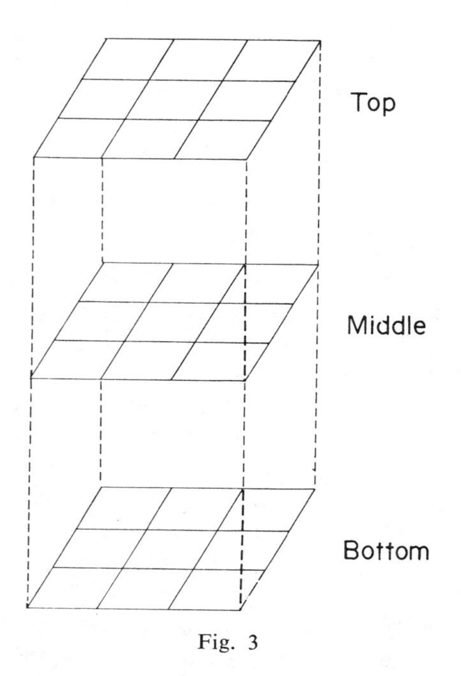

Fig. 3

With a sketch like figure 3 on the board, the pupils and the author went through a number of examples of what was and what was not a winner. For example, in figure 4 the *U*s, *X*s, or *Y*s provide a winner, while the *Z*s and *W*s do not. Some children got the three-dimensional perspective immediately, but others never did. Fortunately members of the latter group were still able to win since a two-dimensional winner, such as with the *U*s in figure 4, was possible.

At this point the idea of repeating numbers came up. The children wanted to be able to use the same number more than once in filling in their squares. They decided to try using repeated numbers but soon found that this led to quick wins with many ties, so they went back to using numbers only once.

The three-dimensional form of the game had the obvious advantage of encouraging spatial visualization and making a three- to two-dimensional "projection." Yet this form had not enriched the numerical experience inherent in the game so the author decided to introduce another modification of the game.

The children were instructed to put 1, 10, and 100 respectively under the three squares on the three-dimensional Multo sheet (see fig. 5) and then to number

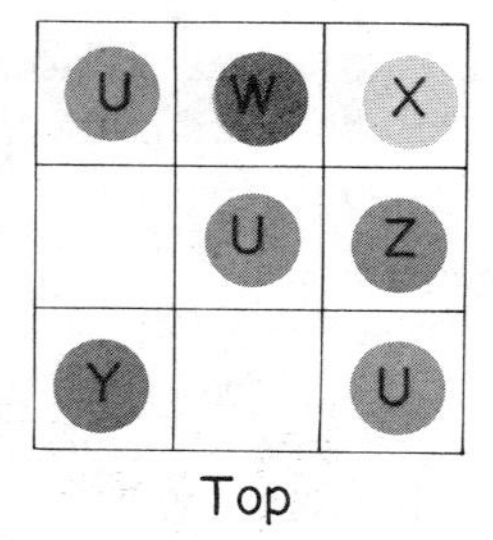

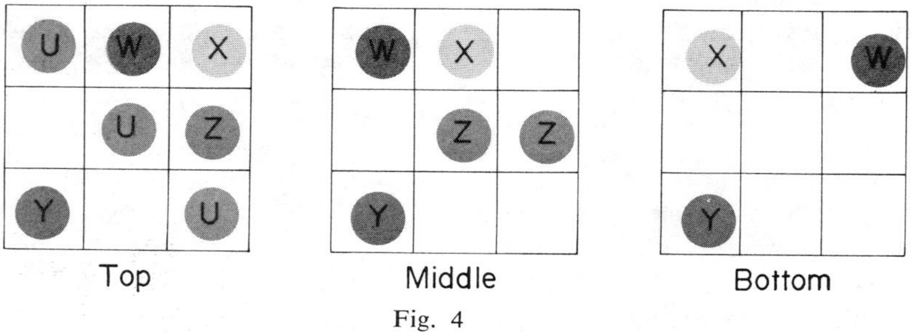

W X Z Z Y

Middle

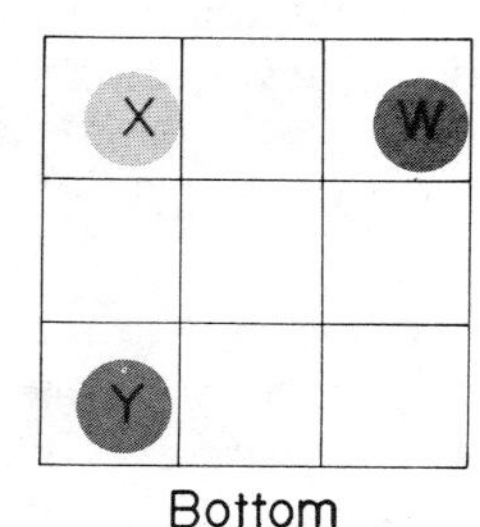

Fig. 4

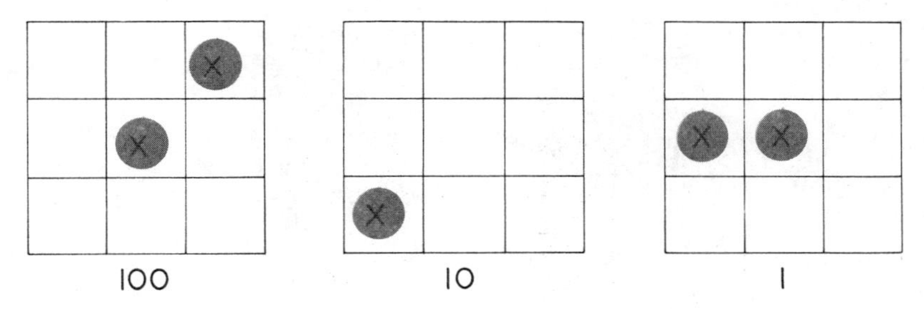

Fig. 5

the squares as before. This time the caller was to call off five numbers (as multiplied pairs). The winner would be the pupil with the highest total, where each number called in a square counted for the number of points *under* the square. This modification, while abandoning the three-dimensional aspect, had the advantage that each pupil was required to add up his score to see if he had won. For example, the pupil who had the *X*s in figure 5 would receive a score of 212.

The pupils evolved a streamlined system for deciding who had won. Any child who felt he had a good score would announce it. Someone whose score was higher would announce his, and so on. Thus the winner was quickly identified. The children also engaged in discussions on which scores were possible and from these the next version of the game arose.

The next and final form of the game emerged from an effort by the author and the pupils to fill in the gaps between the possible scores in the previous form. The small squares in each of the three big squares were given values between 1 and 9, as shown in figure 6. The children then filled in their Multo numbers as before, and the game proceeded in the same fashion. So, for example, the *X*s in figure 6 would yield a score of

$$(4 + 1) \times 100 + (7 + 4) \times 10 + 5 \times 1$$
$$= 6 \times 100 + 1 \times 10 + 5 \times 1$$
$$= 615$$

Quick ways of adding scores to take advantage of the compatibility of the format with the base-ten numeration system were discussed.

There are clearly many additional directions we could have moved in. Other operations and combinations of other operations could have been used for naming points. The squares could have been given values corresponding to other bases—1, 2, 4 or 1, 3, 9. Many more probability and statistics problems could have been pursued. And basic games other than Multo could have been used as a point of departure.

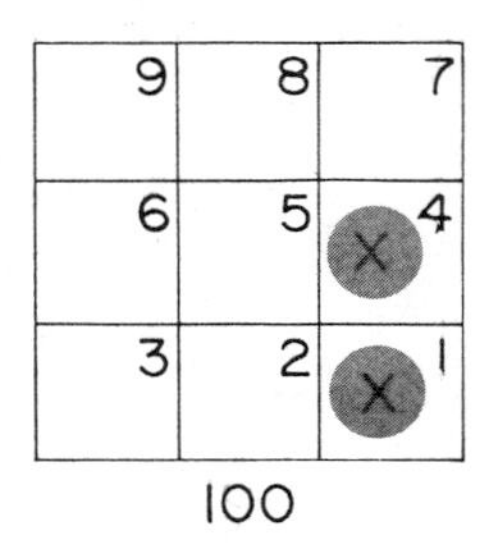

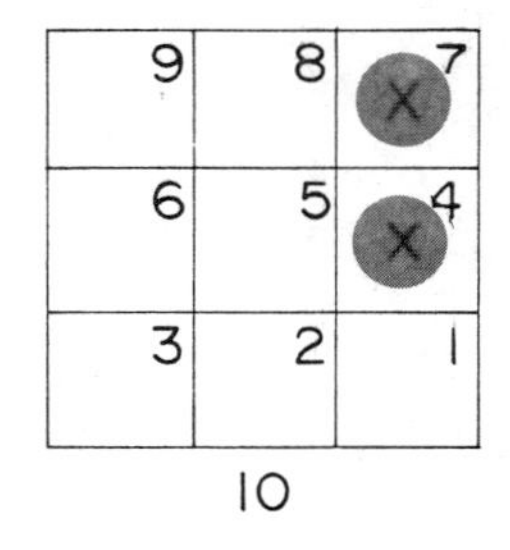

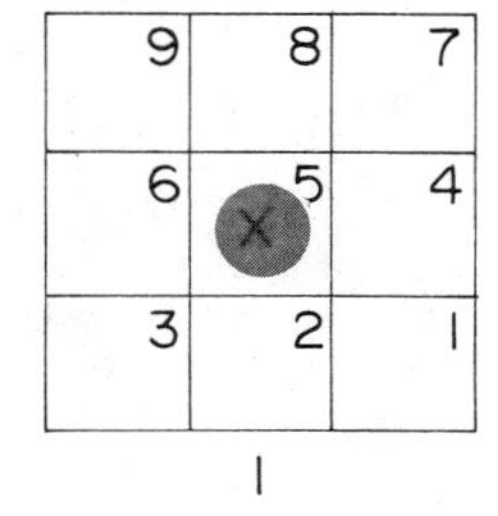

Fig. 6

The following points are significant in the sequence of activities that have been described:

(*a*) the skill reinforcement in the game,

(*b*) the pupil involvement in the evolution of the game,

(*c*) the teacher involvement in providing direction to the evolution and in seeing that many pupils participated, and

(*d*) the introduction of new mathematical concepts such as elementary statistics and probability.

The particular game and its variations are not important as such. More important is the idea of involving children in the evolution of a game and of directing that evolution in order to supplement and extend the regular mathematics program.

AUTHORS NOTE: The author wishes to express his appreciation to teachers Katheryn Vaughan and Beth Williams, and to Principal Edwin Smith for giving him the opportunity to have the experiences described here in fifth- and sixth-grade classes at Templeton School in Bloomington, Indiana.

Sparking interest in the mathematics classroom

DONALD L. BRADFIELD

Donald Bradfield is a classroom teacher in Sand Springs, Oklahoma.

From the glamorous stars of the entertainment world come the clues for sparking interest in the mathematics classroom. Basically, these stars make use of three kinds of entertainment—comedy, mystery, and drama. Through proper use of these methods of entertainment, the mathematics classroom can come alive with diversions and excitement. The intent of this article is to illustrate how to spark interest in the mathematics classroom by utilizing types of recreational mathematics that are categorized under three entertainment media headings. Suggestions will be made as to how to assimilate each type into a regular program of mathematical instruction.

Before a discussion of mathematical entertainment, the need for sparking interest in the mathematics classroom should be stressed. Many teachers of mathematics are unbelievably boring in their approach to teaching. Consequently, many potentially strong students never develop any great liking for mathematics. Weak students are disenchanted quickly with the course, and elect to avoid all courses in mathematics. This loss of interest is an educational waste—both to the individual and to society as a whole. Therefore, a more stimulating atmosphere is needed in most mathematics classrooms. One way to stimulate the atmosphere is to use recreational mathematics materials.

The comedy of mathematics

To entertain is to amuse. To many pupils, there is nothing funny about mathematics. However, in the world of mathematics, catchy problems and tricky problems exemplify types of comedy that portray humorous situations and unexpected solutions which, consequently, are funny. Consider the following two examples:

> Monkey Problem
>
> A monkey is at the bottom of a 30-foot well. Every day, he climbs up three feet and slides back two feet. When does he reach the top?

> Cigarette Problem
>
> If a tramp can make one cigarette from six butts, how many cigarettes can he smoke from 36 butts?

The solution to the monkey problem is 28 days, since on the 28th day he would climb and not slip back. The solution to the cigarette problem is seven cigarettes, since after making six cigarettes and smoking them he would have six more butts from which another cigarette could be made. These comedy problems capture attention and appeal to a sense of humor. Similar problems from the field of recreational mathematics may be labeled as "Rat Problem," "Squirrel Problem," "Coin Problem," "Egg Problem," "Barrel Problem," "Marble Problem," "April Fool Problem," etc. These titles attract the attention of children and ignite sparks of interest within the classroom. By solving these comedy problems, children develop a pleasant feeling toward problem solving that should transfer to the regular curriculum of mathematics.

The mystery of mathematics

To entertain is to mystify. To many pupils, all of mathematics is a great mystery. However, in the field of mathematics, magic problems, puzzle problems, code problems, and progression problems illustrate special types of mysteries that will fascinate many students. The following four examples provide mystery and surprise.

Faded Document Puzzle

```
        · · 8 · ·
12 ) · · · · · · ·
     · · ·
     -----
          · ·
          · ·
          -----
          · · ·
          · · ·
          -----
```

Reproduce the numbers on the document.

Age and Family Magic

Write down your age. Double it. Multiply by 5. Add 15. Add the number of members of your family less than ten. Subtract 15. The first two digits in the result is your age. The last digit is the number in the family.

Pythagoras Coded Message

Use the following key to decipher the message below:

0,1,2,3,4,5,6,7,8,9,△,□,~
a,b,c,d,e,f,g,h,i,j,k,l,m
n,o,p,q,r,s,t,u,v,w,x,y,z

Pythagoras says:

0 7~1 4 4 5

4 7□4

6 7 4

7 0 8 8 4 4 5 4

Salary Progression

If you were offered a salary of one penny the first day, two pennies on the second day, four pennies on the third day and so on for thirty days, would you accept?

Days:	*Salary*
1	.01
2	.02
3	.04
4	.08
5	.16
.	.
.	.
.	.
30	Total ——

What is the total salary?

The quotient in the faded document puzzle is 90809. The coded message of Pythagoras says, "Numbers rule the universe." The total salary in the progression problem is $10,737,418.23. These problems and their solutions provoke amazement, stir curiosity, and challenge the intellect. Similar problems in the field of recreational mathematics may be labeled as "Age and Coin Magic," "Cryptorhythm Puzzle," "Wiggle Operation Puzzle," "Gauss's Secret Message," "Notebook Paper Progression Problem," etc. These kinds of problems catch the eye of children and spark interest in the classroom. By solving these mystery problems, children experience success in problem solving.

The drama of mathematics

To entertain is to dramatize. To many pupils, mathematics is a dull subject and has no dramatic appeal. However, in the area of mathematics, many problems can be presented to children in such a manner as to have dramatic appeal. Consider the following three examples:

Kings Problem

A king suspected his advisors were halfwits. To test the wit of the advisors, he threw them into a dungeon and told them that they would lose their heads if they did not solve the following problem:

```
.  .  .  .
.  .  .  .
.  .  .  .
.  .  .  .
```

The problem is to draw six straight lines through the dots without lifting your pencil or retracing a line. Could you have saved your head?

Josephus Problem

When the Romans invaded their town, Josephus and forty other men took refuge in a cave. Believing that they were about to be captured, the group wanted to commit suicide except Josephus and his friend. Fearing the others in the group and wanting to save himself and his friend, Josephus suggested a plan whereby the killings would take place in an orderly manner. He suggested that the group form a circle and begin killing every third man until only two would be left. These remaining two persons would then supposedly kill themselves. If Josephus selected the first man in the circle from which to begin the counting, where did he place himself and his friend in order that they would keep from getting killed?

Three Jealous Men and Their Wives Problem

Three jealous men and their wives come to a river to be crossed. Using one boat that holds only two people and having no wife ever in the presence of another man unless her own husband is present, how can the group cross the river?

The solution to these problems is left for the reader to find. These kinds of dramatic problems arouse emotions, encourage ingenuity, and test perseverance—as the reader will discover as he attempts to solve these problems. Similar problems in the field of recreational mathematics may be labeled "Railroad Problem," "Christian and Turk Problem," "Bridge Problem," etc. Such problems have emotional connotations for children, sparking interest in the classroom. By solving these dramatic problems, children discover the human-interest factor in mathematics.

To effectively use recreational mathematics in the classroom, the following procedures are helpful. First, make color posters of problems similar to those listed in the previous topics of this article and place them on the walls and bulletin boards within the classroom. These posters create an exciting mathematical environment and can be referred to by the teacher when appropriate. Second, allocate time in the schedule of mathematical instruction to study topics from recreational mathematics. Between units of study and the days before vacations are especially good times to provide pupils with a treat that will leave them with a good feeling toward mathematics and make them more willing to tackle new mathematical topics. Third, add recreational problems to examinations and give extra points for solving them correctly. These problems keep fast students occupied profitably while slower students are finishing the main part of a test; the only danger involved in this procedure is that slower students must be cautioned to finish the main portion of the test before attempting the bonus problems. A variety of other procedures can be devised for using recreational mathematics in the classroom. The teacher is limited only by his own ingenuity and creativity.

EDITORIAL COMMENT.—The author has provided an interesting classification structure for enrichment problems. Almost every teacher could find many opportunities to use a file of puzzles and problems such as these. If you are interested in starting or extending your own set of enrichment activities, you will find many games, puzzles, and problems in the *Arithmetic Teacher*, in the *Mathematics Teacher*, in various NCTM publications, in children's magazines, in the magazines provided by airlines, in the newspaper, in paperback books on mathematics, in books on methods of teaching mathematics, in the teacher's manuals for various series of textbooks, and in many books devoted primarily to games, puzzles, number magic, and so on. A teacher is limited only by his own effort and imagination in finding and creating "interest getters" in mathematics.

Payoff in increased instructional time and enrichment activities

ROBERT E. REA
University of Missouri, St. Louis, Missouri

JAMES FRENCH
Hazelwood Public Schools, Hazelwood, Missouri

Are your pupils below grade level in mathematics achievement? Are you looking for ways to accelerate your pupils' achievement in mathematics? If your answer to either of these questions is yes, you will be interested in Rea and French's answer to the question, "Do regular instructional activities devoted to skill development produce marked increases in achievement scores?"

This article shares some research findings that will interest every mathematics teacher. Anyone looking for evidence that encourages the use of number puzzles and other enrichment activities will take particular delight in these results. We hope that this article will generate discussions among teachers, or spawn other studies of this type. Most important, we hope the article will encourage an examination of the supplemental activities that are currently being used in our classrooms.

Robert E. Reys

A recent research project assessed the merits of oral instruction in mental computation versus pencil and paper enrichment activities for sixth-grade pupils. An outcome of particular interest to teachers of elementary mathematics is that *both* treatments, as measured by standardized achievement-test scores, resulted in a significant growth over a small number of instructional sessions.

Procedure

A sixth-grade class was divided into two treatment groups that were matched on the bases of age, sex, I.Q., and achievement scores. The matching process produced the two groups shown in table 1. The instructional activities for group 1 were based on Kramer's mental computation series (1965), and the activities for group 2 were adapted from Crescimbeni's collection of enrichment activities (1965). Care was taken to ensure that the *type* of computational examples was the same for both groups—that is, if the mental computation exercises involved multiplying a three-digit number by a two-digit number, then the game, puzzle, or activity taken from Crescimbeni was modified to provide similar practice.

SRA achievement-test scores were obtained for both groups on the first and twenty-fifth days of the project, with twenty-four successive instructional periods intervening. Both groups received their regular mathematics period each day, plus fifteen minutes of special activities. The grade-equivalent scores derived from these test administrations are shown in the scattergrams in figures 1 and 2.

In both groups, there were individual

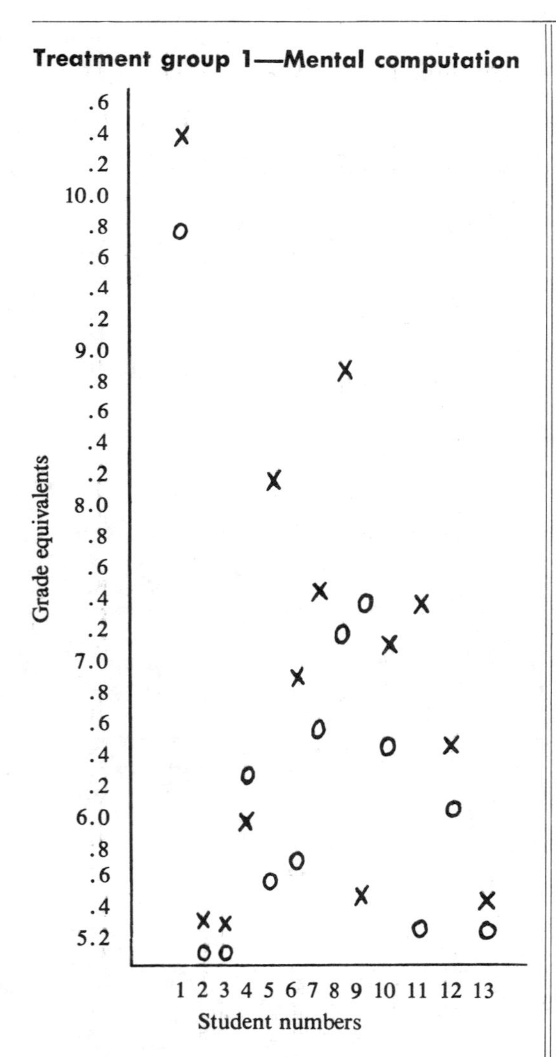

Grade equivalents derived from table accompanying SRA Achievement Series: Arithmetic, Form E, Blue level, 1971.
O = Pretest grade equivalent
X = Posttest grade equivalent

Fig. 1

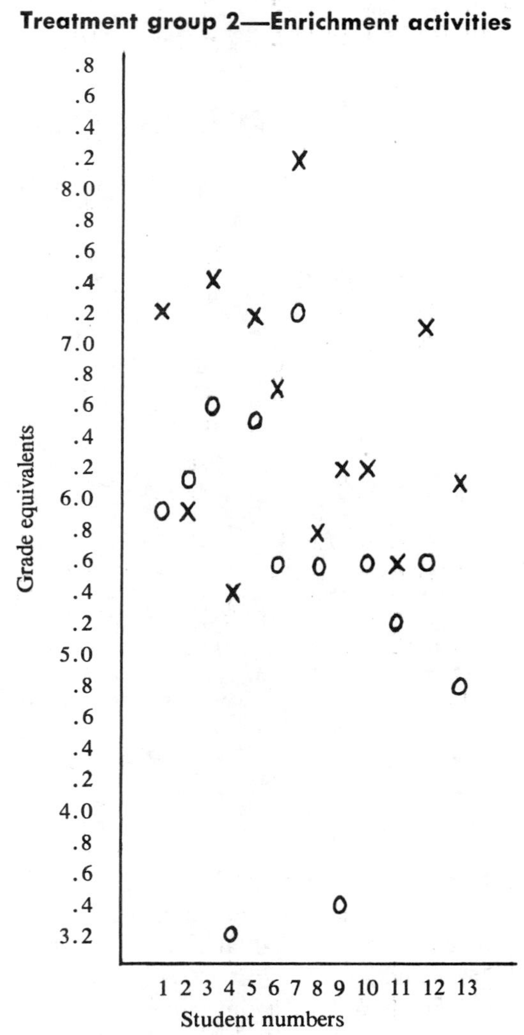

Grade equivalents derived from table accompanying SRA Achievement Series: Arithmetic, Form E, Blue level, 1971.
O = Pretest grade equivalent
X = Posttest grade equivalent

Fig. 2

Table 1
Matched data for treatment groups

	Group 1 N (7 girls, 6 boys)		Group 2 N (7 girls, 6 boys)	
	Range	*Median*	*Range*	*Median*
Age	11.1–12.0	11.65	11.0–12.0	11.5
I.Q.	86–141	102.6	86–133	102.0
Achievement scores*	3.9–9.2	5.2	4.2–6.2	5.2

Data from 5th-grade records, SRA Achievement Series: Arithmetic, Form C, Blue level, 1964. Administered September 1970.

youngsters whose scores increased only slightly (students 3 and 13 in figure 1) and even certain individuals whose scores decreased (students 4 and 9 in figure 1 and student 2 in figure 2). However, the majority of the students in both groups gained so dramatically in achievement scores that the average growth for group 2 was one full year, and for group 1 was eight months.

Discussion

While this project is subject to many limitations and possible sources of error, the increase in achievement scores seems to be worth pursuing (formally in more rigorously designed experiments and informally in classroom situations where a similar set of experiences may be desirable). The following examples of activities and suggestions for additional modifications are offered to elementary teachers who may wish to try a few of them in their own classrooms. Teachers wishing to use a greater variety of activities may refer to the sources listed in the references.

The examples presented here were well received by the youngsters in this project and can easily be modified to most topics or levels of computation. The mental computation and enrichment examples that follow do not involve the same computational skills. However, parallel examples can easily be constructed. In fact, constructing such parallels greatly increases a teacher's understanding and appreciation of the power of these activities in generating computational practice.

Mental computation activities

The information given in figures 3 and 4 was reproduced for use in group instruction. The instruction began with the reading of the first sentence, which contains enough information for the solution of the questions that follow it. At this point, the children were cautioned not to write the problem down on paper, but rather to think of it in the way that was indicated in the instructions. Children were then asked to search for other methods of solving the problem mentally. Following this, a similar set of problems was presented for mental solution by the method that was presented, or by an alternative method that was discovered by the class.

The second example in both figures 3 and 4 presents the inverse of the process in example one. In all lessons, an example of one process was followed immediately by an example of the inverse process. This

MENTAL COMPUTATION ACTIVITY

1. The Deck family traveled 250 miles on the first day of their trip and 160 miles on the second day. How many miles did they travel on the first two days?

$$250 + 160 = \square$$

Think: 160 is 100 + 60
250 plus 100 is how much? (350)
350 plus 60 is how much?

180 + 160 = ____ 160 + 150 = ____
430 + 170 = ____ 140 + 150 = ____
340 + 120 = ____ 290 + 120 = ____

2. Mr. Thomas has a debt of $120. If he pays $70 of it, how large a debt will he have left?

$$120 - 70 = \square$$

100 − 20 = ____ 100 − 20 = ____
110 − 40 = ____ 130 − 40 = ____
140 − 60 = ____ 150 − 70 = ____
110 − 20 = ____ 160 − 40 = ____

Fig. 3

MENTAL COMPUTATION ACTIVITY

1. In an auditorium there are 10 rows of 14 chairs. How many chairs are there in all?

$10 \times 14 = \square$

Think: A number is multiplied by 10 by annexing a zero to the numeral to find the number.

$10 \times 24 =$ ____	$10 \times 19 =$ ____
$10 \times 69 =$ ____	$10 \times 46 =$ ____
$10 \times 95 =$ ____	$10 \times 35 =$ ____
$10 \times 85 =$ ____	$10 \times 58 =$ ____

2. Ten boys had to carry 110 books to the science room. How many books did each boy have to carry?

$110 \div 10 = \square$

$240 \div 10 =$ ____	$720 \div 10 =$ ____
$380 \div 10 =$ ____	$470 \div 10 =$ ____
$540 \div 10 =$ ____	$840 \div 10 =$ ____
$730 \div 10 =$ ____	$640 \div 10 =$ ____

Fig. 4

sequence was followed in the hope of making the inverse relationship easier for the children to grasp.

Enrichment activities

Ladder Arithmetic, Draw-a-Trail, Star-Burst, and *Brain Teasers* are examples of the activities used with group 2. Such activities are a direct attempt to encapsule mathematics reinforcement, practice, and enrichment in forms that are appealing and enjoyable for the children. The purposes of activities of this nature include making mathematics more enticing, and involving students in situations that stimulate curiosity and quantitative thinking.

Ladder Arithmetic has a built-in, self-checking device (see figure 5). The frames are filled with the numbers that make the horizontal sentences true. These numbers provide the digits necessary to solve the vertical sentences. The horizontal sentences can be arranged to give the correct sequence of digits or, if additional complexity is desired, a scrambled order of digits. Scrambled digits further encourage children to perform the self-checking feature of the activity.

In *Draw-a-Trail,* solutions vary from student to student depending on the starting point and the path that is followed to the bottom box. Another dimension may be added to the exercise by asking the children to discover as many paths as possible

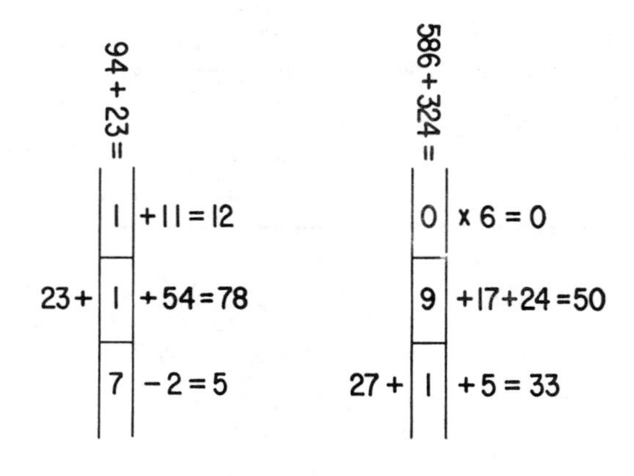

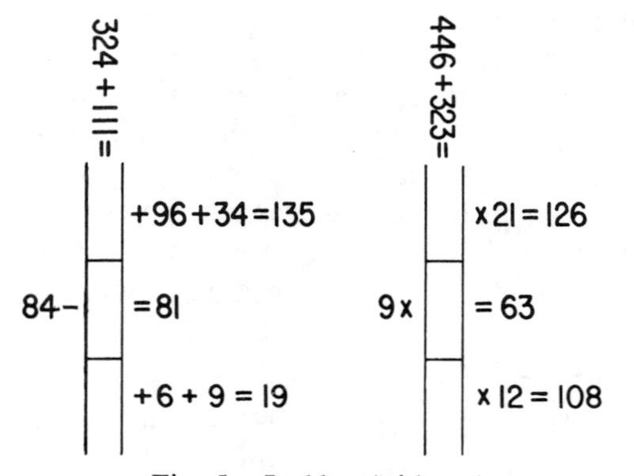

Fig. 5. Ladder Arithmetic

Directions: Fill in the frames with the number that makes the horizontal sentences true; then determine if these digits complete the vertical sentences. The digits may be in order, reversed, or scrambled.

within a given time limit. The teacher can adjust the level of difficulty of this exercise by increasing the size of the numbers, or by varying the operations to be performed. Such variations should be designed to meet the needs of a particular group or individual. (See figure 6).

The examples of *Draw-a-Trail* selected here illustrate variations that might be used —from a simple addition exercise to a more complicated mixed-fraction and whole-number exercise. Once children understand the procedure for finding the solutions to *Draw-a-Trail,* they may be encouraged to construct their own examples. After constructing a trail, pupils can exchange examples and check one another's solutions.

Star-Burst is an activity that provides

1/2	×2	−1/4	3	1/3
×1/3	3/4	1	1/2	−2
1/3	−1	2/3	1/8	1/4
2/3	1/2	−1/3	2	4/4
1/3	3	− 4	×2	1/2
×1/3	3/4	3	1/2	−2
4	−1	2/3	1/8	1/4
		4		

7	6	8	9	5
5	8	9	4	9
6	7	9	8	8
7	4	8	7	9
8	5	6	7	8
9	8	7	9	9
7	8	7	5	5
		61		

Fig. 6. Draw-A-Trail

Directions: The objective is to draw a trail that leads to the bottom box of the puzzle. The trail must be a continuous line connecting numerals in the boxes. It must end at the numeral in the bottom box, and the numbers in the trail must add up to the total in the bottom box. All boxes need not be used, but each box may be used only once.

practice in various operations simply by changing the sign of the number in the middle circle. The level of difficulty of this activity can be adjusted to different groups, and it can also provide for the development of specific computational skills. Two activities are illustrated in figure 7.

Brain Teasers, such as the two that follow, were interspersed with other activities, and the pupils seemed unusually motivated to find the correct solutions.

John was given $2.00 to buy some school supplies for the opening day of school. He went to the store and found that pencils cost 10¢ each; erasers, 15¢ each; paper clips, 25¢ per box; and rubber bands, two for 1¢. He bought 100 things. What did he buy?

A man bought a lawn mower for $20.00. He later sold it for $25.00. Still later, he bought the same lawn mower back for $15.00. Did he lose money or make money on the lawn-mower trade?

These activities represent a most difficult level of word problems, yet all pupils achieved an increased degree of success. Although no specific objective data were

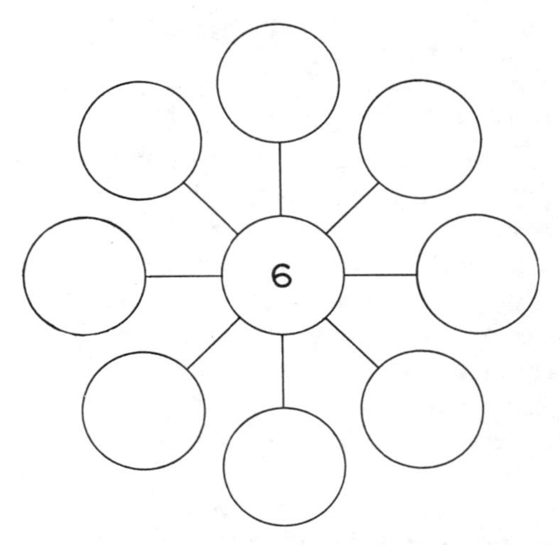

Put the numbers 11 through 18 in the outer ring of circles so that each line totals 35.

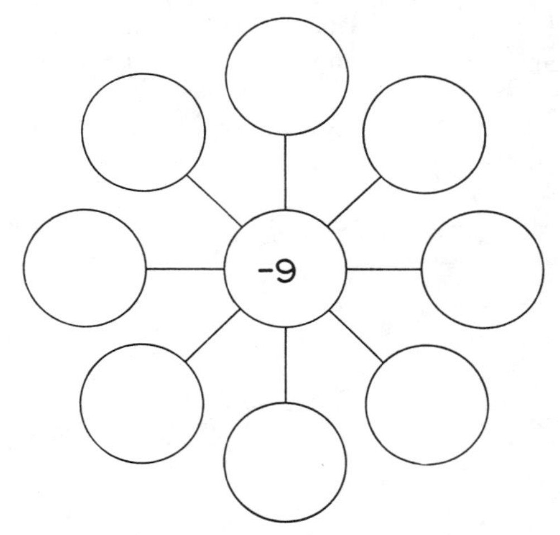

Put the numbers 21 through 28 in the outer ring of circles so that each line totals 40.

Fig. 7. Star-Burst

collected concerning word-problem skills, classroom observation provided ample testimony that such practice was both enjoyable and beneficial.

Conclusion and recommendations

A small-scale study using activities in mental computation and selected enrichment activities produced almost a full year's increase in mathematics achievement scores for a sixth-grade class. This result was obtained in a total of six extra hours of instruction accumulated over a five-week period.

There can be little doubt that these results were influenced by factors other than the instructional activities such as: (*a*) the Hawthorne, or halo effect, (that often accompanies enthusiastic experimentation); (*b*) the instructional variable (since this project was directed toward specific computational skills, instruction was aimed at any test that measured such skills); (*c*) a testing set established during the pretest (even though five weeks elapsed between test administrations, there may have been some retention of test material).

While there is room for further speculation on these results, two recommendations are justifiable here:

1. Additional research is needed to determine more accurately the relationships among the amount of time, nature of the instructional activity, and mathematics achievement.

2. Exploratory use of such activities in elementary mathematics classes where computational performances are lagging seems imperative.

References

Crescimbeni, Joseph. *Arithmetic Enrichment Activities for Elementary School Children*. West Nyack, N.Y.: Parker Publishing Co., 1965.

Kramer, Klaas. *Mental Computation*, bks. A–F, teacher's guide. Chicago: Science Research Associates, 1965.

Science Research Associates. Achievement Series: Form E, Blue Level. Chicago: Science Research Associates, 1971.

NOTE. A more detailed report of the research findings reported in this article is available from the authors.

Whole Numbers

2

A major portion of the elementary school mathematics curriculum is devoted to a study of the whole numbers and their operations. Many of the games selected for this chapter offer a variety of ways to provide the much-needed drill and practice on the basic facts of addition, subtraction, multiplication, and division.

In the opening article, Deans offers several game activities to aid young children in developing their counting skills. This article should stimulate the reader to think of ways to adapt various classroom and playground games to aid in mathematics instruction.

The idea of using pictures to motivate students to complete a drill sheet is suggested by Crouse and Rinehart. Many students will find these "connect the dot" activities to be most enjoyable, and it is possible to create similar activities without any special artistic ability.

The "Secret Number Sentence" game suggested by Swart will provide students with an opportunity to polish up their skill in asking questions while they work with number sentences at an appropriate level of difficulty.

Patterson provides a nice set of directions for constructing drill wheels that may be used in a variety of individual or group activities to practice the basic facts. In the "Go Shopping! . . ." activity suggested by Orans, students will have opportunities to role-play a shopping trip to a candy store. This should be a sweet experience for all!

The three addition games suggested by Heckman are easy to play, and the needed materials can be readily prepared by the classroom teacher for a very small price. She also suggests variations to provide practice in multiplication.

Howard Gosman suggests that dice are "loaded" with activities to help children master basic number facts while enjoying the experience. He then provides instructions for playing at least a half-dozen games using dice. In the next article, Bartel suggests a "computer" game that also uses dice. She provides several "for instances" to demonstrate how her activity can be varied to suit the needs of children at various grade levels. The dice game "Multi-bet" by Miki will provide practice with the "more difficult" basic multiplication facts and appeal to students who have a flair for gambling.

For those who need practice in two- and three-digit addition, Dilley and Rucker suggest an interesting game called "Build the Greatest Sum." This game also provides opportunities for students to formulate strategies that will increase the chances of winning. Several other games are also described in this article. An activity for three-digit subtraction is offered by Smith in his "Witch's Best Game." The students will have many opportunities to gain experience with the various regrouping situations that occur in three-digit subtraction problems.

The card game described by Arnsdorf should provide the players with opportunities to make strategic decisions while they continually do mental computation in the four basic operations. This should provide an effective and enjoyable review of the basic facts.

Broadbent's "Contig" game is simple to make and will provide much practice with the basic facts of addition, subtraction, multiplication, and division. Many students will enjoy playing this game at home with their older brothers, sisters, and parents.

Six games that help the learning and retaining of a variety of mathematical skills and understandings are described by Metzner and Sharp. Each of the activities requires three decks of ordinary playing cards, and several teams of two or three students each may play in a game. In the next article, Ristorcelli suggests "Green Chimneys Poker" as a game to develop proficiency with the multiplication tables through twelve.

"Diffy" is a game in which the student himself initiates the drill activities. The game automatically provides for individual differences and can be productive academic recreation at almost every grade level. Wills provides an excellent set of directions and guidelines for effective use in the classroom.

In his article on "Magic Triangles . . .," Zalewski offers a number of ideas for providing practice in addition while simultaneously stimulating the students to practice the searching, guessing, and thinking that is essential for problem solving. Similar experiences may be developed from the work with magic squares proposed by Cappon.

Games for the early grades

EDWINA DEANS

Games are one means of helping young children develop number concepts.

Good learning situations have been observed as teachers and children engage in games such as the ones described below. It will be noted that some of these are adapted from familiar games with slight variations or adjustments made to highlight the mathematics aspect. Perhaps readers will see similar possibilities in many classroom or playground games that they use frequently in connection with their instructional programs.

Kitty-cat

A game for kindergarten and first grade

The children sit on the floor in a circle. One child is chosen to be Kitty-cat. Kitty-cat sits in the center of the circle. When all are ready to begin the game, one child slowly counts to twenty. The children forming the circle have a designated number of small balls which they roll from one to another very quickly across the circle. As the balls roll across, the Kitty-cat catches as many as he can before the counter reaches twenty. The number of balls caught is counted, and another Kitty-cat and another counter are chosen.

Any number of children can play this game at one time. Children forming the circle sit close together so that no gaps will be created through which balls may roll to the outside.

The number value, of course, comes through the checking.

"How many balls did we have?"

"How many balls did Kitty-cat get?"

"How many did he fail to get?"

The number of balls is changed from time to time. All children should be aware of any change in the number of balls before the game begins.

Tenpins

Commercial tenpins can be used but homemade ones work just as well. Large dowel rods or broom handles sawed into about four-inch lengths make very good ones. Paper cones made from construction paper may also be used. Old golf or tennis balls are very good for rolling. Two to four children play. The tenpins are placed in the formation shown in Figure 1.

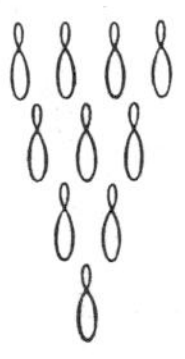

Figure 1

Each child is given two turns to see how many pins he is able to knock down. Scoring can be done according to the method best suited to the ability of the children who are playing. This game may be played with any number of pins. It is often played with five or six at first with an additional pin being added as the children are able to cope with larger numbers.

A similar game, well adapted to the school room, makes use of paper blocks instead of pins. The paper blocks are made from heavy construction paper, using the familiar sixteen square fold. The paper blocks are piled up pyramid fashion (Fig. 2). Each child has one turn and scores according to the number of squares knocked down. Since the paper blocks make very little noise as they fall, this game can be played while group activities are going on without causing undue distraction.

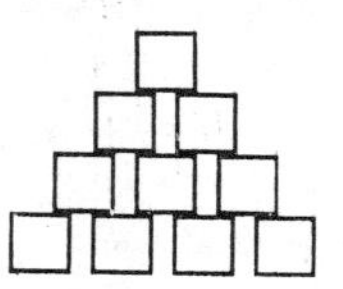

Figure 2

Guessing how many

This simple game is appealing to small children. The only materials needed are a given number of small objects which can be held in one hand such as wooden beads, little pieces of chalk, or pennies.

Three to five children may play. The game starts by showing the objects to the children asking, "How many do you see?" Suppose the children count and discover that there are six. The leader continues, "I have them all in one hand. Now I am going to turn around so you cannot see, and put some in one hand and some in the other. Think about them and see if you can guess how many I have in one hand and how many in the other."

The children take turns guessing the number in each hand—four and two, five and one, six and none. After each child has had his turn at guessing the leader opens his hand and shows the sets. The first child to guess correctly gets the next chance to be leader.

Interest is maintained if the material used and number of objects is changed often.

An interesting variation of this game is to show children the total and also the number in one hand. The children guess how many are in the other hand.

Creative drill with pictures

RICHARD CROUSE

University of Delaware, Newark, Delaware

ELIZABETH RINEHART

Krebs School, Newport, Delaware

The word *drill* has some bad connotations for many mathematics educators—probably generated by the recollections of gross misuses of drill in the past. However, many educators believe that drill as meaningful practice has a definite place in the sequence of teaching activities, and they accept axiomatically that drill should occur after students understand a concept or skill.

A key to effective drill seems to be how it is motivated, and teachers often find this to be a difficult task. The following activity is one that some teachers will find useful in their own classrooms.

The idea is to use pictures to motivate students to complete a drill exercise. Figures 1 and 2 were used in a third-grade class to review simple addition and subtraction combinations, but other skills could easily be substituted and practiced. The directions for this activity are simple:

Place your pencil at the point where the answer is 1. (This point is marked with the star ★. Look for the point that is marked 2, and connect the two points by a line segment. Next, look for the point marked 3, and connect it by a line segment to the point marked 2. Repeat this process as far as you can.

Some teachers may feel that they do not have enough artistic ability to make these drawings. However, the drawings in this article were made from a coloring book. The picture that is wanted can be placed on top of ditto reproduction paper and traced with a pencil, or preferably a ballpoint pen. The rest is filled in freehand.

As an added feature to this activity, pictures could be chosen that would lend themselves to other activities. For example, after the picture of the sheriff was completed, the children could make up a story about the sheriff and the number of men he captured. Many of these pictures could lend themselves to a blending of subject areas and perhaps make teaching and learning more worthwhile and exciting. The teacher is limited only by her imagination.

7+9
15-0
8+9
9+9
9+5
7+6
10+9
7+5
6+5
10+10
25-4
10+0
18-9
1-0
6-1
8-6
5-1
3+3
13-10
4+3
0+8

Fig. 1

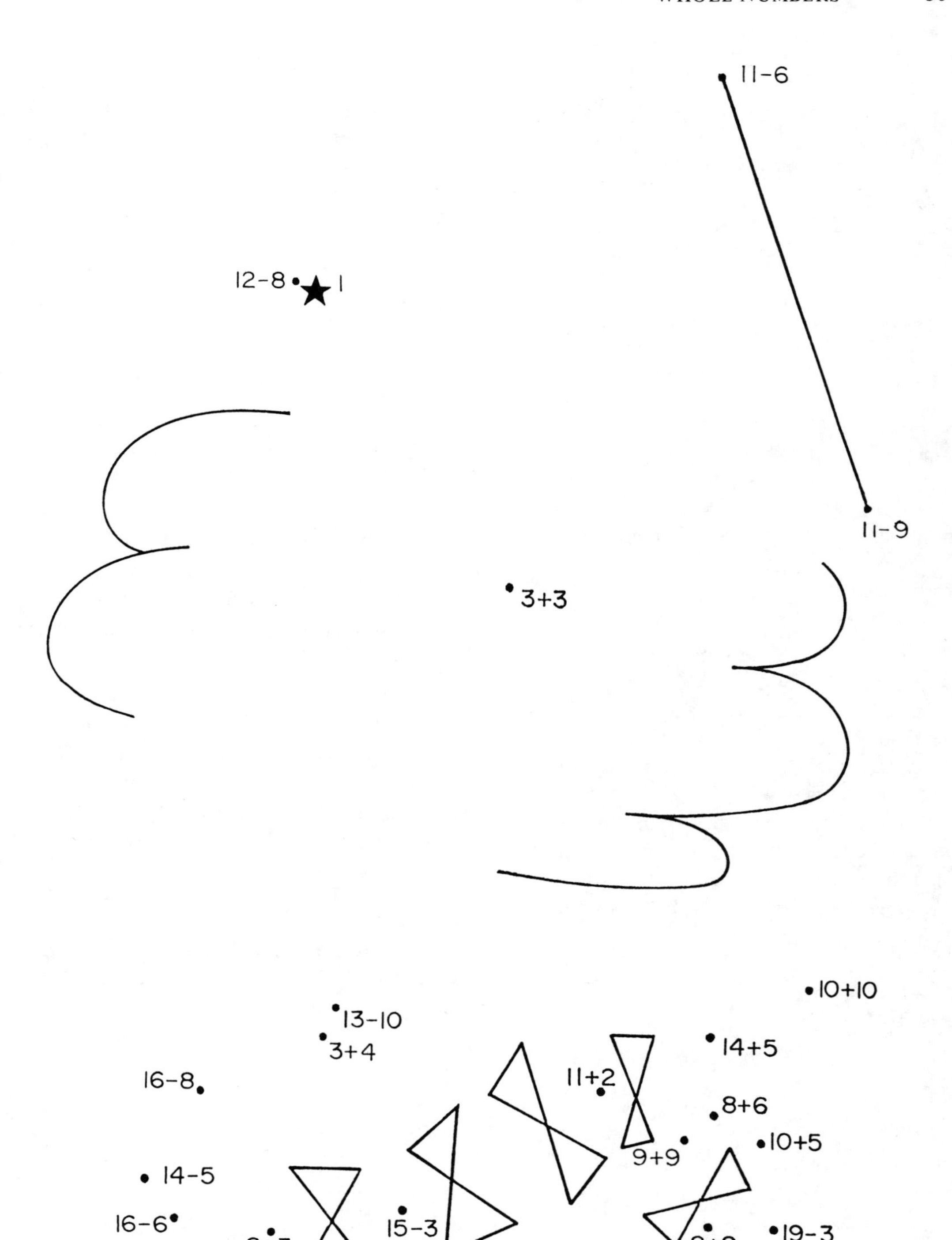
11-6
12-8
1
11-9
3+3
10+10
13-10
3+4
14+5
16-8
11+2
8+6
9+9
10+5
14-5
16-6
15-3
8+3
8+9
19-3

Fig. 2

Secret number sentence

WILLIAM L. SWART
Central Michigan University, Mount Pleasant, Michigan

William Swart is assistant professor of mathematics at Central Michigan University.

Here is a game for primary arithmetic, providing stimulating drill on mathematical sentences and on addition, subtraction, multiplication, and division combinations.

Player A secretly fills in this pattern with a number sentence:

□ ○ △ = ——

Player A must choose his numbers from some agreed-upon set. In Grade 2, a suitable set would be 0, 1, 2, 3, 4, 5, 6, 7, 8, 9. If he selects $6 + 3 = 9$ as his secret number sentence, he replaces the square and the triangle with 6 and 3, the circle with the operational sign +, and the blank with 9. It must also be understood which operations are to be used.

Player B puts the following on his paper:

□	○	△	=	——
0	+	0		0
1	−	1		1
2		2		2
3		3		3
4		4		4
5		5		5
6		6		6
7		7		7
8		8		8
9		9		9

Then Player B proceeds to ask certain questions of Player A until he discovers what Player A's secret sentence is. The object is to discover the secret sentence in as few questions as possible. Player B can only ask the following questions:

1. Is the numeral in the (square, triangle, blank) equal to__?
2. Is the numeral in the (square, triangle, blank) greater than__?
3. Is the numeral in the (square, triangle, blank) less than__?
4. Is the operation__? (addition, subtraction, etc.)

Each time Player B asks a question about a position, he is able to eliminate one or more numerals under that position. For the secret sentence $6 + 3 = 9$, the question, "Is the numeral in the square equal to 4?" would eliminate 4 as a possibility for the square. And the question, "Is the numeral in the square greater than 7?" would be answered "No," revealing that 8 and 9 are not possible. After these two questions, Player B's paper should look like this:

□	○	△	=	——
0	+	0		0
1	−	1		1
2		2		2
3		3		3
~~4~~		4		4
5		5		5
6		6		6
7		7		7
~~8~~		8		8
~~9~~		9		9

If his next question were, "Is the numeral in the square less than 5?" he would get a response of "No," and cross off 0, 1, 2, and 3. Then one or two more questions will reveal that 6 is the numeral in the square.

As the game is played, the pupil should become aware that "greater than" and "less than" questions usually yield more information than "equal to" questions.

When Player B succeeds in discovering Player A's secret sentence, the roles are reversed and Player A becomes the questioner. The player who discovers the other player's secret with the fewest questions wins.

It is advisable to start the game with the teacher in the role of Player A and the children asking questions, and to proceed in this manner until the children know how to play. However, while the game is highly stimulating as a group activity, it is doubtful that the group approach will result in significant gain for the child who needs it most. The kind of thinking promoted by the game requires that the child have time to ponder—to decide which questions to ask, and what information an answer provides him.

In the upper grades the game can be made more challenging by including all four fundamental operations and by allowing the relationship to include "greater than" and "less than" as well as "equals," so that secret sentences such as $4 \times 8 < 380$ may be used.

EDITORIAL COMMENT.—This is one of many game activities that provide students with opportunities to gain experience in several important areas. First, the game will provide desirable practice with the basic facts of arithmetic. Second, and perhaps even more important, the activity will provide a framework in which students may learn how to ask the "right questions" to secure needed information.

There are many types of games that can aid in the development of questioning skill. For instance, one student can select a whole number from some specified interval, such as *less than 30,* and the second student can try to identify the number in five questions or fewer. Points can be awarded on the basis of the results of the questioning as follows:

Points scored	*If number is identified in*
5	1 question
4	2 questions
3	3 questions
2	4 questions
1	5 questions
0	6 questions or more

If the students are ready to work with negative integers, the scoring could be continued as follows:

−1 point	7 questions
−2 points	8 questions
−3 points	9 questions
.	.
.	.
.	.

The game could be continued for a specified amount of time, or until one student gets a prescribed number of points.

Making drill more interesting

W. H. PATTERSON, JR.

University of Southwestern Louisiana,
Lafayette, Louisiana

Looking for ways to make drill more interesting, entertaining, and even fun? Isn't everyone? I have developed a little device that may help you provide more interesting drill sessions for your students. I call it a drill wheel.

The drill wheel is easy and inexpensive to make. It consists of a board on which are mounted two or more spinners. Cardboard rings with numerals printed on them are fitted over the spinners. The spinners are then twirled and the students add (or subtract, or multiply, . . .) the numbers indicated by the spinners. A drill wheel with two spinners is pictured in figure 1.

Although simple, the device has several advantages. It is different and will add fun to drill sessions. Also, a few rings will provide a large supply of drill problems for a teacher. For example, 25 numerals printed on each of two wheels will result in 625 problems!

To make a drill wheel first cut out the mounts for the spinners, using posterboard or cardboard. Paste or glue these mounts to the rectangular base, which can also be made from posterboard or cardboard. Next, punch holes through each mount and, using a fastener, attach the spinners. The spinners can be made of any convenient, stiff material, such as posterboard, cardboard, or an old photo negative. The number rings can be cut easily from cardboard and labeled with felt pens. Care should be taken so that the hole in the ring fits over the spinner mount. Whole numbers have been printed on the rings shown in figure 1, but other types of numbers (negative integers, proper and improper fractions, and so on) can be used. An operation indicator (see fig. 1) is a nice addition to the drill wheel.

I have made drill wheel models as large as 4 feet wide and as small as a regular piece of notebook paper. A teacher can build up a file of number rings and, with a large drill wheel hanging in the front of the classroom, be ready for many types of drill. For example, if a class needs drill in adding improper fractions, the teacher simply selects two (or more) rings with improper fractions on them, sets the operation indicator on +, and goes to work.

A large supply of rings for a classroom wheel can be obtained quickly by simply letting the students, either individually or in small groups, make them.

While larger drill wheels can be used for regular class drill, they can also be effectively used in other ways. For example, a class might be divided into two teams that compete in working problems with the teacher (or team captains) doing the spinning. To add interest, some numerals on the rings could be printed in different colors and bonus points given when these numbers turn up.

Smaller models of the drill wheel can be

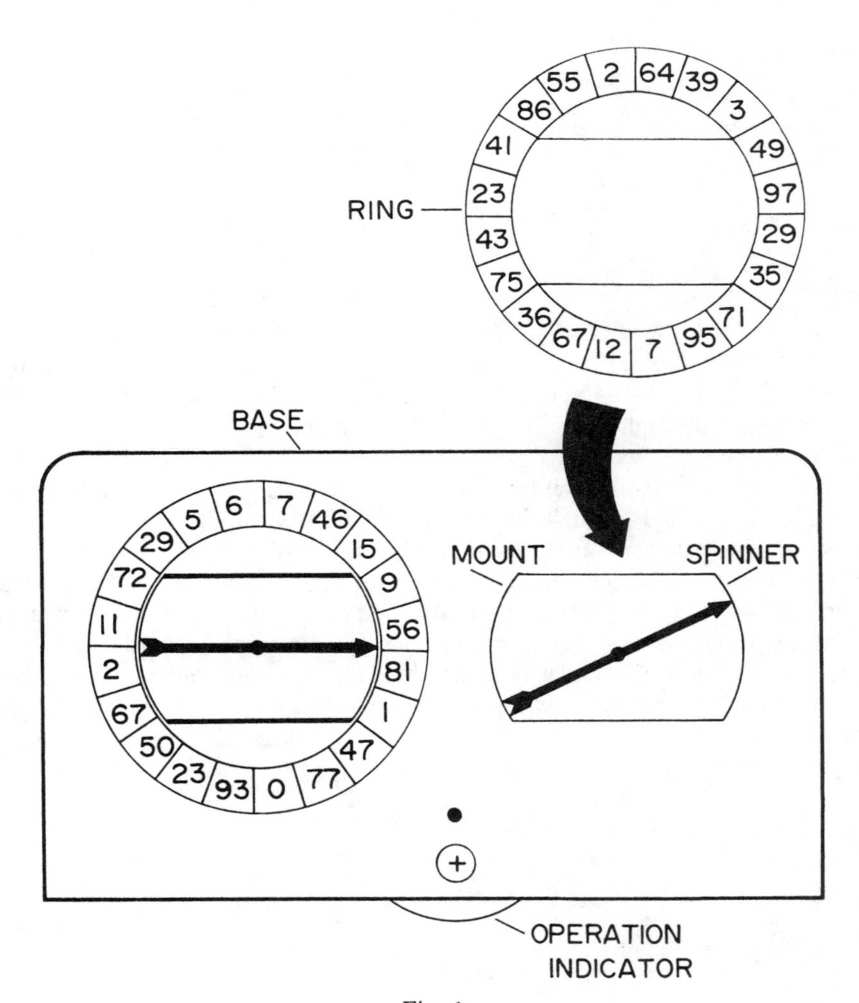

Fig. 1

made and used individually by students. Those students with any artistic inclinations will probably enjoy making their own spinner wheel and rings. In fact, an interested student can make a "drill wheel kit" with a pocket for the rings, that will fit in his notebook.

Consider using the drill wheel in your classes. I think that you, too, will find that it is an inexpensive, economical device that can give you an opportunity to vary your instructional activities, add interest to your classes, and make your students more enthusiastic about drill.

EDITORIAL COMMENT.—Another activity that you can use to make drill more interesting is called "Pencil Point Facts." First you need to construct some practice cards for the related facts you want the students to work on. For example, a multiplication/division practice card would appear as follows:

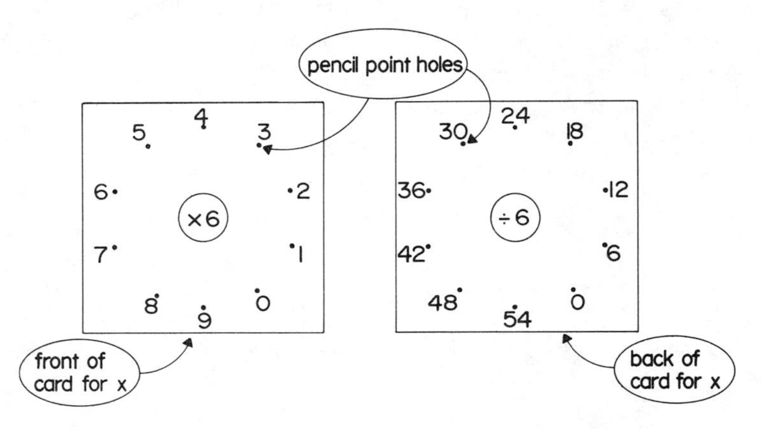

Students may work together in pairs to practice their facts. The holder of the card inserts his pencil point from the back of the card through the hole for 8. The other student must give the product for 6 × 8. The holder of the card can check the answer (48) on the back of the card beside the hole through which he inserted his pencil point.

For division, the card is reversed so that "÷6" shows for the student computing the answers. The holder inserts his pencil from the back side of the card as previously described. If the pencil point is inserted through the hole for 18, the student gives the quotient for 18 ÷ 6. The holder can check the answer (3) on the back of the card.

Cards may be constructed in a similar manner for related addition/subtraction facts.

The cards may be used in game activities for pairs of students by setting goals and awarding points for correct answers. For example, one student may pick a card (from the set identified by the teacher according to need) and challenge his opponent with five "pencil point facts" on the card. One point is awarded for each correct answer; then the two students reverse roles.

Go shopping! Problem-solving activities for the primary grades with provisions for individualization

SYLVIA ORANS *Parkway School, Plainview, New York*

Materials

Real money (one dime, two nickels, ten pennies) brought from home by each child

Real candy (large candy gum drops, several colors)

Construction paper for price tags

Clear plastic bags (sandwich size)

The approach

Problem situations requiring a minimal reading vocabulary of ten words, using real materials and real money in a gamelike atmosphere of playing store.

The procedure

Our candy-store activities were initiated after much exploration with simple games using real money. Our bulletin board, on which real candies were displayed, was our "store" (see fig. 1). Several candies of one color were placed in each plastic bag. The bags, containing orange, yellow, red, green, and black candies, were tacked to the bulletin board. Price tags, made by the children, were changed often to vary the level of difficulty and/or to lend interest.

NOTE: Color names may be omitted. Color of candy is visible.

FIGURE 1

At the beginning, store activities were teacher-directed. All the children used the coin or coins designated at their desks. One child was selected to "go shopping" for each problem situation. Another child was the storekeeper, and gave change under teacher supervision. As each problem was solved, the shopper was reimbursed with his original coin or coins. The children who were seated at their desks were involved in checking out each problem. The children used their coins; the candy, however, remained on the bulletin board (!) with the promise of its distribution to all when the store closed, at the end of the week. Some oral problems follow.

1. Go shopping.
   You have a dime.
   How many green candies can you buy? (green—5 cents each)
2. Go shopping.
   You have a dime.
   How many candies do you want to buy?
   What must you pay?
   How much change will you get?
3. Go shopping.
   You have a nickel.
   Can you buy black candy? (black—6 cents each)
   How much more money do you need?
4. You bought five candies of one color.
   You had a dime.
   What color candy did you buy?
5. You bought one candy.
   You had a dime.
   You got four cents change.
   What color candy did you buy?
6. You bought two candies.
   You had a dime.
   You got three cents change.
   Were your candies the same color?
   Could they be the same color?

Then we went shopping on the chalkboard. The candies were still on display, to refer to, and for "show" appeal. A chart was set up (see figs. 2a, 2b, and 2c).

Written work followed when the children showed a readiness to apply the techniques. Problems suggested in the charts, "Variation" and "Another Variation," were used for written work. The pupils went

You have	You buy	Each candy costs	You must pay	Your change is
5¢	2 green	2¢	4¢	___¢
5¢	1 red	4¢	___¢	___¢
10¢	3 yellow	3¢	___¢	___¢
20¢	2 black	5¢	___¢	___¢

FIGURE 2a

Variation

You have	You buy	Each candy costs	You must pay	Your change is
10¢	_ yellow	3¢	___¢	4¢
10¢	_ ____	_¢	___¢	5¢
25¢	_ ____	_¢	___¢	10¢

FIGURE 2b

Another variation—open-ended problems

You have	You buy	Each candy costs	You must pay	Your change is
10¢	_ yellow	3¢	___¢	___¢
20¢	_ black	6¢	___¢	___¢
20¢	_ red	5¢	___¢	___¢
* 25¢	_ green	8¢	___¢	___¢
* 50¢	_ orange	5¢	___¢	___¢
* 25¢	_ ____	_¢	___¢	___¢

* These problems were elective.

FIGURE 2c

"shopping" on paper. They copied the information from the charts on a ditto that had five columns and headings to correspond to the chart. The problems had become more difficult (* items were elective), yet the problems were still within the range of the slowest pupil, because he could choose to buy only one of each candy. The more capable pupil enjoyed the challenge of choosing candy that was priced at a figure that enabled him to spend all of his money or to get the least amount of change.

It was possible to involve all the pupils, regardless of reading level, in these problem-solving activities. Because of the variables of price, number of candies, and other variables in the chart, these store activities provided individualization for a broad range of abilities and challenged each pupil's unique mathematical potential.

They all add up

M. JANE HECKMAN

Now principal of the Pembroke Elementary School, Birmingham Public Schools, Troy, Michigan, Jane Heckman has also served as an elementary school teacher and a mathematics consultant for grades kindergarten through six.

The games described here have provided practice in addition for many children in numerous classes. The games are easy to play and the needed materials can be readily prepared by a classroom teacher for a small price. They all add up to addition 'n fun.

Roll the sum

Number of players: 2

Game equipment:

Two cubes with the numerals 0, 10, 9, 8, 7, 6 in red on the faces of the cubes.

Two cubes with the numerals 0, 1, 2, 3, 4, 5 in green on the faces of the cubes.

Two ROLL THE SUM gameboards. (Fig. 1.)

Markers—beans, corn, buttons, disks (at least 40).

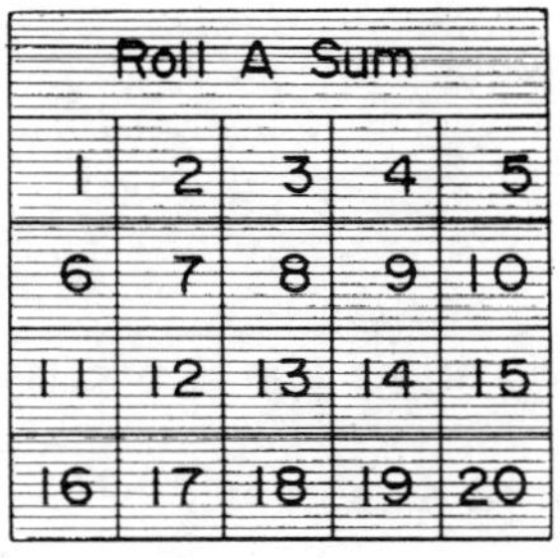

Roll A Sum				
1	2	3	4	5
6	7	8	9	10
11	12	13	14	15
16	17	18	19	20

Fig. 1

Directions:

1. To start the game, each player rolls one of the *red* dice. The player with the largest number appearing on the top face of his thrown die plays first.
2. On his turn, a player selects two of the four dice and rolls them on the playing area. The *sum* of the two numbers appearing on the top faces of the thrown dice is covered on the player's gameboard. If the sum is already covered, the player loses a chance to cover a square.
3. Players take turns rolling the dice.
4. Play continues in this manner until one player has either four markers in a vertical row or five markers in a horizontal row.
5. The game can be varied by having players—
   - cover the entire board with markers.
   - cover four corners.
   - cover a "T" (top horizontal row and middle vertical row).

Author's note. By giving each player a chance to pick his dice, this game allows for a choice of possible addends. Thus a player must have some knowledge of which numbers can and cannot be added to give the sum he wants to cover. Another game, ROLL THE PRODUCT, provides practice in multiplication and requires the same kind of knowledge of products and their factors.

To play the multiplication game, two ROLL THE PRODUCT gameboards (fig. 2) are needed and more markers.

Roll The Product					
1	50	60	70	80	90
0	2	3	4	5	6
7	8	9	10	12	14
15	16	18	20	21	24
25	27	28	30	32	35
36	40	42	45	48	49
54	56	63	64	72	81

Fig. 2

Five sums in a row

Number of players: 2

Game equipment:

Two cubes with faces marked 0, 1, 2, 3, 4, and 5.

Two cubes with faces marked 2, 3, 6, 7, 8, and 9.

One playing board. (Fig. 3.)

Markers—beans, corn, buttons, disks.

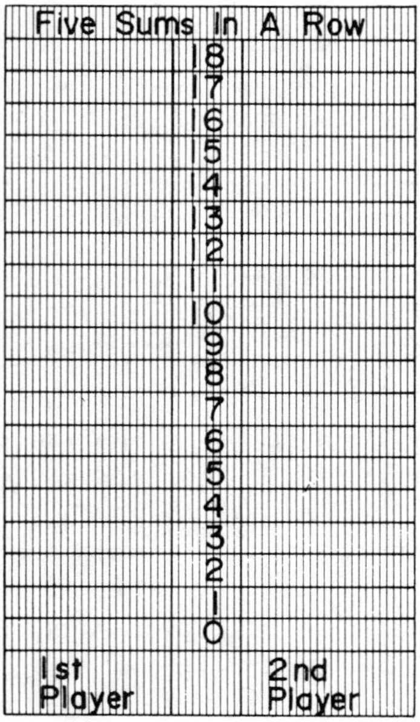

Five Sums In A Row		
	18	
	17	
	16	
	15	
	14	
	13	
	12	
	11	
	10	
	9	
	8	
	7	
	6	
	5	
	4	
	3	
	2	
	1	
	0	
1st Player		2nd Player

Fig. 3

Directions

1. Each player in turn rolls the four dice and adds the numbers appearing on the top faces. The player with the greatest total starts the game.

2. On his turn, a player rolls the four dice. He selects two of the numbers appearing on the top faces to form a sum on the playing board. He places one of his markers opposite that sum on his side of the playing board.

3. Players take turns rolling the dice.

4. Play continues in this manner until one player has placed his markers opposite *five sums in a row* and wins the game.

Author's note. Players may test two or three sums before they make the selection for their play. This game can be varied by making sums with three dice instead of two.

Kumpoz

Number of players: 2 to 4

Game equipment:

Eighty-one cards, each 4½" by 2", with nine cards of each of the following: 1, 2, 3, 4, 5, 6, 7, 8, 9. (Fig. 4.)

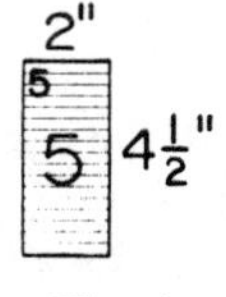

Fig. 4

Directions:

1. Cards are placed face down on playing area.

2. To begin the game, each player turns over one card. The player turning over the highest number starts the game. (In case of ties, players in question turn over new cards.) Play then proceeds clockwise. Drawn cards are returned face down to the playing area.

3. The starting player selects one number from the set {10, 11, 12, . . . , 18}. The selected number will be used for this round of play. That is, *the starting player selects the number.*

4. Each player draws 15 cards from the playing area.

5. The first player puts down two or more cards. The numbers on the played cards must add up to the number selected in direction 3. For example, suppose the selected number is 16 and the first player puts down the three cards shown in figure 5. This is called a run. The run must total the selected number, in this case 16.

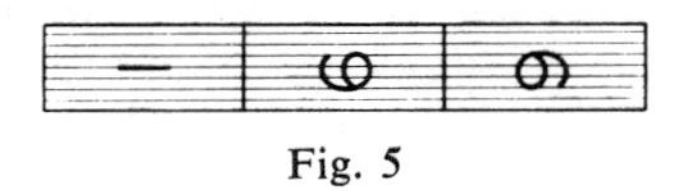

Fig. 5

6. The second player then puts down cards to make a new run. The new run may be connected to the 1, the 6, or the 9 card of the first run. The new run must also total 16. (See fig. 6.)

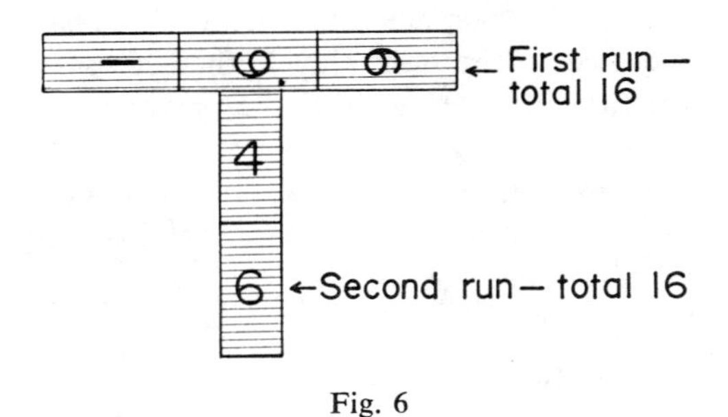

Fig. 6

7. Play continues clockwise around the table with each player in turn putting down cards to make new runs. All runs in a game must total the selected number. Possible third and fourth plays in the sample game are shown in figures 7 and 8.

8. Play continues in this manner until one player has used all of his cards, or until no player can make a play. The player who has played all his cards is declared the

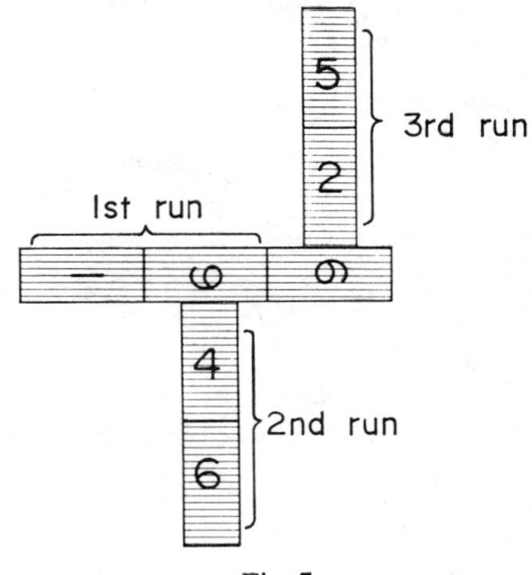

Fig. 7

winner, or if no player can play all his cards, the player having the least number of cards when play stops is declared the winner.

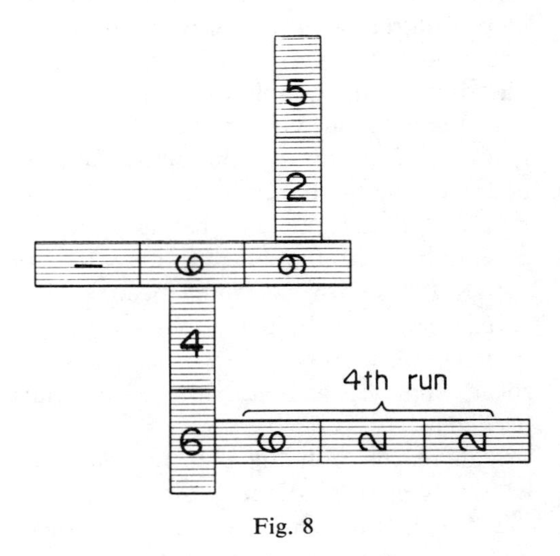

Fig. 8

Author's note. In this game players have practice in adding two or more numbers to make a given sum and on each turn they must test several combinations.

Mastering the basic facts with dice

HOWARD Y. GOSMAN

Mathematics coordinator and teacher at P.S. 165 in New York City, Howard Gosman is also doing additional graduate work at Teachers College of Columbia University.

The dice are "loaded" with activities to help children master basic number facts and enjoy the experience. "Double Trouble" is a good example of an activity that can be used in any grade. This game helps children master the basic facts in addition, subtraction, multiplication, and division. The game can be used in any classroom, and the teacher may modify it to suit his own classroom needs.

The game can be played with one, two, three, or four players. Two dice, a score sheet, and a pencil are all the materials needed.

The players roll the dice to determine who will go first. Each player then rolls the dice in turn. His score is the sum of the numbers that come up on the two dice. In the simplest form of the game (called "Trouble"), the first player to score 100 points (or a lower score in the lower grades) is the winner. A simple score sheet should be prepared by the teacher so that the children can keep a running count of the score.

In Double Trouble each player may roll the dice one or more times during his turn. After each roll the player must decide to either pass or play. His score for the round is the total number of points scored. If the player rolls a double before he passes the dice to the next player, then his score for that round is zero. If the player passes without rolling a double, then he adds the score for the round to his total score. Once a player passes, he cannot lose any of his points.

Variations

Double-Double Trouble, Triple-Double Trouble, and other Multiplication-Double Troubles. The players roll the dice as usual. Their score on each round is the number on the dice multiplied by two, by three, or by any number selected before the game begins. Of course, the total number of points needed to win the game would also be multiplied by the same number.

Addition-Double Trouble. The players roll the dice as usual. Their score on each round is the number rolled on the dice plus a number selected before the game begins.

Multiplication-Addition Double Trouble. The players roll the dice. The number on the dice is multiplied by a preselected number, and a second number is added to the resulting product. For example, the number on the dice is multiplied by five, and eight is added to the product. ($5 \times$ number thrown on dice $+ 8 =$ score.)

Triple Trouble. Three dice are used instead of two. The player's score for the round is reduced to zero if two of the three dice thrown make a double. The player's total score is reduced to zero if all three dice show the same number. Double-Triple Trouble would mean doubling the score on the three dice.

Other variations can be developed using subtraction, division, and fractions. The children can also make up their own games.

Double Trouble (or any of its variations) is a fun way to learn the basic number facts. The game is completely under the control of the children. They must roll the dice, do the appropriate computation, keep track of their scores, and decide to pass or play after each throw of the dice. Children can learn their basic number facts quickly without losing interest in numbers.

EDITORIAL COMMENT.—Another interesting activity with dice was suggested by Carol Stephens in the April 1967 issue of the *Arithmetic Teacher*. The game is called "Yahoo," and the basic information is as follows:

Equipment: 2 score sheets
2 dice
a multiplication fact chart
a set of bonus cards from colored construction paper with the following facts written on them:
7×2 through 7×9
8×2 through 8×9
9×2 through 9×9

Number of players: 2

Directions: Each person rolls the dice in turn. The two numbers shown on top indicate the factors. If the player knows the product, he records it on the chart. Except for the doubles facts, each product may be recorded in one of two possible places. If a player rolls a double, he gives the product and writes the answer on the chart. If correct, he selects a bonus card from the pile, gives its product, and records the answer on the chart. When a player completes a row (vertical, horizontal, or diagonal), he earns a *yahoo*. A game consists of twenty turns for each player, and ten points are awarded for each yahoo. The objective is to earn the most points in the game. A yahoo may also be earned by completing any three consecutive bonus squares.

x	1	2	3	4	5	6	7	8	9
1									
2									
3									
4									
5									
6									
7									
8									
9									

Multiplication Fact Chart

Turn	
1	
2	
3	
4	
5	
6	
7	
8	
9	
10	
11	
12	
13	

Score Sheet

An obvious variation of this game would be to adapt it to drill with addition facts.

Let's play computer

ELAINE V. BARTEL

An associate professor of education at the University of Wisconsin—Milwaukee, Elaine Bartel is director of the Intern Teaching Program for the School of Education on that campus.

Despite the burgeoning market of manipulative devices and games to enhance pupils' understanding of basic mathematics concepts, classroom teachers and paraprofessionals are still exhausting every available resource in their search for ideas and materials that meet the following criteria:

1. Relatively inexpensive
2. Easily adaptable to various levels of understanding
3. Related to identified problem areas of the mathematics curriculum
4. Simple enough to be introduced and used with a minimum of teacher guidance or supervision.

One idea that has passed the test of all of these criteria and has produced very positive feedback from preservice and inservice teachers, tutors, and paraprofessionals, is the game that I have titled "Computer." The only materials needed are several pairs of dice of varying colors. Since the game has been designed specifically to aid in pupils' grasp and memorization of basic addition, subtraction, and multiplication facts, it speaks directly to an identified problem area in a fresh, game approach. Since "rules" for the game (programming the computer) are established by the group using it at the time, the game can be used effectively with pupils at any age or level of understanding.

Rather than attempt to explain the many possibilities of the game in lengthy detail, it might suffice to go through a few "for instances" at various grade levels to give the reader a general idea of how "Computer" can be used with students.

Imagine, if you will, that our materials include three pairs of dice; one pair each of red, green, and white. At the primary grade level, the teacher could choose two youngsters, a small group, or a group divided evenly into two teams. Initially the teacher would "program" the computer: "$R + r$" or "$R - r$" or "$R + r + G + g$." (The letters refer to the color of the dice. The capital letter always denotes the larger of the two numbers on each pair of dice, and the small letter the smaller of the two numbers, unless they are the same.) Eventually the children will "program" the computer themselves, writing the problem on the chalkboard or on a large piece of paper on the table where the game is being played. When the children set up a problem like "$R + W - G$," they will face the possibility of encountering a negative number for their answer. This should cause no difficulty, especially if the classroom numberline extends to the left beyond zero.

In the intermediate grades pupils will tend to set up problems that match their ability to handle the basic addition and multiplication facts accurately and rapidly. This will also serve to encourage pupils to work at the mastery of these facts, since the need to know them well will be clearly demonstrated when pupils participate in the game. Such problems as

$$(R + r + W + w) - (G + g)$$

or

$$\frac{(W + G) \times (w + g)}{R - r}$$

will be used often at this level, but it will not be long before students will attempt to set up much more sophisticated problems.

At the junior high level students will spend a great deal of time and mental energy while competing with each other on a problem like the following:

$$\frac{R^r + w^W}{G \times g}$$

For example, suppose six dice are rolled so that $R = 4$, $r = 3$, $W = 6$, $w = 2$, $G = 5$, and $g = 4$. Then, assuming that it is understood that the answer must be in the simplest form and expressed as a mixed number when possible, the correct response would be $6\frac{2}{5}$.

For each game session, it is strongly suggested that children be permitted to establish their own "rules" for that session. For the example just given, students might conceivably establish the following rules:

1. A referee will be selected who will in turn choose the number and color of dice to be used, and select someone to "program" the computer (establish the problem).

2. The referee will be responsible for tossing the dice, checking answers, and keeping score.

3. No participant may touch or move the dice while computation is taking place.

4. The referee will jot down the first answer given, wait a few moments, then work through the problem orally to determine the correct answer.

5. Participants may use scratch paper for computation.

6. Scores will be determined as follows: If the first answer called out is correct, the person (or his team) gets that number of points added to his score. If the answer is under 5 or is a negative number, the answer (to the nearest whole number) is squared to determine his score. If the answer is incorrect, the correct answer is deducted from his score.

Needless to say, the possibilities for this game are endless; and if it does happen that children establish problems that are for the moment beyond their ability to solve, this can be a valid learning experience for them, and may be just the encouragement they need to increase their own level of expertise in mental computation.

Some of the obvious advantages of using this game approach for mastery of basic skills are that (*a*) it clearly demonstrates to students the need to have the addition and multiplication facts "at their finger tips," (*b*) it encourages the utilization of mental arithmetic, (*c*) it stresses accuracy and speed in a computation situation where the student must determine the balance between accuracy and speed, and (*d*) it places students in competition with partners of similar ability.

Rather than attempt to persuade you to introduce this game to your students, I would like to suggest that you find a willing colleague, set up a few problems, and play the game yourself. This will serve two purposes: It will demonstrate to you the applicability and feasibility of the game for many and varied levels, and it will quickly dispel any preconceived notions that it is boring or will lose its appeal for mathematics students. And remember, once you have introduced it to your students, they might on occasion ask you to serve as referee. You would be in a much better position to serve in that capacity if you feel comfortable about matching your own skills with those of your students. I can't guarantee that you will always win, but I can guarantee some highly tense moments and the joy of seeing pupils really involved in their own learning.

Multi-bet

ARTHUR K. MIKI

Wayoata Elementary School, Transcona, Manitoba, Canada

The game "Multi-Bet" provides practice with the more difficult multiplication combinations and has proved to be one of the most popular games in the fourth grade.

Multi-Bet uses two specially marked dice—each die has the numbers 4, 5, 6, 7, 8, and 9; a supply of counters—poker chips or bottle caps are excellent; and a chart like the one pictured in figure 1. The numbers on the chart represent all the possible products that may result from rolling the two dice. The chart also shows the various combinations of numbers that can be played and the odds to correspond with the different combinations. A student may bet on one number, two adjacent numbers, four adjacent numbers or four numbers in a row, eight numbers, or ten numbers. Placement of the counters for the different combinations is indicated on the chart.

Each player begins with ten to fifteen counters. In each round players place their counters on the chosen numbers or groups of numbers. One player then rolls both dice and calls out the product of the two numbers turned up. The product is the winning number for that round. When a

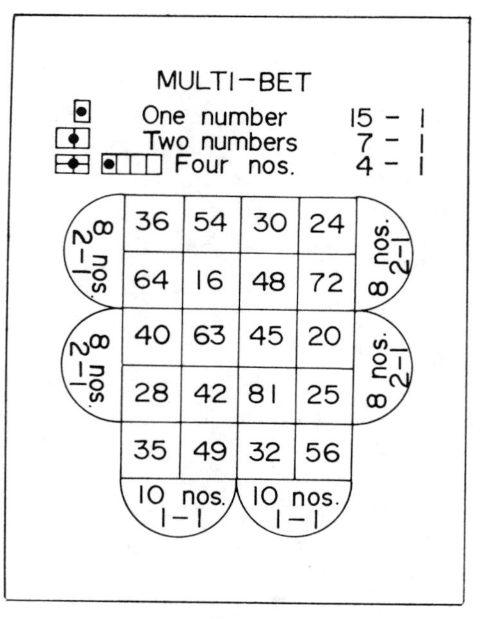

Fig. 1

player has played two numbers, if either number is the winning product the student gains seven counters from the house pot. Similarly, if a player has played eight numbers, he gets two counters from the house pot if any of his numbers is the winning product. When the number or players played are not winners, the counters are lost to the house pot. The players then place their bets for the next round and the next player in turn will roll the dice and call out the product.

The house pot would be all the counters left after each player had received the designated amount at the beginning of the game. A student who has mastered the multiplication facts, or a teacher could act as the house man.

The game can be played for a designated time period. If a player loses all his counters before the game is over, then he is out of the game. At the end of the time period the person with the most counters is the winner.

The game can also be used in the study of simple probability. The chart itself can be used to introduce the subject: Which numbers are the most likely to occur? How were the various odds chosen? Discussions could stimulate experimental work on the different combinations that are most likely to occur when rolling two dice.

For students who are just beginning to learn basic multiplication facts, the dice can be marked from 1 to 6. The game can also be applied to addition facts.

Arithmetic games

CLYDE A. DILLEY and
WALTER E. RUCKER

Urbana, Illinois

Often there is little variation in the setting in which children practice arithmetic skills (basic combinations and algorithms). In many cases the practice is unchallenging and unrewarding. An appropriate game can provide children with both a challenging and rewarding experience and an opportunity to discover mathematical concepts. Some games that have been used successfully with children at all levels are described below.

The first game, "Build the Greatest Sum," can be played as soon as children have been introduced to the addition algorithm. Ahead of time, you should make up digit cards by writing each of the ten digits (0–9) on a three-by-five card. The game is played on a table like that shown in figure 1. Each player can draw

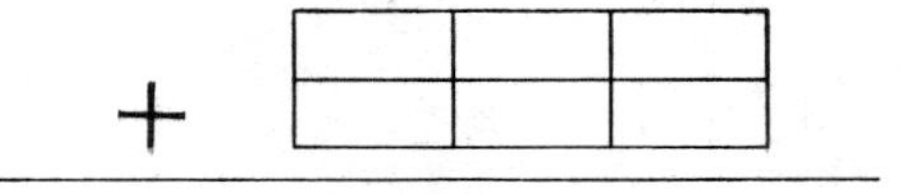

Fig. 1

his own tables, or you can duplicate copies to distribute to the players. To play the game you shuffle the digit cards and select one at random. Tell the children to copy the digit that is on the card in any of the boxes of the table. Replace the card with the others, reshuffle, and pick another card. Tell the children to copy that digit in one of the five remaining boxes. Repeat until a total of six digits have been picked and recorded in the tables. Then ask the children to add the two three-digit numbers. The player with the greatest sum wins.

Since the cards are selected at random, chance is a factor, especially at first. However, after the game has been played a few times students will begin to formulate strategies that will increase their chances of winning. The opportunity for developing such strategies is probably the most important facet of this particular game. (The other objectives are to provide reinforcement of place-value concepts—a digit in one column is worth more or less than the same digit in another column—and to provide practice using the addition algorithm.) These strategies involve an intuitive use of probability concepts because to decide where to write a given digit the child must consider the chances of getting a greater or lesser number on a later draw. Such a strategy does not, of course, guarantee a win, but the child who uses it has a definite advantage over the child who fills in the squares at random.

There are many obvious variations of this game. For example, one variation is *not to* replace each digit card before the next is drawn. In this case there will be no repetition of digits. Another variation is to extend the table so as to provide work with greater numbers. Still another is to have students aim for the least sum instead of the greatest sum. Other variations involve subtraction, multiplication, and division. In fact, one of the most interesting of such games is the "Build the Least Difference" game.

Another type of game that children enjoy is called "Target Number." Here is an example of a Target Number game. Ask the children to copy a sentence like the one shown in figure 2. Then, as digits are picked at random as in the above games, children record them as they choose in the frames of the sentence. When all the frames have been filled, students are to simplify the expression they have built. The player who has built the number nearest the target number, in this case 24, wins the game. A winning strategy for this game is not so obvious, but children who have "number sense" will have a distinct advantage.

Fig. 2

Such a game provides an interesting setting for the practice of many arithmetic skills. Of course, the pattern sentence can be deliberately designed to cultivate those skills in which the students lack proficiency.

Two more pattern sentences that might be of interest are shown in figure 3. (Of course, 0 should not occur in the denominators.) You can make up pattern sentences of your own, and children should be encouraged to suggest others.

Our students have always enjoyed these games, and we think that yours will too.

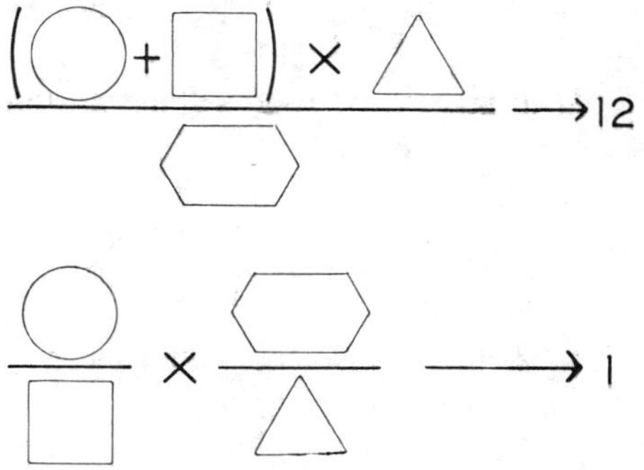

Fig. 3

The Witch's Best Game

C. WINSTON SMITH, JR.
College of Education, Wayne State University, Detroit, Michigan

As students encounter three-digit subtraction, many who previously have "mastered" two-digit subtraction falter, not only because of the complex regrouping required in some problems, but also because of the wider selection of regroupings to consider. While two-digit subtraction requires the student to choose between two alternatives —no regrouping or regrouping from tens to ones—three-digit subtraction can call for (1) no regrouping, (2) regrouping from tens to ones, (3) regrouping from hundreds to tens, or (4) regrouping from hundreds to tens *and* from tens to ones. This range of choices, when linked with the understanding and use of basic subtraction facts and regrouping itself, creates for some a task that is very perplexing.

An activity that encourages each child to search for the *best renaming,* while providing a gamelike atmosphere, is one that the students call "The Witch's Best Game" (Which Is Best). To begin the game the teacher chooses a three-digit number. After recording this on the board, she writes the four names used most often when subtracting from this number.

$$\begin{array}{r} 245 \\ -\boxed{129} \\ \hline \end{array} \qquad \begin{array}{l} (1)\ 200 + \ \ 40 + \ \ 5 \\ (2)\ 200 + \ \ 30 + 15 \\ (3)\ 100 + 140 + \ \ 5 \\ (4)\ 100 + 130 + 15 \end{array}$$

In some classes students assist by suggesting names and illustrating their choices on the twenty-bead abacus. Using this number as the sum (minuend), the teacher then writes an addend (subtrahend) and asks the class to choose the name from the board that they think would be the most helpful when subtracting. After surveying the four choices and deciding upon the best renaming, each student indicates his choice by raising one, two, three, or four fingers. After the class has shown its response to the first addend (subtrahend), the teacher may substitute another addend (subtrahend) and ask the class to indicate their choices for this new problem in the same manner. (For example, 162 may be substituted for 129, and a new example, 245 − 162, presented.) After five or six problems, the teacher may wish to change the sum (minuend) to vary the problem situation.

As the students raise their fingers indicating their choices, the teacher will have an opportunity to give immediate reinforcement with a nod, a smile, or a word of praise. With a perplexed look the teacher can encourage others to reconsider their answers. Observation of the total class response will aid the teacher in locating those students who are in need of individual help and will also point out areas which require reteaching to the entire class.

It is important throughout for the teacher to stress that there are times when two or more names may be appropriate—therefore the class is looking for a response that is *best* or *most helpful,* not one that is "right" or "wrong." Students who choose names that do not appear to be helpful to the rest of the class should be given an opportunity to explain their choices.

EDITORIAL COMMENT.—To provide additional reinforcement with three-digit subtraction problems, create a "Pick Your Problem" game as follows: Cover three shoe boxes with decorative paper and mark them *A*, *B*, and *C*. In box *A*, place cards that have only three-digit subtraction problems requiring no renaming. In box *B*, place cards that have three-digit subtraction exercises requiring one renaming. In box *C*, place cards that have three-digit subtraction examples requiring double renaming. Two to six players may participate at one time. Each student in turn spins a spinner marked *A*, *B*, and *C*, as shown. He then selects a card from the box indicated by the spinner and works the problem. If worked correctly, problems from *A* are worth 2 points, problems from *B* are worth 5 points, and problems from *C* are worth 10 points. The student who accumulates the most points in a specified number of turns or times is the winner. You may wish to have an answer key available for use if the students disagree on answers.

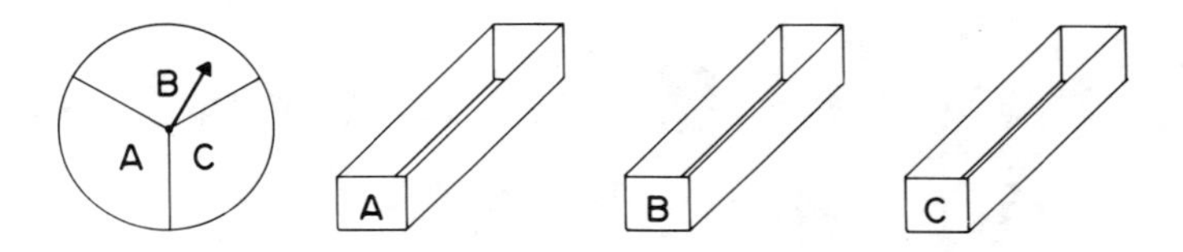

Beanbag-toss games may also be created to provide practice. For example, the following game board may be made from posterboard or plywood:

326	598	200	141	475	Sorry, Lose 5 Points
94	Bonus 5 Points	100	783	249	505
609	723	670	Bonus 10 Points	184	393
542	177	316	489	320	152
210	Sorry, Lose 3 Points	425	206	588	600

Have each student throw the two beanbags on the game board and then add (or subtract) the two numbers indicated by the beanbags. One point may be awarded for each correct answer. If a bag lands on a "Bonus" or a "Sorry" block, the player records the points accordingly and takes another turn. Encourage the students to work their problems on the chalkboard or overhead projector so the results may be verified.

A game for reviewing basic facts of arithmetic

EDWARD E. ARNSDORF

Sacramento State College, Sacramento, California

A delightful game to use in reviewing addition-subtraction and multiplication-division facts for whole numbers has been developed by a Sacramento State College education student, Mary Dunlap.

Materials for the game consist of forty-five cards cut from tagboard. Three-by-two-inch cards work well. The tagboard cards are numbered as shown in figure 1. Two different colors should be used for the numerals on the cards. With all top numerals one color and all bottom numerals another color, the cards can be arranged easily, right side up.

The game may be played by two, three, or four persons. To begin play, the deck is shuffled and five cards are dealt to each player. One card is turned face up and the remaining undealt cards are placed in a stack face down. The person at the immediate left of the dealer plays first.

Suppose the (6, 2) card (the one with the 6 on top and the 2 on the bottom) is the one that has been placed face up. The numbers on this card may represent $6 + 2 = 8$, $6 - 2 = 4$, $6 \times 2 = 12$, or $6 \div 2 = 3$. If the first player has a card containing two numbers whose sum, or difference, or product, or quotient is 8, or 4, or 12, or 3 respectively, he may play the card from his hand by placing it face up on the (6, 2) card. If he does not have such a card in his hand, he must draw the top card from the pack and then determine if it can be played. If the drawn card cannot be played, it is added to the player's hand and he draws again. The process of drawing from the pack continues until an 8, or 4, or 12, or 3 is found. When a card can be played, it is placed on the face-up pile and the next player takes his turn. For example, assume that the first person plays

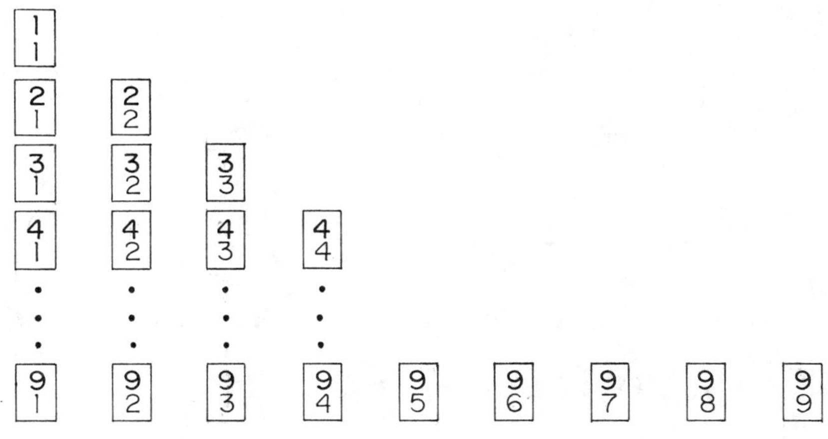

Fig. 1

the (8, 2) card, representing $8 \div 2 = 4$, on the opening (6, 2) card. The second player must now determine the new numbers that can be played. He may use any one of the four operations to name one of the new numbers, in this instance either 10 or 6 or 16 or 4.

Any time a card is played, another player may challenge the play by asking why that particular card can indeed be played. The game is won by the first person to play all his cards. If all the cards have been drawn and the game has not been won, all but the top card on the face-up, or discard, pile are reshuffled and the game continues.

Notice that the two numbers on some of the forty-five cards cannot be used to name four different whole numbers. The sum and the product may name the same number; or the product and the quotient may name the same number; or the operation of division may not yield a whole number. Since some cards are more likely to be playable than are others, there are opportunities to make strategic decisions.

Players of this game are continually mentally performing the operations of addition, subtraction, multiplication, and division. Hence this game can be both an effective and an enjoyable way to review basic facts of arithmetic.

EDITORIAL COMMENT.—Many familiar card games, such as Old Maid, War, Fish, and so on, may be adapted to serve particular needs for practice experience. For example, the Old Maid game could be renamed "Mathlete" and a set of cards constructed as follows for addition and subtraction facts.

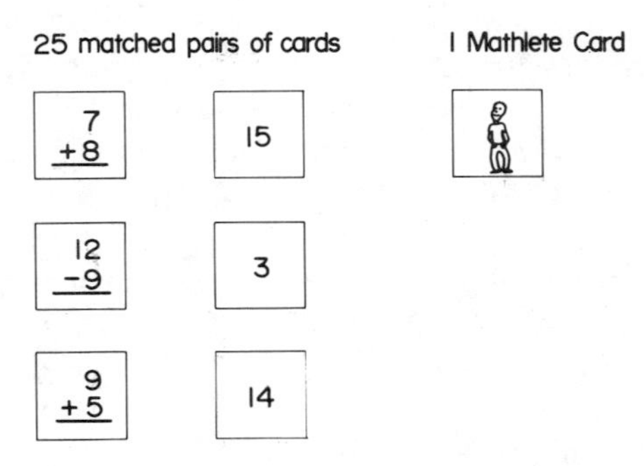

Each game could be played with 4, 5, or 6 players. All cards are dealt to the players, and the first player chooses a card from the hand of the player on his right. The second player chooses a card from the player on his right, and so on. Whenever a player gets a matched pair, he lays it down on the table. If correct, he earns a point; if not, he loses a point. The game continues until all matched pairs are on the table and a player is left with the "Mathlete" card. The player who is left with the "Mathlete" card loses a specified number of points (or is awarded a specified bonus).

"Contig": a game to practice and sharpen skills and facts in the four fundamental operations

FRANK W. BROADBENT

Syracuse University, Syracuse, New York

"Contig" is a game that intermediate-grade children love to play and with which they can challenge their older brothers, sisters, or parents. Four things are required—three dice, a score pad, markers, and a Contig board. The dice, pad, and markers are readily available, and a supply of boards can be made by reproducing the sample provided at the end of this article on sheets of paper.

Rules of the Game

1. Two to five players may play Contig.
2. To begin play, each player in turn rolls all three dice and determines the sum of the three numbers showing. The player with the *smallest* sum begins play. Play then progresses from left to right.
3. The first player rolls the three dice. He must use one or two operations on the three numbers shown on the dice. He is then allowed to cover the resulting number on the board with a marker. When he has finished his turn, he passes the dice to the player on his right. A player may *not* cover a number that has been previously covered.
4. To score in Contig, a player must cover a number on the board which is adjacent vertically, horizontally, or diagonally to another *covered* number. One point is scored for each *adjacent covered number.*
5. When a player rolls the dice and is unable to produce a number that has not already been covered, he must pass the dice to the next player. If he incorrectly passes the dice, believing he has no play when in fact he does have a play, any of the other players may call out the mistake. The first player to call attention to the error may place his marker on the proper uncovered number. This *does not* affect the turn of the player citing the error.
6. A cumulative score is kept for each player. A player is eliminated from further play in a game when he fails *in three successive turns* to produce a number that can be covered. When *all* players have experienced three successive failures to produce a coverable number, the game ends. The player with the highest cumulative score wins.

Variations of Contig

1. Use a one-minute egg timer to time the

CONTIG

1	2	3	4	5	6	7	8
9	10	11	12	13	14	15	16
17	18	19	20	21	22	23	24
25	26	27	28	29	30	31	32
33	34	35	36	37	38	39	40
41	42	44	45	48	50	54	55
60	64	66	72	75	80	90	96
100	108	120	125	144	150	180	216

turn of each player. This will tend to speed up the game.

2. Allow any player to challenge an opponent if the opponent does not choose the number that will score the maximum number of points. The challenger should then receive the difference between the number of points scored by the chosen number and the greater number of points that could have been scored.
3. For a faster game, allow only five turns for each player. The player with the highest score at the end of the fifth round would be the winner.
4. Allow students to play it as a solitaire game and attempt to score as many points as possible before experiencing three successive unsuccessful rolls of the dice.
5. To make the game easier, use a four-by-five array and the numbers 1, 2, 3, 4, 5, 6, 7, 8, 9, 10, 11, 12, 15, 16, 18, 20, 24, 25, 30, and 36. In this case, students use only *two* dice but play by the rules above.

Sample play (four players)

Suppose four players roll the dice in their respective turns as shown in table 1. Player 1 covers the 9 on the playing board ($2 + 3 + 4 = 9$) but does not score because no other numbers have been covered. Player 2 covers the 10 on the board ($[4 \div 2] \times 5 = 10$) and scores 1 point because the 10 is adjacent horizontally to the already covered 9. Player 3 covers the 11 ($[1 \times 5] + 6 = 11$) and scores 1 point because the adjacent 10 has already been covered. Player 4 covers the 2 ($4 - [6 \div 3] = 2$) and scores 3 points, one for the vertically adjacent 10 and one each for the diagonally adjacent 9 and 11.

Table 1

	Roll of dice	*Number covered*	*Points scored*
Player 1	2, 3, 4	9	0
Player 2	2, 4, 5	10	1
Player 3	1, 5, 6	11	1
Player 4	3, 4, 6	2	3

After the children have played the game for a while, questions like the following might be explored:

a) How were the numbers used in Contig selected?

b) What numbers would you use if you had two dice or four dice?

c) How many ways can you cover each number in Contig?

d) What is the highest score (without challenges) a player can make in a game?

e) Why are some numbers between 1 and 216 left off the Contig board?

f) Would it be possible to use all the numbers from 1 to 216 on a Contig board if the dice went from 1 to 10?

g) What is the probability of being able to cover 216 on your first throw?

h) Why is part of the Contig board still uncovered when everyone has passed three times?

EDITORIAL COMMENT.—You may wish to have each student write a number sentence to illustrate his answer after each play. This will give him some of the desired written practice and also make it possible for the other players to check his work more easily.

The questions suggested at the end of the article are excellent and should stimulate the students to do some analytical thinking.

A variation of the game could be created by using spinners instead of dice to generate the numbers. Another variation could be created by using stacks of cards and having a player draw a card from each stack to start play.

Cardematics I—using playing cards as reinforcers and motivators in basic operations

SEYMOUR METZNER and
RICHARD M. SHARP
California State University, Northridge, California

Materials for pupil activity-learning in all areas, and especially mathematics, are difficult to find even though more educators are advocating this approach to learning. Some teachers have used considerable initiative and ingenuity to develop and adapt inexpensive materials for pupil activity-learning. One of the most inexpensive, easily obtained, and adaptable items that can be used is an ordinary deck of playing cards.

Six games or activities that have relevance to the learning and retention of basic mathematical skills and understandings are described here. The games motivate students and add a certain "pizzaz" to the mathematics program. Each of the activities requires three decks of playing cards. All nonface cards have their numerical face value, with an ace as 1. Play can be limited to the nonface cards, or if face cards are used, the jack, queen, and king can be assigned the numerical values of 11, 12, and 13 respectively. Other numerical values can be given the face cards, if desired.

All of the described activities suggest teams of students. Although any size team could play the game, there is greater student participation if the teams are made up of no more than two or three students. Small teams can better discuss possible answers to problems.

All games begin the same way. After the teams have been determined and the decks of cards are shuffled, each team draws three cards. Then the teacher draws a card (or cards) and shows it to all of the teams. The procedures that follow then vary with the games, but whatever the procedure, it is repeated a predetermined number of times. In each game, the team with the largest "bank" at the end of play is declared the winner. And in every game, as soon as a team "banks" cards, it draws to replace the banked cards. A team always begins play with three cards in its hand.

These activities are self-correcting if each team is paired with another team to verify each other's answers before being allowed to bank cards or draw additional cards. The teacher may act as a court of last resort in questionable decisions.

Prime-o

Objective

Identification of prime numbers, addition and subtraction practice

Procedure

Each team, by adding or subtracting one of their cards from that displayed by the teacher, should try to reach a prime number. It is possible that more than one of a team's cards can fill this function, but the cards must be used separately. All cards that are used successfully to reach a prime number are added to the team's bank.

Round-up

Objective

Practice with basic number operations

Procedure

Each team tries to use as many of its cards as possible to reach the number drawn by the teacher. The team may use any one or any combination of the basic operations. (For example, if the teacher's card is a 4 and a team has a 5, an 8, and a 10, then the team could use all three cards by showing that $5 \times 8 = 40$ and $40 \div 10 = 4$.) All cards that are used are added to the team's bank.

Magic number

Objective

Addition and subtraction practice

Procedure

The teacher announces a magic number and then draws a card. The teams add or subtract one of their cards from the card drawn by the teacher to reach the magic number. (For example, if the magic number is 8 and the card drawn by the teacher is a 5, then a team would need a 3 to add to 5 to get 8.) More than one card can be used by a team as long as each card is considered separately. All cards that are used are added to the team's bank.

Divvy-do

Objective

Division review with remainders

Procedure

The teacher draws *two* cards from the deck; the students then divide the larger number by the smaller. Each team must then see if they can get the same quotient and remainder by dividing any two of the cards in their hand. A pair of cards that can be used to get the appropriate quotient and remainder are added to the team's bank.

Even-steven

Objective

Division practice with no remainders

Procedure

The teacher draws a card from the deck. Each team then tries to combine one of its cards with the card displayed by the teacher to make a division problem with no remainder. The card drawn by the teacher can be used either as a divisor or a dividend. (For example, if the teacher draws a 4, a team could use 2, since $4 \div 2 = 2$, and an 8, since $8 \div 4 = 2$.) A team may use more than one card as long as each card is used separately. All cards that are used are added to the team's bank.

Multi-match

Objective

Multiplication review

Procedure

The teacher draws *two* cards from the deck. The students calculate the product of the numbers on the drawn cards and then teams try to use any two or three of their cards to get the same product. (For example, if the teacher draws a 7 and an 8, a team could use a 2, 4, and 7 to get the same product.) All cards that are used are added to the team's bank.

This game can be varied by having teams compete to see which team can get *closest* to the product of the cards drawn by the teacher. The team that gets the nearest product banks the cards it uses.

Conclusion

All of these activities should be viewed simply as starting points. Teachers can adapt these games or activities to their own particular situation or to class needs and interests. The general format can be extended to areas not covered by these half-dozen activities.

Green Chimneys poker

T. RISTORCELLI
Green Chimneys School, Brewster, New York

This game can be played by two, three, four, or more players using an ordinary deck of cards. Six cards are dealt to each player and the remaining cards, placed face down, form a drawing pile. The idea is for each player to try to rid himself of all his cards before any of the competing players by playing his cards one at a time consistent with the rules of the game.

The first player (clockwise from the dealer) leads a card face up on the table, thus starting a playing pile. Each succeeding player in his turn plays a card on the playing pile, but when he plays the card he must multiply the number on his own card by the number on the card that is face up on the playing pile. If he multiplies correctly, the player will have gotten rid of one card and the play goes to the next turn. If he makes a mistake in the multiplication, he must take back the card he played and pick up an extra card from the drawing pile.

All number cards stand for the number that is written on them and face cards are valued as follows: a jack is 11, a queen is 12, and an ace is 1. The kings and jokers are free discards at the beginning of a player's turn and both have zero point value. Jokers are wild. Three or more of a kind and a run of three or more cards of the same suit are also free discards at the beginning of a player's turn.

The player who first disposes of all his cards wins the hand. Points can be scored in one of two different ways. The winner of the hand can get credit for all the points held by the other players when the hand ends, in which case a high score is the goal of the game. Or, when a player goes out, the other players can be penalized the number of points they hold in their hands. In the latter case a low score is the goal of the game.

For players who are not yet up to the multiplication tables of 12, 11, 10, 9, or others, the corresponding cards are simply removed from the deck before starting the game.

DIFFY

HERBERT WILLS

An associate professor of mathematics education at Florida State University in Tallahassee, Herbert Wills has directed NSF summer institutes in computers, and last year was director of an NSF Academic Year Institute at Florida State University.

The game DIFFY is an academic game that provides intrinsically interesting drill experiences and allows for individual differences. Moreover, teachers need not construct any of the exercises. The student himself initiates the drill, and the nature of the game provides for variety in the numbers encountered. Thus the activity promotes self-generated drill. This feature is commendable, since students won't run out of material and they become personally involved in their work. Besides automatically adjusting to students of diverse abilities within a given classroom, DIFFY also provides productive academic recreation at every grade level. This flexibility stems from the fact that the game may be played with a variety of numbers: whole numbers, nonnegative rationals, integers, all rationals, or all real numbers.

To introduce a class to DIFFY, give each student a supply of game sheets, illustrated in figure 1, then play a sample game on the chalkboard or with the use of an

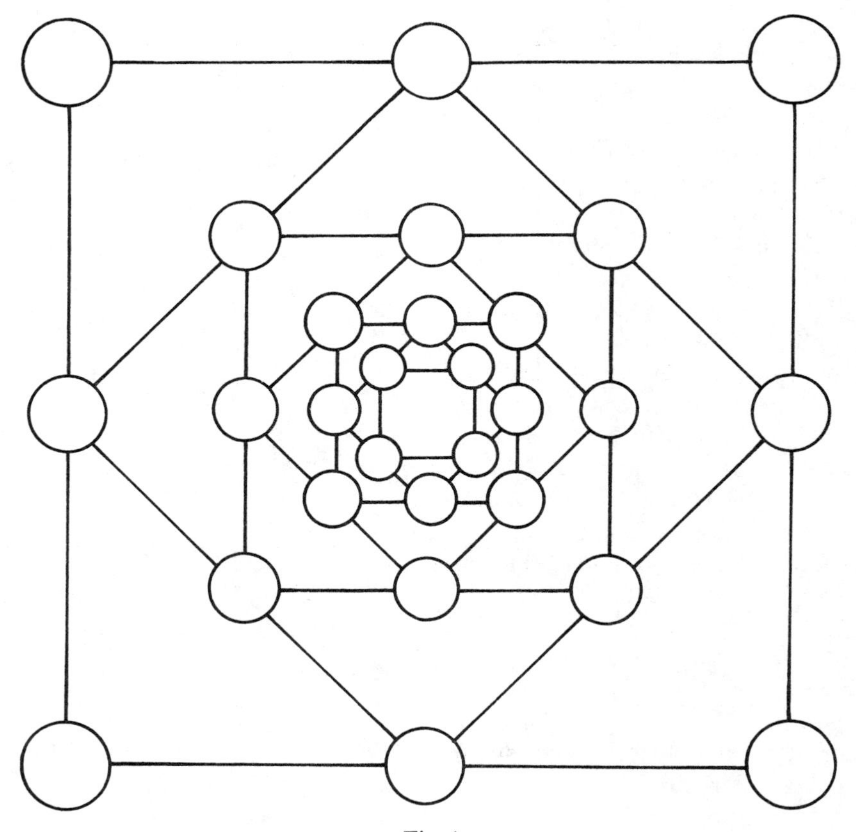

Fig. 1

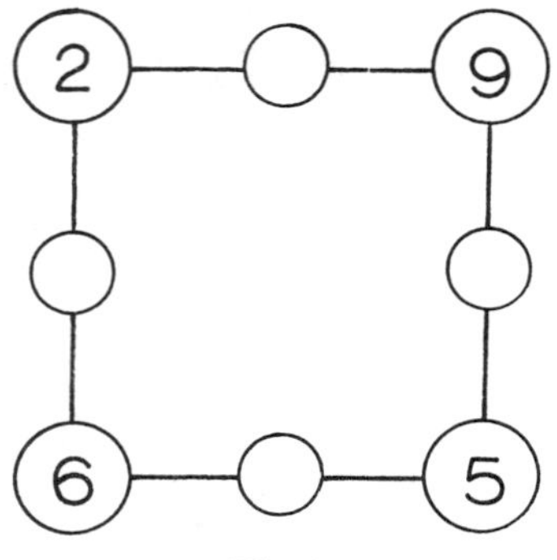

Fig. 2

overhead projector. This provides each student with the information needed to proceed independently and stimulates "Let me try it myself" interest. The game is started by placing numbers in the four outermost corner cells. Players may choose any numbers they wish for this purpose. Let us suppose that students chose the numbers shown in figure 2 to start a game.

The adjacent corner cells determine the number to be written between them. In this particular game three of the numbers generated have been entered in figure 3. See if you can guess the remaining entry.

Once these middle cells have been filled, we have determined a new four-cornered diagram (fig. 4) which, in turn, has its middle cells vacant.

Players fill these cells the same way that the previous ones were filled. The only rule in DIFFY, as you probably have guessed, prohibits subtracting a larger number from a smaller one. Younger students, who have not learned to work with negative numbers, will need only to have this rule stated. More mature students can be introduced to the absolute-value function and apply it to the differences. The game being considered ends at the next move, since it yields a zero in *each* middle cell, as seen in figure 5.

This game took only two moves before zero was written in each middle cell. If zero appears in some of the middle cells but not all of them, the game continues. The game sheet distributed to the students has room for six moves (see fig. 1). Should a player reach the innermost set of cells without writing zero in each of them, he wins.

It is gratifying when a student introduces numbers in the game that produce problems that challenge him. One is also pleased with the care players take in doing their calculations or checking their work. This care assures them that a declared win is truly a success and not a fluke resulting from miscalculation. Also, it is not uncommon to see students checking the work of others, attempting to find an error that will refute a proposed win. This checking is essentially additional drill, and carefully done drill at that! Compare this with the lack of purpose and motivation of regular drill for drill's sake.

Besides providing a vehicle for self-generated drill in subtraction, DIFFY may be used to introduce investigations that are more intellectually stimulating. For example:

> Start a sample game that requires several moves but stop short of its culmination. Instruct the pupils to continue independently. They will find by themselves that zero is the only number that gets listed beyond some move. Moreover, they can be encouraged to try additional games allowing them to discover for themselves that a set of zeros eventually results. This may encourage some to keep trying again and again in an effort to elude an arrival at zeros.
>
> Does the order in which the numbers are listed in the corner cells affect the number of moves in a game?

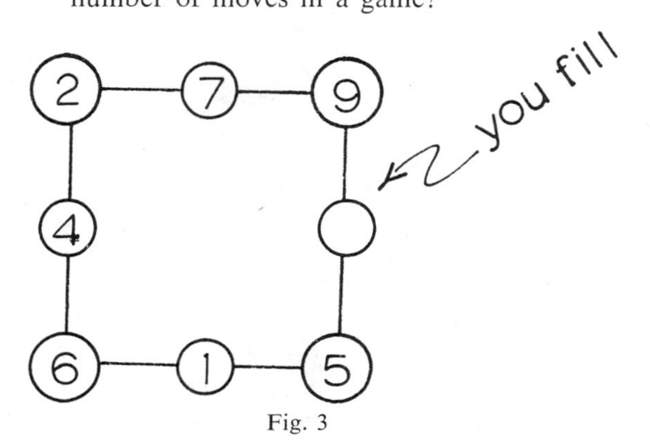

Fig. 3

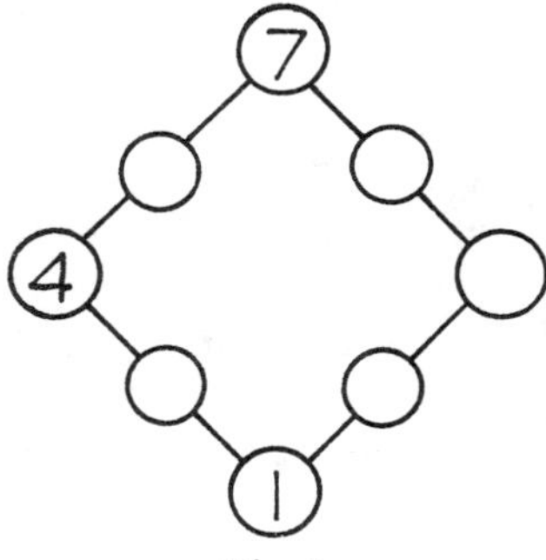

Fig. 4

Once a player has won the game a few times (six moves without all zeros) have him try for seven moves, eight moves, nine moves, and so forth. A continuing contest can be held to see who can come up with the longest game. Others can then try to do at least one better. This will always be possible, since there is no "longest game."

Find a winner using only single digits in the starting cells. Find another.

How does the number of moves of one game compare with the number of moves in a second game whose starting cells are entered with doubles of the entries of the first game? With triples? With other multiples?

Another time add 5 or some other number to each of the initial entries and compare the number of moves.

For younger students, exhibit a completed six-move game and ask them to come up with one that takes exactly five moves. Then four moves, and so forth.

Let older students try fractions.

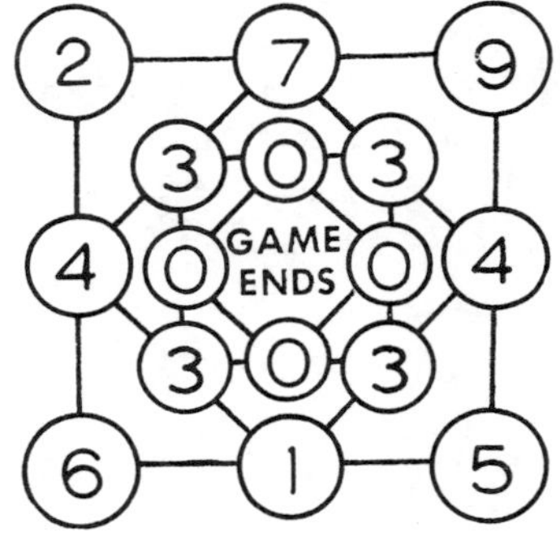

Fig. 5

DIFFY not only provides a good vehicle for drill in subtracting rational numbers but also gives practice in comparing them, since the rule does not permit a larger number to be subtracted from a smaller. Other skills drawn on in this activity involve multiplying, reducing fractions, and recognizing equivalent fractions.

Try DIFFY with any quadruple of integers—positive, negative, or zero.

Try π, e, $\sqrt{91}$, 19. Before you start, consider whether you think it will ever end. Observe the manipulative skills required in this activity.

One need not stop here, as there are many more intriguing aspects about DIFFY. Not the least of these is the fact that it involves an iterative process and so lends itself very well to the use of a computer. Many students who are familiar with BASIC or other programming languages may enjoy putting DIFFY on the computer. For this purpose, though, a change in format proves useful. We can show this by appealing once more to a sample game; this one is shown in figure 6.

```
   5 − 2
2  9  6  5
 7  3  1  3
  4  2  2  4
   2  0  2  0
    2  2  2  2
     0  0  0  0
```

Fig. 6

In this format the starting numbers are listed horizontally. The "distance" between neighboring numbers is listed directly below the space between them. The last entry in each row other than the first row is acquired by finding the distance between the first and last entries in the preceding row. This format also has obvious advantages in a search for long games, since one need not concern himself with expanding the original game chart.

The computer may be programmed to print out each intermediate step or just the number of steps required. One may wish to program the game so that the player "inputs" the number of steps to beat before he plays the game. The computer would then inform the player whether he won or not. The variations available provide opportunities to apply programming skills at many levels.

Since DIFFY proves so useful for drill in subtraction, might we not try to get additional mileage out of it by considering another operation?

Indeed we might. We can change our operation and still use the same playing chart. This time, instead of subtracting, we shall divide. Furthermore, as in DIFFY we never subtracted a larger number from a smaller one, in DIVVY we shall never divide a smaller number by a larger. This time instead of ending with zero in the last set of cells we shall end with something else. Can you guess what it will be? Go ahead and guess; then check your guess by trying a game of DIVVY. Of course, in playing DIVVY, be sure not to enter any zeros. You wouldn't get very far, since we can't divide by zero.

DIFFY and DIVVY are but two academic games that stimulate self-generated drill that is purposeful to the student. There are several other such games having the same fine attributes, among which is that of providing for individual differences. The slower learners can try smaller numbers and proceed at their own pace—and win in time—while the more clever student may choose numbers as challenging as his heart desires and try for a win involving more moves than his previous one.

In my opinion, it is important that such activities become more widely known, since we are beginning to hear cries that "what our children need today is old-fashioned drill." True, they need drill, but not the barren set of exercise after exercise devoid of intellectual purpose. Sometimes it's better to give a child his medicine with a sugar coating. Shouldn't we give children proper drill and maintain their interest in mathematics at the same time?

EDITORIAL COMMENT.—There are many ways to provide interesting practice in elementary school mathematics. For example, have the students look at a calendar for a given month and select any four adjacent dates in a square array, such as

9	10
16	17

Lead the students to observe that 9 + 17 = 16 + 10. Have the students check to see if they can find similar arrangements on a calendar that do not work in this way. Test the idea on a three-by-three arrangement, such as

12	13	14
19	20	21
26	27	28

Does 12 + 20 + 28 = 26 + 20 + 14?

You can also develop coin puzzles such as the following:

A box contains 3 coins. The total value of the coins is 20¢. Name the coins in the box.

A box contains 6 coins. There are 3 different kinds of coins in the box. The total value of the coins is 32¢. Name the coins.

Magic triangles—practice in skills and thinking

DONALD L. ZALEWSKI

Presently an assistant professor of education at the University of Nebraska at Omaha, Donald Zalewski teaches mathematics methods and supervises student teachers. The article was written during his investigation of mathematical problem solving while he was at the University of Wisconsin in Madison.

Can you place each of the numbers 1 through 6 in the triangular chain of circles in figure 1 so that each side has the same sum? If you are successful, congratulations, you have just created one example of a magic triangle. These simple figures may not have as many "angles" as the well publicized magic squares, but they possess a few powers that may be interesting to both you and your students. Obviously, important practice in addition skills is built into the puzzle and it is possible that you can at the same time stimulate the formation of problem-solving strategies.

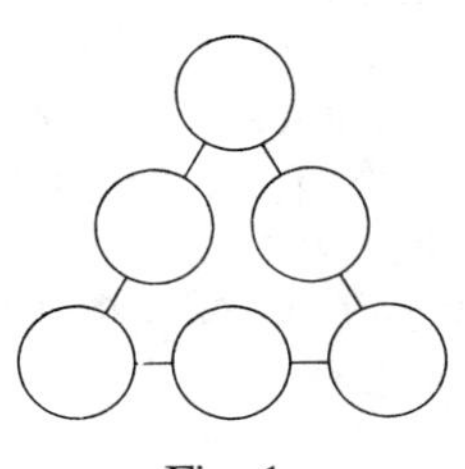

Fig. 1

A *magic triangle* is a triangular chain of circles (an equal number of circles on each side) containing numbers such that the sum on one side is equal to the sum on any other side. "Level one" magic triangles contain six circles in the chain. If only the first six positive whole numbers are inserted in the circles, the configuration will be referred to as a *basic magic triangle*. Although magic squares and magic triangles both provide practice in computation and problem solving, the triangle problem holds two distinct advantages for the beginner: (1) There are fewer sums to find and check, and (2) the simple design offers children an easier opportunity to use elementary common sense and logical thinking to solve these puzzles.

Now, back to the problem. Don't feel too smug if you found one correct answer—there are four common sums possible using the numbers 1 through 6. Can you find the others? The four solutions are given in figure 2. Since each solution can be written in six possible forms, any forms that are related by a rotation or flip will be considered the same solution. Figure 3 gives the six possible representations for solution *A* from figure 2.

When posing this problem to elementary school children, it might be better to indicate which sum (9, 10, 11, or 12) they should seek. For instance, you might ask them to try to get 9 as a common sum. If a child incidentally finds one of the other possible sums, he should certainly be commended for his efforts and you should take the opportunity to challenge the children to search for other solutions.

Sometimes, hints on possible strategies for filling in the circles may be necessary to help children get started or to help them organize nonproductive trials. One helpful suggestion is, "Pick three numbers and put them in the corner circles (vertices), then try to arrange the remaining three numbers to get equal sums." A more direct cue might have the children choose 1, 2, and 3 as the vertices and then arrange 4, 5, and 6 as interior points until common sums are found. You might also suggest that the children keep a record of the numbers they have tried as vertices and interior points to avoid duplication of effort.

You may wonder if there is a method to this madness, or if trial and error is the only procedure that can be used. Inspection of the trios of numbers that formed the vertices and interior points of the solutions in figure 2 produces identifiable patterns. If we horizontally order the numbers 1 through 6 from smallest to largest, we see that solution *A* has the first three numbers as vertices and the last three as interior points. (Fig. 4) Solution *D* has the roles of these two trios interchanged. Solutions *B* and *C* share a common partitioning of the trios into the odd and even numbers. (Fig. 5) Some children may notice these regularities, but most may need to be asked if they see any patterns that might help them solve the puzzle.

After identifying the patterns of number trios, the child has an opportunity to apply mathematical reasoning when trying to form magic triangles. Assume that three numbers have been selected for the vertices. Then the interior points can be identified by inspecting the incomplete sums. For

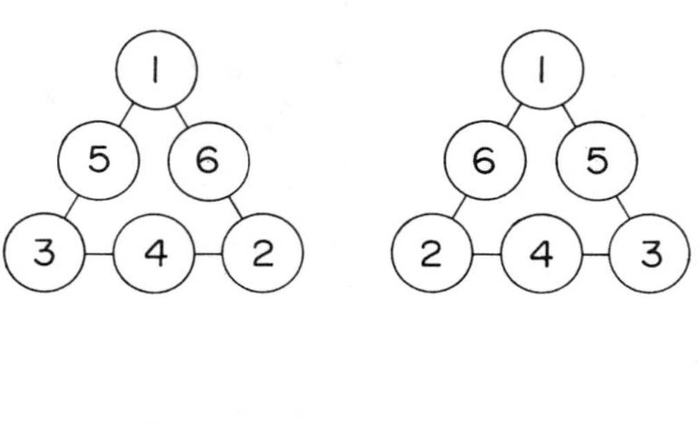

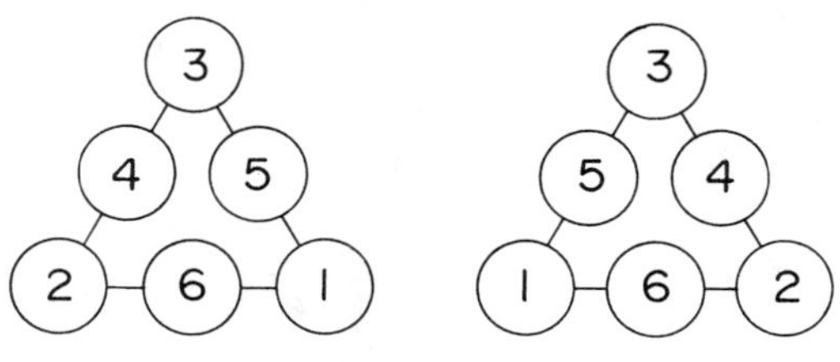

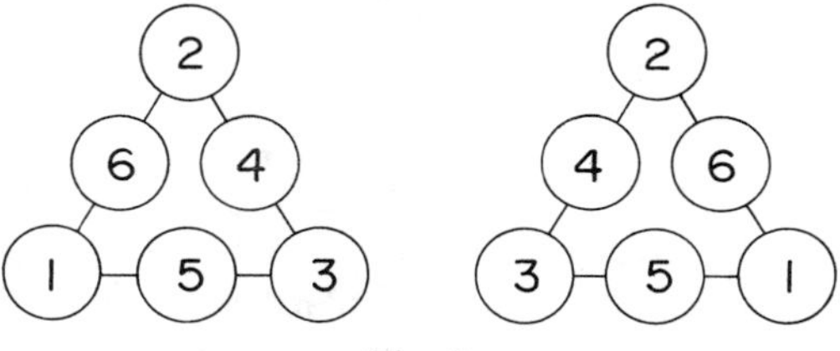

Fig. 3

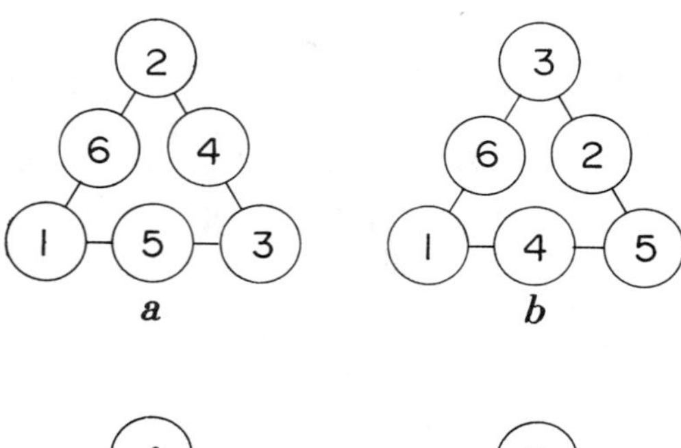

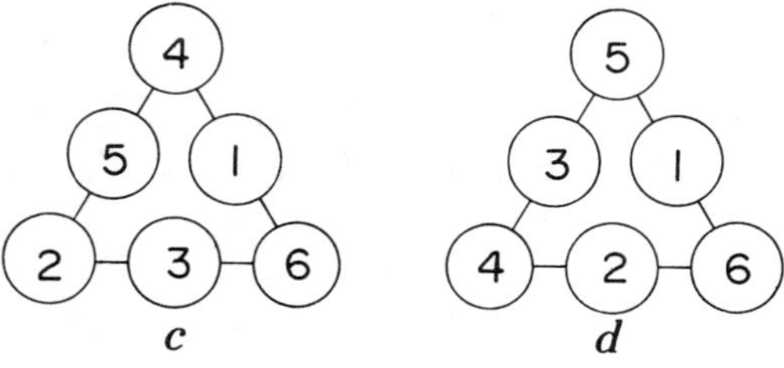

Fig. 2

example, let 1, 2, and 3 be selected as vertices. (Fig. 6) Since 2 and 3 form the largest sum thus far, that side must receive the smallest of the remaining numbers, namely, 4. Similarly, since 1 and 2 form the smallest incomplete sum, that side must get the largest of the three interior numbers, namely, 6. The third side receives the remaining number, 5.

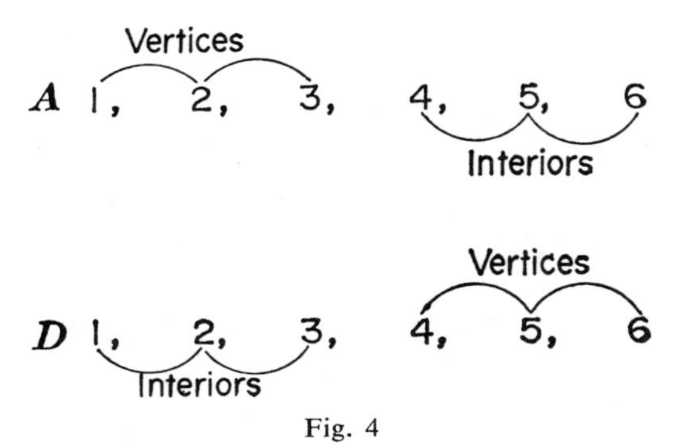

Fig. 4

When an incorrect trio is chosen for the vertices, the same reasoning reduces the number of trials necessary to eliminate it. For example, if 1, 2, and 4 were chosen for vertices (fig. 7), then 3 must be inserted between 2 and 4, and 6 must be put between 1 and 2. The two completed

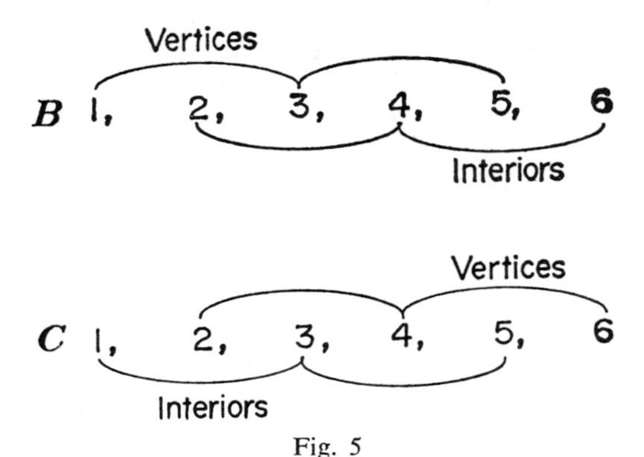

Fig. 5

sides form a common sum of 9, but putting 5 between 1 and 4 gives a different sum of 10. Since no other arrangement of 3, 5, and 6 as interior points is logically possible, we must go back and try a new trio of numbers for the vertices.

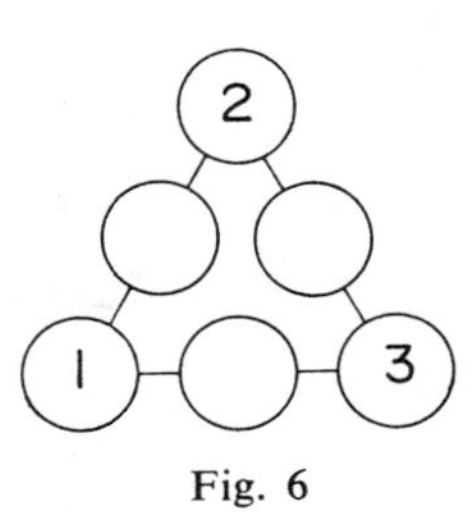

Fig. 6

If you have children who enjoy this type of mathematical problem, you need not stop at the basic magic triangles in figure 2. For enrichment, two or three digit numbers can be used. You may devise your own combination of numbers, or better yet, have children find some. There are an unlimited number of combinations, but any six consecutive whole numbers will suffice.

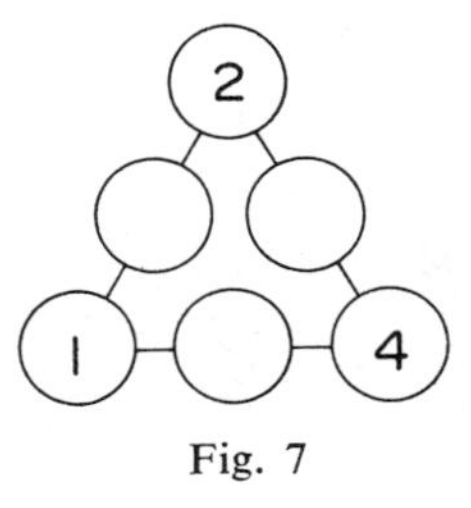

Fig. 7

Furthermore, the use of six consecutive integers permits the children to check the hypotheses formed and patterns observed in the basic magic triangles. (You might notice that new magic triangles can be built systematically by adding or subtracting the same number to each entry of a basic magic triangle.)

Additional challenge can be added to this problem by looking at a "level two" magic triangle consisting of nine circles. (Fig. 8) The numbers 1 through 9 would be used in attempting to form the equal sums of basic level-two magic triangles. You can decide whether to consider answers with the same vertices but with different arrangements of interior points (fig. 9) as the same or separate solutions. The same questions about possible sums, patterns, and strategies can again be raised.

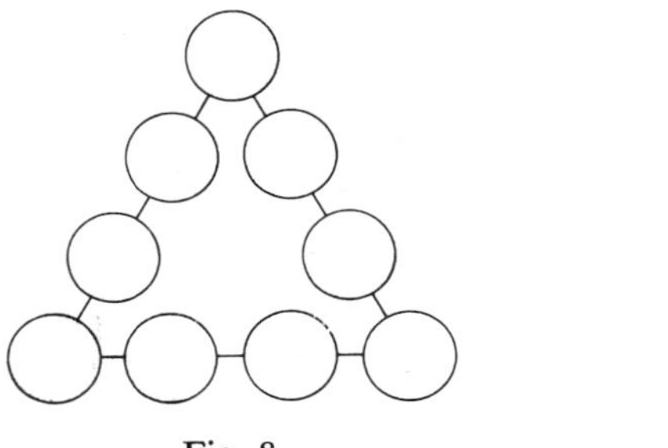

Fig. 8

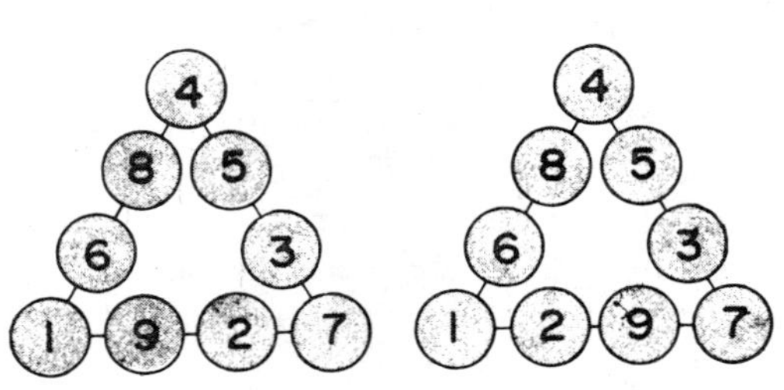

Fig. 9

Children need much practice in skill development and maintenance. Building magic triangles can be an interesting respite from routine drill. It can also provide opportunities for children to practice the searching, guessing, and thinking that is essential for problem solving.

References

Brown, G. "Magic Squares, More than Recreations." *School Science and Mathematics* 66 (1966): 23–28.

Freitag, H. T., and A. H. Freitag. "Magic of a Square." MATHEMATICS TEACHER 63 (January 1970): 5–14.

Gorts, J. "Magic Square Patterns." ARITHMETIC TEACHER 16 (April 1969): 314–16.

Keepes, B. D. "Logic in the Construction of Magic Squares." ARITHMETIC TEACHER 12 (November 1965): 560–62.

May, L. J. "Enrichment Games Get Pupils to Think." *Grade Teacher* 83 (1966): 53–54.

McCombs, W. E. "Four-by-four Magic Square for the New Year." ARITHMETIC TEACHER 17 (January 1970): 79.

Munger, Ralph. "An Algebraic Treatment of Magic Squares." MATHEMATICS TEACHER 66 (February 1973): 101–9.

EDITORIAL COMMENT.—Puzzles are excellent devices to stimulate and interest many students. As a variation of the "magic triangle," you may wish to give a student nine square pieces of blank paper and instruct him to number the paper squares in order, beginning with any whole number (you may wish to specify a beginning number in order to have the student practice at a specific level of difficulty). After the student has numbered the nine paper squares, instruct him to arrange the squares in a 3×3 array so that—

1. he gets the same sum in each row (horizontally);
2. he gets the same sum in each column (vertically);
3. he gets the same sum in each diagonal;
4. he gets the same sum simultaneously in each of the three directions, horizontally, vertically, and diagonally.

The difficulty level of this activity could be decreased by using only four paper squares (2×2 array) and the difficulty could be increased by using sixteen (4×4 array) or twenty-five (5×5 array) paper squares.

Easy construction of magic squares for classroom use

JOHN CAPPON, SR. *Willowdale, Ontario, Canada*

Mr. Cappon is a retired elementary school principal from Amsterdam, Holland. He also taught arithmetic at a private teachers college in Winschoten, Holland. In Canada he is an adviser to a private school.

In the December, 1963, issue of THE ARITHMETIC TEACHER Bryce E. Adkins gave six rules for construction of odd-cell magic squares.[1] His result for a five-by-five magic square is given in Figure 1. It is

17	24	1	8	15
23	5	7	14	16
4	6	13	20	22
10	12	19	21	3
11	18	25	2	9

Figure 1

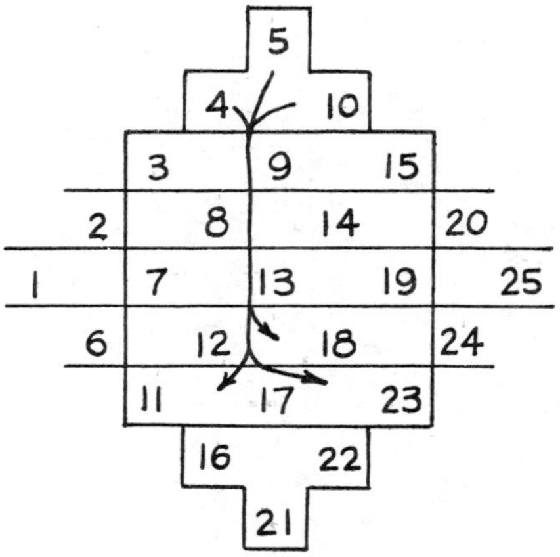

				5				
			4		10			
		3		9		15		
	2		8		14		20	
1		7		13		19		25
	6		12		18		24	
		11		17		23		
			16		22			
				21				

Figure 2

noticed that all horizontal rows, vertical columns, and diagonals have the total of 65.

Actually, the construction of this scheme can be simplified considerably and is easier to remember when certain numerals are replaced in symmetric positions outside the borders of the square as is customary, for example, in the calculation of determinants of the third rank.

A European method proceeds as follows: Cells are added outside the square by building a pyramid on each border line and placing the digits in the order illustrated in Figure 2 for a five-by-five square. Then, the extra cells of the pyramids are used to fill the empty cells within the square along the opposite border. The result is shown in Figure 3.

3	16	9	22	15
20	8	21	14	2
7	25	13	1	19
24	12	5	18	6
11	4	17	10	23

Figure 3

Although the integers 1–25 were used in the example, this is not necessary. Any sequence of numbers which is part of an arithmetical progression will suffice. Moreover, as Frances Hewitt[2] has shown for a

[1] Bryce E. Adkins, "Adapting magic squares to classroom use," THE ARITHMETIC TEACHER, X (December, 1963), 498–500.

[2] Frances Hewitt, "4×4 magic squares," THE ARITHMETIC TEACHER, IX (November, 1962), 392–395.

four-by-four square, there can be gaps in the progression between the parts of the rows. This holds also for all odd-cell magic squares; e.g., the sequence of numbers for a three-by-three square can be chosen as 2, 5, 8 (11, 14), 17, 20, 23 (26, 29), 32, 35, 38, where the numbers in parentheses are omitted (Figs. 4a and 4b).

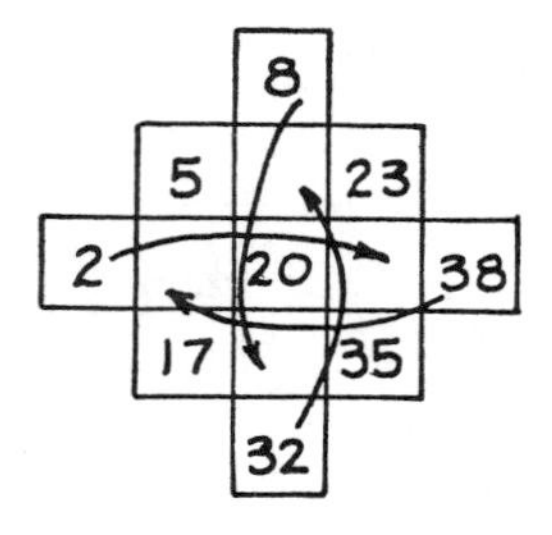

Figure 4a

5	32	23
38	20	2
17	8	35

Figure 4b

Furthermore, the gaps can be taken as any number, positive or negative, if they

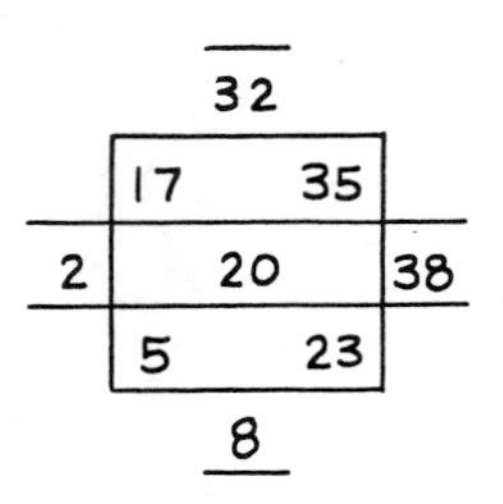

Figure 5a

17	8	35
38	20	2
5	32	23

Figure 5b. *This is the reverse of* 4b!

are all the same, e.g., 2, 17, 32; 5, 20, 35; 8, 23, 38. The gap here is 27 (Figs. 5a and 5b).

All these difficult cases can be summarized in one general rule: The sequence of numbers in the "pyramidally extended square" is formed by two arithmetical progressions, one determining the numbers from the most left-placed cell to the upper cell, and another one determining the numbers from the most left-placed cell to

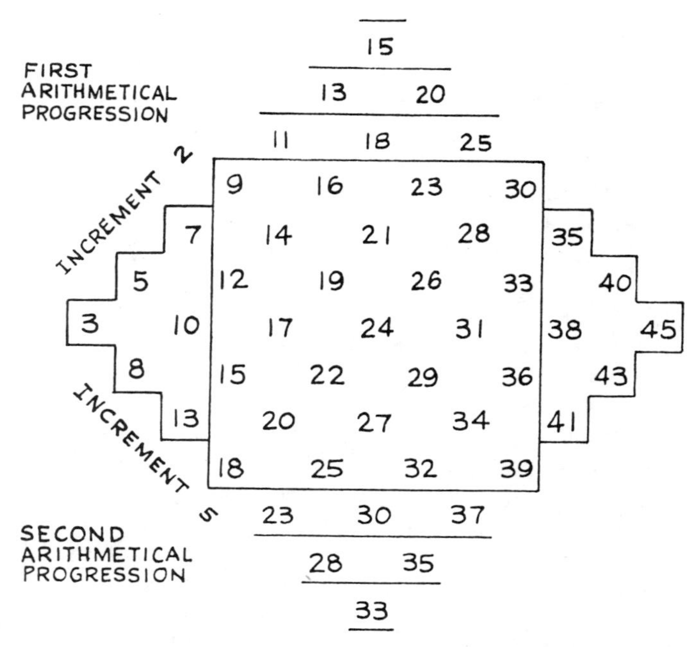

Figure 6a

9	23	16	30	23	37	30
35	14	28	21	35	28	7
12	40	19	33	26	5	33
38	17	45	24	3	31	10
15	43	22	15	29	8	36
41	20	13	27	20	34	13
18	11	25	18	32	25	39

Figure 6b. *The magic square formed by Figure* 6a.

the bottom cell. Both arithmetical progressions can be taken arbitrarily. (See Figs. 6a and 6b.)

The rule given by Frances Hewitt, that a new magic square can be formed from a given square by adding or multiplying the digits with a certain number or fraction, follows immediately from our general rule.

Adkins gave an easy rule for a four-by-four magic square formed from the numbers 1–16 placed in four rows of four digits. The series of the numerals on each diagonal must be reversed. Again, it is not necessary to restrict oneself to the numbers 1–16. In fact, we can again build up the sequence of numbers from two arithmetical progressions, one denoting the order in the first row and the other the order in the first column. (See Fig. 7a.)

Adkins' rule also applies to a square of these numerals; the square of Figure 7a is converted to a magic square by reversing the diagonals. (See Fig. 7b.)

Practical applications

For adopting magic squares to classroom use it is desirable to have an easy method that enables us to construct a magic square for any given number. Most easily one may employ a diagonal for the construction of the auxiliary figures of either the odd-cell squares or the four-by-four squares. The elements of this diagonal are always part of an arithmetical progression.

In the odd-cell squares (see Fig. 6a) the increment of the horizontal diagonal (7) equals the sum of the increments of the border rows (2 and 5 respectively in the figure), while the increment of the vertical diagonal (3) equals the difference of the border row increments. The same applies to the oblique diagonals of the four-by-four squares (see Fig. 7a).

Furthermore, the sum of the first and last elements of either diagonal is an important quantity. For the odd-cell squares this sum equals twice the term in the middle, which in turn equals the magic sum divided by the rank of the square (see Fig. 6a: $[3+10+17+24+31+38+45]\div 7=24$). In the four-cell squares the sum of the first and the last elements of the diagonal equals half the total of the four elements. With the knowledge of this it is possible to construct any magic square with a given number as totals of rows, columns, or diagonals.

Columbus discovered America in the year 1492; let us make a four-by-four magic square with totals 1492. Starting with the diagonal of the auxiliary figure, the first and the last numbers of the diagonal must equal $1492\div 2=746$. The last number of the diagonal equals the first one plus three increments, from which it follows that the difference of the num-

INCREMENT 5 (across); INCREMENT 3 (down)

2	7	12	17
5	10	15	20
8	13	18	23
11	16	21	26

Figure 7a

26	7	12	11
5	18	13	20
8	15	10	23
17	16	21	2

Figure 7b

bers must be divisible by three (if we want to avoid fractions). The numbers 100 and 646 suffice. The increment is then $(646-100)\div 3=182$. The entire diagonal is therefore 100, 282, 464, and 646. Further, we split this increment in two parts, e.g., 82 and 100, and use these parts to construct the first row and the first column. Then the rest is known. The example is demonstrated in Figures 8a and 8b.

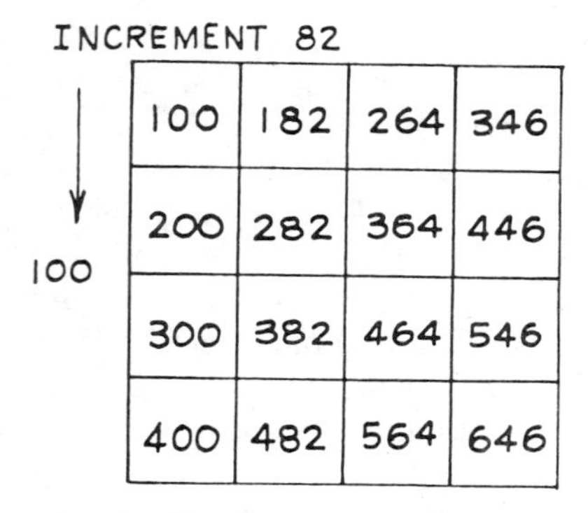

Figure 8a. Auxiliary square.

646	182	264	400
200	464	382	446
300	364	282	546
346	882	564	100

Figure 8b. Magic square after reversing diagonals.

After the teacher has made up the square the children check the results by performing the fifteen possible additions: 4 rows, 4 columns, 2 diagonals, 4 two-by-two squares in the corners, and 1 two-by-two square in the middle. By the choices of the first and last numbers of the diagonal and the increment, the teacher can vary the difficulty of the arithmetic calculation according to the progress of the class.

The year 1964 consists of 19 and 64. We shall make a magic square for 19 and 64. The number 19 is a prime number. Hence, we must avail ourselves of fractions. In the auxiliary figure for the five-by-five square, one places the number $19\div 5=3.8$ in the middle. (See Fig. 9a.)

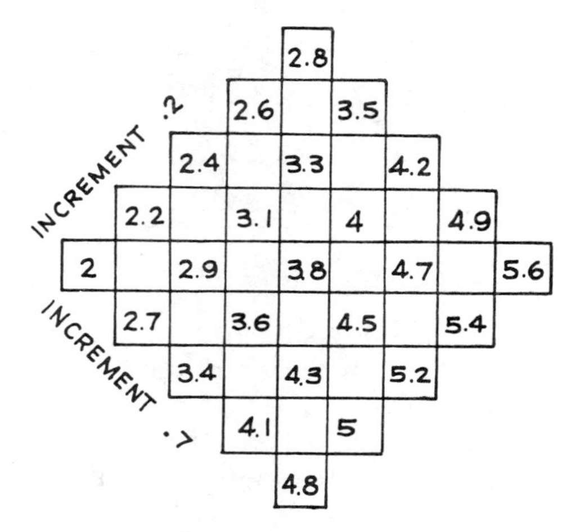

Figure 9a. Auxiliary figure for magic number 19.

The left-cell of the horizontal diagonal is less than 3.8, e.g., 2. The increment of the diagonal is then fixed, viz. $(3.8-2)\div 2=0.9$. The complete diagonal is now known. Then the increment 0.9 is split into two numbers, e.g., 0.2 and 0.7, which determines the border rows. We chose these numbers so that the entire square contains all different numbers. The resulting magic square is Figure 9b.

2.4	4.1	3.3	5	4.2
4.9	3.1	4.8	4	2.2
2.9	5.6	3.8	2	4.7
5.4	3.6	2.8	4.5	2.7
3.4	2.6	4.3	3.5	5.2

Figure 9b. The magic square of 19.

The magic square for 64 is made as the one for 1492. The diagonal is 4, 12, 20, and 28. The increment (8) we split in 7 and 1 (Figs. 10a and 10b).

If interested in the magic squares for 32 or 16, one can utilize the magic square of 64 by dividing all terms by 2 and 4 respectively. (See Figs. 10c and 10d.)

INCREMENT 1 → 7

4	11	18	25
5	12	19	26
6	13	20	27
7	14	21	28

Figure 10a. *Auxiliary square for* 64.

28	11	18	7
5	20	13	26
6	19	12	27
25	14	21	4

Figure 10b. *Magic square after diagonal reverse.*

14	$5\frac{1}{2}$	9	$3\frac{1}{2}$
$2\frac{1}{2}$	10	$6\frac{1}{2}$	13
3	$9\frac{1}{2}$	6	$13\frac{1}{2}$
$12\frac{1}{2}$	7	$10\frac{1}{2}$	2

Figure 10c. *Magic square for* 32.

7	$2\frac{3}{4}$	$4\frac{1}{2}$	$1\frac{3}{4}$
$1\frac{1}{4}$	5	$3\frac{1}{4}$	$6\frac{1}{2}$
$1\frac{1}{2}$	$4\frac{3}{4}$	3	$6\frac{3}{4}$
$6\frac{1}{4}$	$3\frac{1}{2}$	$5\frac{1}{4}$	1

Figure 10d. *Magic square for* 16.

It is not always possible to split the increment of the diagonal in such a manner that no two numbers occur twice, particularly in squares with a low magic sum as might be used in lower grades. Suppose a child in the third grade has his ninth birthday, for which occasion the teacher plans to make a magic square with totals equal to nine. The middle term in a three-by-three square is then $9 \div 3 = 3$; the increment of the diagonal is $3-1=2$. This is divided in 1 and 1 (if no fractions are to be used). (See Figs. 11a and 11b.)

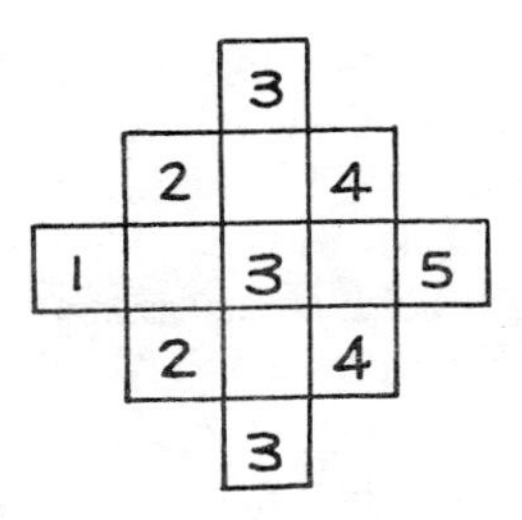

Figure 11a. *Auxiliary square for* 9.

2	3	4
5	3	1
2	3	4

Figure 11b. *Magic square for* 9.

The magic square with totals 10 can be made with a four-by-four square. The diagonal is 1, 2, 3, and 4. The fraction $\frac{1}{2}$ is chosen for both increments. (See Figs. 12a and 12b.)

1	$1\frac{1}{2}$	2	$2\frac{1}{2}$
$1\frac{1}{2}$	2	$2\frac{1}{2}$	3
2	$2\frac{1}{2}$	3	$3\frac{1}{2}$
$2\frac{1}{2}$	3	$3\frac{1}{2}$	4

Figure 12a. *Auxiliary square for* 10.

4	$1\frac{1}{2}$	2	$2\frac{1}{2}$
$1\frac{1}{2}$	3	$2\frac{1}{2}$	3
2	$2\frac{1}{2}$	2	$3\frac{1}{2}$
$2\frac{1}{2}$	3	$3\frac{1}{2}$	1

Figure 12b. *The magic square for* 10.

The magic square for 11 is easily built on a three-by-three square. The middle number is

$$11 \div 3 = 3\tfrac{2}{3}.$$

Choose $\frac{1}{3}$ for the left cell in pyramid.

Magic squares for 12 and 13 are obtained from Figure 4b and Figure 3 respectively. The numbers in the cells are divided by five.

The magic square for 14 can be obtained by adding one to each cell of Figure 12b, by constructing a four-by-four square (diagonal 2, 3, 4, and 5; increments $\frac{1}{6}$ and $\frac{5}{6}$), or by constructing a seven-by-seven square (in the middle $14 \div 7 = 2$; left-cell $\frac{1}{2}$; increments $\frac{1}{20}$ and $\frac{9}{20}$).

These examples are suitable for all elementary school grades from the third grade up, according to the age of the pupils, providing decimal fractions can be used in seventh- and eighth-grade arithmetic.

The high school teacher can use negative numerals; for example, he can subtract 10 from each cell in Figure 1. He can also change the numerals to another numeration system.

EDITORIAL COMMENT.—Magic squares can provide many opportunities for students to learn to think mathematically. However, this will not occur if the students' experience is limited to "plugging in" values in certain places in accordance with rules set forth by the teacher. For example, let's examine the nine-cell magic square in which we are to arrange the numerals 1 through 9 so that the numbers represented in each row, column, and main diagonal have a sum of 15.

a	b	c
d	e	f
g	h	i

First, have the students experiment with the task on a trial-and-error basis. Then focus class attention on the nine cells and ask for ideas about the desired arrangement. Don't tell or show the class how to arrange the numerals. Some class members should begin to observe that certain cells involved will be associated in three different combinations of three numbers, others in two combinations, and so on.

Encourage each class member to make a list of his ideas about the arrangement of the numerals in the nine cells. When each student has completed and organized his list of ideas on the basis of his experiments, develop a class summary.

It is hoped that the students will observe that cell a is associated with three sets of three numbers each [(a, b, c), (a, d, g), and (a, e, i)] and that each corner cell is similarly associated. Other observations should include the fact that cells b, d, f, and h are each associated with two sets of three numbers (for example, b, e, h and b, a, c).

Next, encourage the students to focus on the goal of creating a sum of 15 for each row, column, and main diagonal. Again, it may be easier to tell the students that they should look for various sets of three numbers that total 15. However, far more learning will take place if you allow the students time to come to this conclusion on their own.

Students should be encouraged to do some mathematical thinking for themselves, and magic squares provide excellent opportunities. You may wish to have your students try to form nine-cell magic squares with each of the following sets of numbers. Lead them to look for patterns and to make note of their observations.

1. {2, 4, 6, 8, 10, 12, 14, 16, 18}
2. {3, 6, 9, 12, 15, 18, 21, 24, 27}
3. {1, 3, 5, 7, 9, 11, 13, 15, 17}
4. {5, 10, 15, 20, 25, 30, 35, 40, 45}
5. {10, 20, 30, 40, 50, 60, 70, 80, 90}

3

Numeration

The ideas of base and place value are most fundamental to the understanding of our numeration system and the four fundamental operations on whole numbers. Children need many concrete experiences in grouping single objects in sets of ten and then in grouping sets of ten in sets of ten tens to form a hundred. It is possible to provide much of this practice through games.

The first article in this section describes a game that may be used in the early grades to provide some of the desired experience with place value. MacRae provides a clear set of instructions for constructing and playing her game, and many variations are possible. The Placo game created by Calvo should generate many kinesthetic place-value experiences for children in a most enjoyable atmosphere.

Several games and puzzles involving nondecimal numeration are suggested in the last few articles of this chapter. Shurlow suggests a ring-toss game for developing an understanding of a base-five numeration system. His ideas could easily be adapted for use with other bases, including our own base-ten system.

Alfonso, Balzer, and Hartung provide ideas for several mathematical puzzles relating to base-two and base-three numeration. More mathematical magic relating to the binary system is suggested by Niman, and Ranucci tantalizes the reader with several puzzles built around the base-three system.

A Place-Value Game for First Graders[1]

A Teacher-Made Device

IRENE R. MACRAE

MANY TEACHERS FIND that place value is one of the more difficult number concepts to explain to small children. The following game was devised to help reinforce the understanding of place value after it has been introduced by the use of the *Hundreds*, *Tens* and *Ones* Chart. This chart is familiar to primary teachers. The game uses the same pocket device used in the chart. Thus children can reinforce their initial learning by practice under conditions similar to those in which the learning took place.

The game can be used in the first grade, but can also be adapted to the second and third grades merely by increasing the size of the numbers set as goals. If the goal is set from 25 to 50 for first graders, it is recommended that 2 to 4 children play, depending on the time available. The more players, the longer the game will take. If the goal is set at 100 for second graders, and 1000 for third graders, perhaps two players would be more practical. The game is easily and quickly made, so that a teacher could provide several games for a class. Children could even make the game themselves.

[1] A Chicago Teachers College Student Project.

Directions

OBJECT OF THE GAME:

To score a predetermined number of points.
For first graders: 25–50 points
For second graders: 100 points
For third graders: 1000 points

NUMBER OF PLAYERS:

First grade: 2–4
Second grade: 2
Third grade: 2
Spin to determine who plays first and play to the left.

PROCEDURE:

Each player places the counters in piles in front of him, arranging the colors in the same order as they are placed in the pockets: blue at extreme

The Game in Pictures

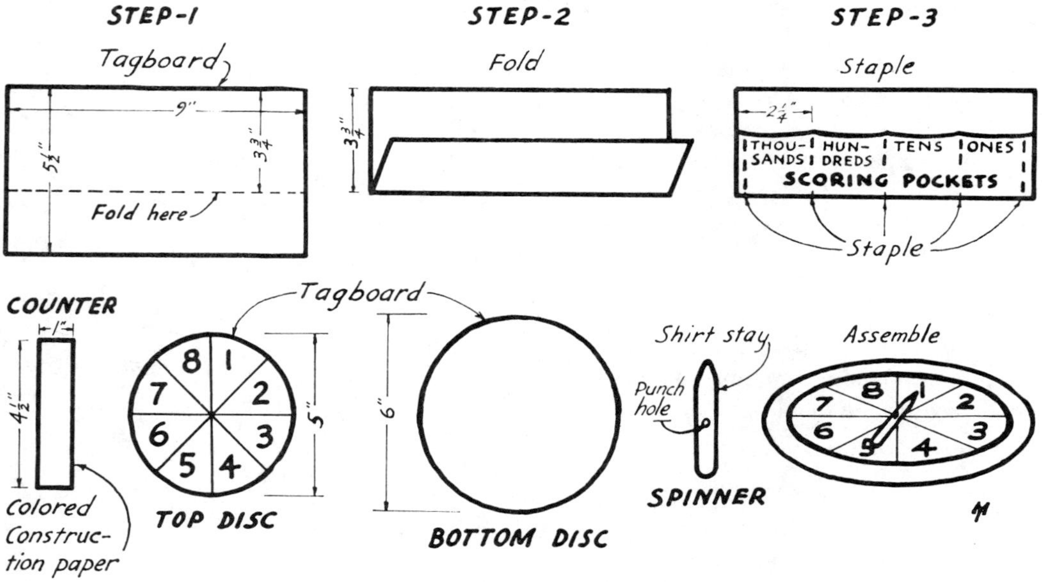

right and proceeding to the left, green, red and yellow.

First player spins and places in the extreme right pocket a number of blue counters equal to the number to which the spinner points. The other players do the same. On the second spin, if a player picks up enough more blue counters to equal ten or more when added to what he already has, he "bundles up" ten blue counters, placing them with the pile of blue counters in front of him, and takes one green counter which he places in the next pocket to the left. This continues until some player has accumulated ten green counters in the ten's pocket which he then removes and puts back in the pile in front of him. He then may place one red counter in the hundreds pocket. The first player to do this wins the game unless the goal is set at 1000.

Since the game is one of chance rather than skill, a fast learner may be paired with a slow learner without doing injustice to the fast learner. He can help the slower child and at the same time reinforce his own learning as well as get fun from the game.

There are other values to be derived from this game in addition to the understanding of place value. When used in the first grade, it will serve to give practice in number value as a result of spinning to determine who plays first. It will provide occasion for children to reproduce abstract numbers with concrete counters. To those beyond the immature level of counting, there will be practice in addition by adding the number which the player spins to the number of counters he has accumulated on his previous turns. When this sum exceeds 10, there will be practice in subtraction by removing the group of ten, leaving the remainder in the ones pocket. The ten ones are bundled and replaced by one counter, representing a group of ten, and moved one place to the left in the scoring pocket. This provides practice in transformation.

The game can be given added interest by providing a pack of penalty and bonus cards, and painting two or three numbers on the spinner dial in colors. When the spinner stops on a color, the player turns up a card from the pack, which is face down, and proceeds to do what the card directs, such as, *lose your next turn*, *take an extra turn*, *add 3 ones*, *you lose three of your ones*, etc.

How to Make the Game

The following detailed instructions are merely suggestions. The teacher can use any materials she has at hand.

1. To make the scoring pockets, use pieces of tagboard, $5\frac{1}{2}''$ by $9''$.
2. Fold to $3\frac{3}{4}''$ by $9''$ so that there is a $1\frac{3}{4}''$ pocket across the bottom.
3. Place a strip of brightly colored mystic tape along each edge to hold pocket in place. The color helps make the game attractive, but the learning values would not be lost if the pocket were held in place by merely stapling.
4. To divide the pockets into four sections representing the place positions, black metal rivets were used which were snapped together but not spread with a hammer. This method allows the pockets to stand out a little so the counters can be easily inserted. If rivets are not readily available, strips of mystic tape might be used, or simple stapling would serve the purpose. The words *Scoring Pocket* were manuscripted across the pocket.
5. For counters, construction paper in four colors is needed. Cut strips $1''$ by $4\frac{1}{2}''$. At least 18 counters are needed for all colors except for the color representing the 1000 position. Only one of these is needed. (Only nine counters are really necessary for the 10 and 100 positions.)
6. For spinner dials, cut four identical circles from tagboard. Mark two into eight equal parts, and write the numbers from 1 to 8 in the sections.
7. The spinners in this particular instance were made of the stays from men's sport shirts by punching a hole in the center with a hand punch and inserting part of a rivet. The other half of the rivet was placed at the back of the dial and the two snapped together, but not spread. The blank circles of tagboard were then glued to the backs. To make the game more attractive in appearance, another larger circle was cut from red poster board to match the red mystic tape used on the pockets. This was glued to the backs to make red borders. (Two spinners were made, but one would be enough if only two children were playing.)

The game can be kept together by placing in a box covered with gay paper or simply marked so that the children will recognize its contents. Keep the counters, divided equally according to color, in the individual pockets to make it easier for the next players who use the game.

⟶

EDITORAL COMMENT.—Many variations of this game are possible. For example, change the numbers represented on the spinner face to 23, 32, 12, 21, 45, 54, 18, and 81 (or any others you wish to use). Then have each student in turn spin the spinner and create sets of tens and ones with the counters to match the number indicated by the spinner.

Placo—a number-place game

ROBERT C. CALVO *Woodland Hills, California*

This game emphasizes fun *and* the learning of positional and place-value aspects of our number system. *Placo* (pronounced play-so) is a number game that boys and girls can play to good advantage in the classroom. The children blindfold each other, one at a time, and then attempt to fit the rings on the highest place-value pole (see diagram). One value of the game is the manipulative aspect. The children enjoy handling the rings, fitting them on the dowels, and then computing to find the amount they "win."

While Placo is chiefly a place-value game, in actual classroom practice it contributes to other operations in arithmetic as well. It provides stimulating practice in computation and enables children to have fun while they play the game. It can be used with success in Grades 2 through 6 and can be refined to challenge even the brightest students. It may be played by as few as two children or as many as the whole class.

Three colored dowels are glued on a base, as shown. These dowels are stationary. The dowel to the left is highest in value, the one to the right is least in value. They may be painted blue, red, and white in that order. They are labeled 100's, 10's, and 1's. Thus 100's (in a whole-number game) are blue, 10's are red, and 1's white.

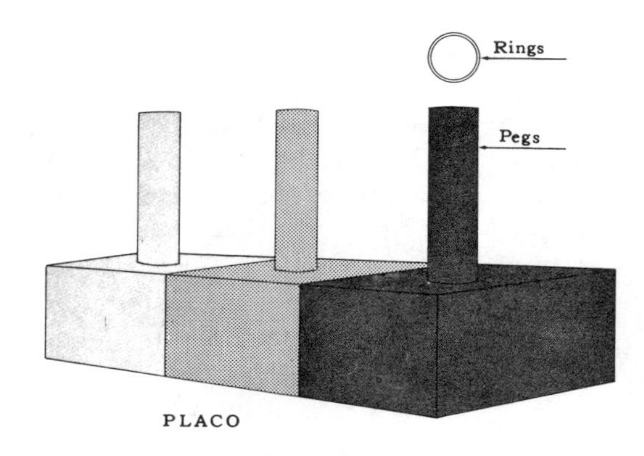

PLACO

One player is blindfolded and given some rings (9–18) of which he may put not more than nine on any one dowel. This blindfolded player then tries to place the rings on the dowels. The other players (split in teams) watch in glee. The object of the game is to get the largest score by putting the rings on the highest-value dowel. After all rings have been placed, the blindfold is removed, and the whole class or a small group computes the amount. The other players then take their turn. Players can play with a goal of 10,000 or with a time limit, the highest score winning. In playing decimal-fraction Placo, the blue dowel represents ones; the red dowel, tenths; and the white dowel, hundredths.

Placo cards may be developed to use with the game. These cards are used as follows: After each turn, a player selects one of the cards and follows the directions given. They say, "Double your score" or "Cut your score in half" or "Subtract three tens," etc. This serves to provide variety in the game. These cards can be developed to reinforce specific skills needed by the class or to provide the class with needed drill. Cards might give directions to add, subtract, multiply, or divide. In addition to dramatizing and creating added interest, they serve to equalize scores of fast and slow learners without penalizing the fast ones.

A word here about the use of colored dowels. While some experts say that place-value instruction should not depend on color, we can't rule it out completely at early levels of learning. In place value, it is important that the child knows that "this ring is in the place to the right" or "that peg is in the units column," but that doesn't mean color can't be brought in to augment these instructions. It merely means color should not be used as a sole means of identifying number-place. If color can help the children gain a concept or can make a game more attractive, then its use is defensible.

Other uses can be found for this game. One possibility is to use it with sight-saving classes by cutting niches in the base to identify the corresponding number-value. Computation could be done mentally. Another potential use would be to roll special dice and compute the score in whatever number base the dice show when rolled. The dice could be made so only certain number bases would result.

Materials list for Placo

1. Piece of wood 12 inches long, 4 inches high, and 6 inches wide

2. Doweling (also known as "closet poling"), $1\frac{3}{8}$ inches in diameter, three pieces of 7 inches each

3. Drapery "cafe curtain" rings, $1\frac{1}{2}$ inches inside diameter, 12 to 20

4. Cards and tacks

5. Sandpaper

6. Paint—white, red, and blue high gloss

7. Carpentry tools, including drill and saw

8. For variation noted, 3-by-5 cards

9. Large handkerchief or clean rag for blinder

10. Oak tag for power chart

Building the game

1. Drill three holes equally distant from each other in the 12-by-6-by-4 inch piece of wood, making these holes 1 inch deep and $1\frac{7}{16}$ inches in diameter.

2. Sand one end of each 7-inch piece of doweling smooth. Put some white glue on the bottom inch of the dowels and the inside of the drilled holes. Slip the dowels in and allow to dry overnight.

3. Purchase cafe curtain rings of the size specified in the materials list.

4. Paint the game three different colors, as mentioned previously, making the dowels and the base of matching colors.

5. You may want to put two small nails about 3 inches from the top of each dowel (exact placement would depend on the thickness of the curtain rings). In this way, players could put only 9 rings on each peg—a 10th would have nothing to keep it from falling off. This style of counting uses only 9 units. When 1 more is added, it becomes the number on the left.

EDITORIAL COMMENT.—Two other variations of Placo will provide excellent kinesthetic learning experiences for children. In the first variation, identify a whole number less than 1000 and have the blindfolded student try to represent the number correctly on the three pegs with rings. The number may be identified by drawing a card, by spinning a spinner, or simply by having someone name it orally. In the second variation, have one student write a three-digit numeral on the chalkboard and then put rings on the three pegs to represent the number.

→

Then have a blindfolded student try to count the rings and name the number. In each variation, it is possible to devise a scoring system and conduct the activity as a game.

The game of five

HAROLD J. SHURLOW *Columbus Public Schools, Columbus, Ohio*

Purpose

It has been stated many times that the key concepts needed by elementary teachers in order to teach arithmetic meaningfully include the distinction between number and numeral, the structure of our numeration system, and the concept of place value. This game is designed to provide teachers with an interesting way of reviewing the important concept of place value and of working into a study of numeration systems that have different bases.

To a class:

"Would you like to learn to play a new game? Of course! Now there are two things you have to know before you can learn this game. First, you must know how to count to five. Well now, you all know that. Then you must know how to count by 5's to 25. Let's try that. 5–10–15–20–25. Next you must know how to count by 25's to 125. Shall we try that too? 25–50–75–100–125. This game is a ring-toss game. How many know what a ring-toss game is?"

(Most children know, and at this point classroom interest picks up.)

"Let's draw a picture of some pegs for our ring-toss game."

(See Figure 1. For lower grades, actual pegs with standards would be more effective than mere drawings, but for older students drawings serve adequately since most students are acquainted with peg games.)

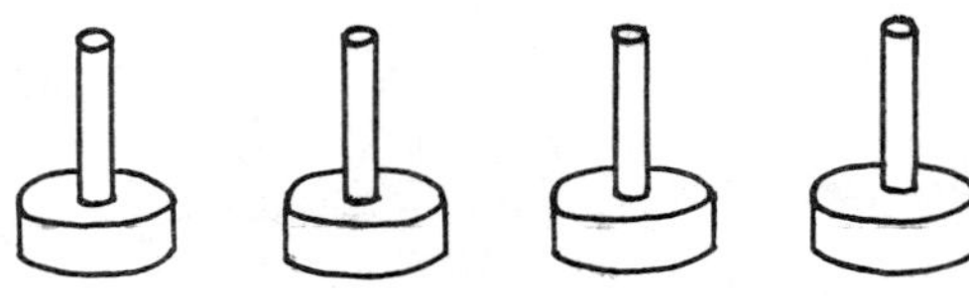

Figure 1

"Now in a ring-toss game there is usually a score, so we have to get that into our game. Let's say that if a ring is on the peg *to the right,* it counts 1 point. If a ring is on the next peg, it counts 5 points; on the next, 25 points, and on the next, 125 points. (See Figure 2.) So now we know what makes our score."

(At this point teachers may pretend to become visibly shaken at the prospect of rings flying all over the room.)

"Now comes a change in this game. In most ring-toss games you toss the rings and count up the score. Not so here. In this game you know the score ahead of time, and it will be your job to *put the rings on the pegs in the right places.*

"Let's think about a score of 13. One way of placing the rings would be to put 13 rings on the peg to the right. But there is more to this game. You have not won this game unless you have used the *fewest number of rings* possible. So with a score of 13, there would be two rings on the peg marked 5 and three rings on the peg to the right."

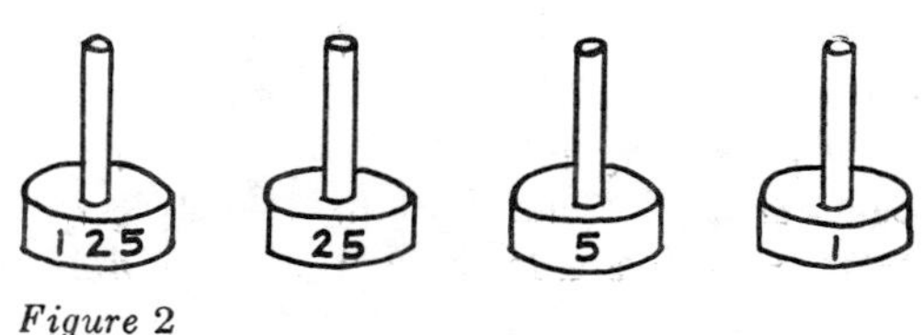

Figure 2

To the teacher

At this point it is best to go over a few more examples, and then let your pupils try the game during rainy-day recess or at other convenient times. The game seems to work best if three or four pupils play together. One pupil can assign a score, and then the other pupils can try to place the rings, using pencil and paper to record where they have placed them. Contests of speed may result. However, the purpose here is for pupils to become quite familiar with the game before they proceed to the next step.

You might next suggest that it would be nice to have an orderly way of recording what we were doing in the game—something like the following:

A SCORE OF		125	25	5	1
13	LEADS TO			2	3
39	" "		1	2	4
53	" "		2	0	3

Figure 3

(Notice if there is no ring we indicate the absence with a zero.)

After this orderly recording has been done for a while, you can draw your pupils' attention to the fact that the recorded numerals—23, 124, and 203—look suspiciously like some old friends. Here ask the question, "In our game is 124 really one hundred twenty-four? No? Then what is it?"

If pupils have been playing this game, they will tell you, "1 twenty-five, 2 fives, and 4 ones." As teachers, you can see that here is the chance to tell pupils that a numeral's place value is important.

It is a simple matter then to show your pupils that 39 and 124 really represent the same score *if we know the value of the places.* Now we need a way to show the value of the places. Here we introduce $39_{(\text{ten})} = 124_{(\text{five})}$. So we have two pictures (numerals) for the same score (number).

I must warn that in using this game the teacher must proceed slowly in order that the students assimilate each idea as it comes: first the game and its rules, then recording the results, then comparing place values, and finally recording numerals by means of different bases.

Here are some good questions for teachers to ask along the way:

1 In the "game of five" what is the greatest number of rings that will ever go on a single peg?

2 The peg on the right has a value of 1, and the next peg to the left of it has a value of 5. How do you get the value of the next peg, the next, etc.?

3 Suppose we wanted to play the "game of 4" (or "game of __"), what would the value on each peg (place) be? What would be the greatest number of rings on a single peg?

4 In base 7, what is the largest numeral used in any single place? Why?

Many of our elementary teachers have told me that they had often wanted to try some work on numeration systems that have bases other than 10, but they did not know a way to introduce such work to children in an interesting manner. "The game of five" seemed to satisfy this need. I hope that you, too, will find it useful.

EDITORAL COMMENT.—Although the author suggests that the activity be conducted without actually playing a ring-toss game, it is possible in many situations to apply his ideas to a real ring-toss game. For example, each student may be given an opportunity to toss four rings on the base-five pegs to establish some base-five numeral. He may then calculate the number of points in our decimal system and score this amount if his work is done correctly.

For students who need more work in our base-ten system, you may wish to try the ring-toss activity with the pegs named ones, tens, hundreds, and so on.

From second base to third base

MICHAEL ALFONSO
Carol City High School, Carol City, Florida

RICHARD BALZER
Hackettstown High School, Hackettstown, New Jersey

PAUL HARTUNG
Bloomsburg State College, Bloomsburg, Pennsylvania

You may have seen the game consisting of four cards—call them A, B, C, and D, for convenience—on which sets of eight numbers are written. (See fig. 1.) Using these four cards, Mr. Rhee asks a student to pick a number from one to fifteen and to tell him which cards the number appears on—whereupon Mr. Rhee quickly announces the number to the amazed student.

How does Mr. Rhee do it? When the student tells him which cards contain his number, Mr. Rhee quickly adds the first number on each of those cards together;

A	B	C	D
1	2	4	8
3	3	5	9
5	6	6	10
7	7	7	11
9	10	12	12
11	11	13	13
13	14	14	14
15	15	15	15

Fig. 1

the sum is the number the student chose. As an example, suppose the number appeared on cards A, C, and D. Then Mr. Rhee adds 1 + 4 + 8 to find the number is 13.

To see why this happens, we write the numbers from one to fifteen in binary notation. (See fig. 2.) Now you can readily see how the cards are developed.

1	0001
2	0010
3	0011
4	0100
5	0101
6	0110
7	0111
8	1000
9	1001
10	1010
11	1011
12	1100
13	1101
14	1110
15	1111

Fig. 2

Card A contains all the numbers having a "1" in the ones place in the base two notation. Card B contains all the numbers having a "1" in the twos place (the second digit from the right in the base two notation). Card C contains all the numbers having a "1" in the fours place (the third digit from the right in the base two notation); and Card D, all the numbers with a "1" in eights place (the fourth place from the right in base two notation). (See fig. 3.)

1	0001	A			
2	0010		B		
3	0011	A	B		
4	0100			C	
5	0101	A		C	
6	0110		B	C	
7	0111	A	B	C	
8	1000				D
9	1001	A			D
10	1010		B		D
11	1011	A	B		D
12	1100			C	D
13	1101	A		C	D
14	1110		B	C	D
15	1111	A	B	C	D

Fig. 3

It is possible to develop another set of cards with which the same game can be played. For example, a new set can be constructed using base three notation. To obtain the cards, we write the numbers from 1 to 26 in base three notation. (See fig. 4.) The first card, Card A, will consist of those numbers whose base three notation has a "1" in the ones place.

1	001
2	002
3	010
4	011
5	012
6	020
7	021
8	022
9	100
10	101
11	102
.	.
.	.
.	.
25	221
26	222

Fig. 4

Card B will include the numbers having "2" in ones place. Card C will include the numbers with a "1" in threes place (the second digit from the right in the base three notation). Card D will have all the numbers having a "2" in threes place. Card E will have all the numbers with a "1" in the nines place. And Card F will have the numbers with a "2" in nines place. (See fig. 5.)

A	B	C
1	2	3
4	5	4
7	8	5
10	11	12
13	14	13
16	17	14
19	20	21
22	23	22
25	26	23

D	E	F
6	9	18
7	10	19
8	11	20
15	12	21
16	13	22
17	14	23
24	15	24
25	16	25
26	17	26

Fig. 5

A	B	C	D	E	F
1	2	3	4	8	12
5	6	7	5	9	13
9	10	11	6	10	14
13	14	15	7	11	15

Fig. 6

Now to play with these cards. Suppose someone tells us his secret number appears on cards B, D, and E. Then the secret number is 2 + 6 + 9 = 17.

Base four could also be used for this game. A set of cards using base four would look like those in figure 6.

The stopping point in any base is relatively unimportant. However, it will always be of the form

$$b^n - 1,$$

where b is the base you're working with and n is a natural number. Examining the stopping points for bases 2, 3, and 4 respectively, we see that

$$15 = 2^4 - 1,$$

$$26 = 3^3 - 1,$$

and

$$15 = 4^2 - 1.$$

Other stopping points could be chosen so that each student would pick a different number, thereby encouraging total participation.

The easiest way to determine the number of cards needed is to add the digits of the largest number, written in whatever base you are working with, as if the digits were in base 10. For example,

$$15 = 1111_2 = 1 + 1 + 1 + 1 = 4 \text{ cards},$$

and

$$26 = 222_3 = 2 + 2 + 2 = 6 \text{ cards}.$$

You will notice that the place values are reversed when we set up the cards. When someone picks a number and tells you what card it's on, the number in that base is converted to base 10. That is, for the secret number 13 (Mr. Rhee's student's number), the base two numeral is 1101—one 1, one 4, and one 8. And 1 + 4 + 8 = 13, the secret of the game.

EDITORIAL COMMENT.—Students who have had considerable experience in looking for patterns may quickly discover that the answer is determined by finding the sum of the first numbers on the identified cards. This will, of course, provide an excellent basis for leading the students to discover why this technique works.

If you wish to create a system in which your secret is more difficult to discover, arrange the numbers on the cards in random order and remember that you must use the smallest number on each identified card when you calculate the desired sum. For example:

A	*B*	*C*	*D*
9	6	15	12
15	11	7	8
3	3	12	14
7	10	14	9
5	2	6	13
13	15	4	11
1	7	5	15
11	14	13	10

If the number is represented on card *A*, you would use 1, which is the smallest number on card *A*; if the number is represented on card *B*, you would use 2, which is the smallest number on card *B*, and so on.

A game introduction to the binary numeration system

JOHN NIMAN

An assistant professor at Hunter College of the City University of New York, John Niman teaches courses in mathematics and mathematics education. His research interests include applied mathematics and curricula development.

Have you ever encountered a teasing child who would only answer you with yes or no? If you ever do, tell him that you are going to read his mind through his revealing yeses and noes—and all you are going to do is ask him the same question several times! Tell him to choose a number between 1 and 256. Then pose the question "Is the number odd?" If he says yes, tell him to subtract one and divide the remainder by two. If he says no, simply tell him to divide his number by two. Follow this procedure until the question has been asked at most eight times. The game stops when the number one is reached.

In order to play the game, the child has to know how to subtract and to divide by 2. He must also know what odd and even numbers are. Suppose, as an example, that your friend chose the number 98. The dialogue would go as follows:

Is it odd?	(98)	NO	Divide by 2.
Is it odd?	(49)	YES	Subtract 1 and divide by 2.
Is it odd?	(24)	NO	Divide by 2.
Is it odd?	(12)	NO	Divide by 2.
Is it odd?	(6)	NO	Divide by 2.
Is it odd?	(3)	YES	Subtract 1 and divide by 2.
Is it odd?	(1)	YES	

Now you have all the information you need—the seven yeses and noes. Starting with the *last* answer, replace each no with a "0" and each yes with a "1." Such a process transforms the result into 1100010. This is the number you are looking for! However, it is expressed in the binary numeration system. The number expressed by 1100010 in the binary system is expressed by 98 in the decimal numeration system.

To convert an expression from the binary to the decimal system, start with the last digit on the right and multiply it by 2^0. (Note: Any number greater than zero raised to the zero power is defined as one.) Then multiply the next digit by 2^1, the next by 2^2, the next by 2^3, and so on. Finally, sum the results.

In our example, 1100010 becomes

$$\begin{aligned}(0\times 2^0)+(1\times 2^1)+(0\times 2^2)+(0\times 2^3)\\ +(0\times 2^4)+(1\times 2^5)+(1\times 2^6)\\ =0+2+0+0+0+32+64\\ =98.\end{aligned}$$

The game is summarized in figure 1.

The choice for the number does not have to be restricted to numbers between 1 and 256. Generally, no restrictions are necessary. The number of questions needed depends on the number to be guessed. For example, consider how many questions are needed for numbers from 1 to 1000. We

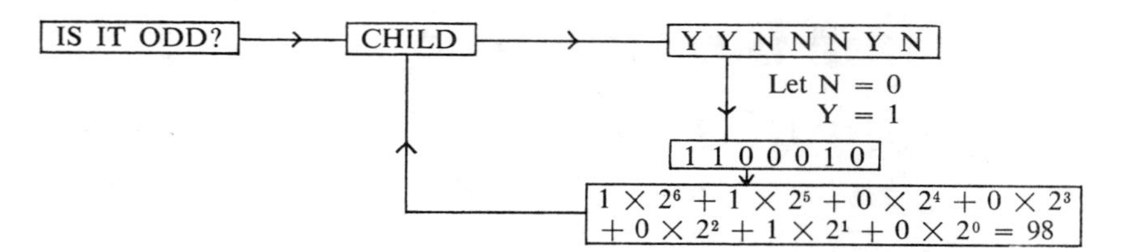

Fig. 1. "Guessing" a number

can solve the problem by finding out how many times it is necessary to divide successively by two to end up finally with one. For numbers from 1 to 256, at most eight divisions are needed, since $2^8 = 256$. Similarly, at most nine divisions are needed for numbers from 1 to 512 ($2^9 = 512$), and at most ten divisions are necessary for numbers from 1 to 1024 ($2^{10} = 1024$). Consequently, to guess a number between 1 and 1000, at most ten divisions are needed. The above leads to the result that the greatest number of questions needed to guess a particular number from 1 to N is equal to one plus the minimum exponent of 2 such that when 2 is raised to that exponent, the resulting number is equal to or greater than N.

The game could be used in a variety of ways to determine such characteristics as age, height, and weight. The fact that one does not have to ask his subject the final result after the sequence of questions makes the game quite intriguing. Furthermore, it would serve as an effective tool for acquiring practice in the basic operations of the arithmetic of counting numbers, and an explanation of how the game works could serve as an introduction to the base-two numeration system.

EDITORIAL COMMENT.—This is one of many mathematical games and puzzles that appear to be "number magic." Such activities can serve to stimulate student curiosity and create interest in mathematics if used effectively. However, the student should ultimately investigate the mathematics that explains the *why* of the "number magic."

Tantalizing ternary

ERNEST R. RANUCCI
State University of New York at Albany, Albany, New York

Dr. Ranucci is professor of mathematics education. Last summer he was in Colombia, South America, for three months on a Fulbright grant. His work there was with teacher education for both elementary and secondary levels.

Arithmetic computation in bases other than ten is a fertile field to explore. Especially interesting are certain problems that arise in the base three (ternary) system.

Place value in the ternary system is based upon powers of three. The units digit has a value of 1, the base digit a value of 3, the base-two digit a value of 9, the base-three digit a value of 27, etc. The number 12012 in base three has a value of 140 in base ten.

$$(12012_{\text{three}} = 140_{\text{ten}})$$

1	2	0	1	2	
3^4	3^3	3^2	3^1	1	. . . Place value as a power of three
81	27	9	3	1	. . . Calculated place value

$$\begin{aligned} 12012_{\text{three}} &= 1 \times 81 + 2 \times 27 + 0 \times 9 + 1 \times 3 + 2 \times 1 \\ &= 81 \quad 54 \quad 0 \quad 3 \quad 2 \\ &= 140_{\text{ten}}. \end{aligned}$$

In the ternary system only three digits are needed. We shall use 0, 1, and 2.

Many intriguing puzzles are based on the ternary system. For the exploration of several of these, we need to convert the decimal numbers from 1 to 40 into base-three numerals. (See Table 1.) Note that

$$40 = 1 + 3 + 9 + 27.$$

This fact will prove to be of extreme significance later on.

Let us first consider the "butcher problem." A certain Scottish butcher finds that he can get along surprisingly well with nothing but a simple balance scale and individual weights of one, three, nine, and twenty-seven pounds. As long as no customer asks for meat in a quantity requiring the use of half-pounds, the butcher finds that he can weigh any integral number of pounds of meat from one to forty. He can, what's more, do this in one weighing (providing the customer is not so fastidious as to object to the placing of weights on the same pan with the cut of meat.) To weigh out five pounds of chopped beef, for example, he places a weight of nine pounds on one pan. He then places weights of one and three pounds on the other pan and adds meat until the scales

Table 1

Decimal notation	*Ternary notation*	*Decimal notation*	*Ternary notation*
1	1	21	210
2	2	22	211
3	10	23	212
4	11	24	220
5	12	25	221
6	20	26	222
7	21	27	1000
8	22	28	1001
9	100	29	1002
10	101	30	1010
11	102	31	1011
12	110	32	1012
13	111	33	1020
14	112	34	1021
15	120	35	1022
16	121	36	1100
17	122	37	1101
18	200	38	1102
19	201	39	1110
20	202	40	1111

balance (Fig. 1). This leads to the simple equation:

$$M + 1 + 3 = 9; M = 5.$$

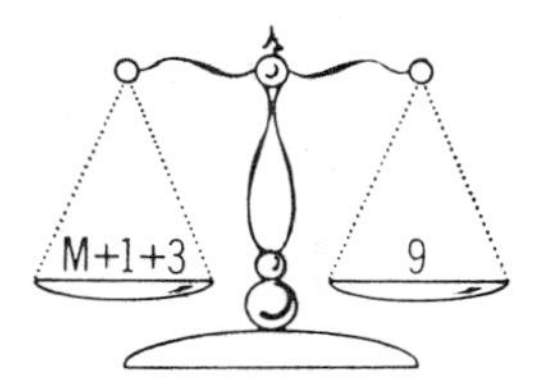

FIGURE 1

Weighing ten pounds of meat is easy. All this takes is the placing of weights of one and nine pounds on one pan and the addition of meat to the other pan until they balance (Fig. 2). The equation here is

$$M = 1 + 9; M = 10.$$

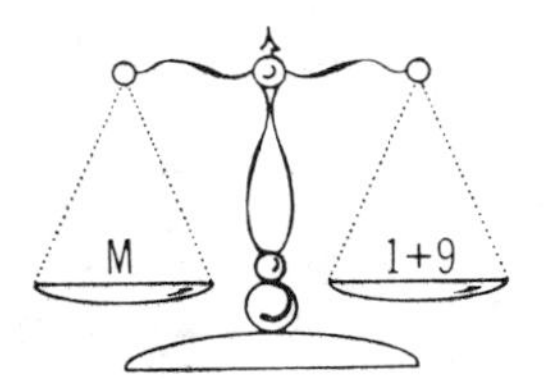

FIGURE 2

Curiously, each total, from one to forty, can be achieved by only one combination of the various weights available. Table 2, based on the ternary system, indicates the unique structure behind the "butcher problem." Remember that there is but one weight of each size.

The table is used as follows: All positive weights are placed on one pan; all negative weights are placed on the other pan. Meat is always added to the pan on which the negative weights have been placed. In weighing out 20 pounds, for example, weights of 27 and 3 are placed on one pan; weights of 9 and 1 are placed on the other. Meat is added to the 9, 1, side until a balance is secured. With the addition of another weight of 81 pounds to the set, every integral weight from 1 to 121 may be achieved. The "butcher" approach to equation solving can be handled by the average sixth grader if the approach is gradual.

Martin Gardner, in an article in the May 1964 issue of the *Scientific American*, offers an unusual approach to the "butcher problem." This starts with the value of the desired weight expressed in the ternary notation. To weigh out 33 pounds, for example, first consider the ternary numeral 1020_{three}. Add a 1 and subtract a 1 in the column where a 2 occurs. It is advantageous to rewrite as follows:

Table 2

1	1	21	27 − 9 + 3
2	3 − 1	22	27 − 9 + 3 + 1
3	3	23	27 − 3 − 1
4	3 + 1	24	27 − 3
5	9 − 3 − 1	25	27 − 3 + 1
6	9 − 3	26	27 − 1
7	9 − 3 + 1	27	27
8	9 − 1	28	27 + 1
9	9	29	27 + 3 − 1
10	9 + 1	30	27 + 3
11	9 + 3 − 1	31	27 + 3 + 1
12	9 + 3	32	27 + 9 − 3 − 1
13	9 + 3 + 1	33	27 + 9 − 3
14	27 − 9 − 3 − 1	34	27 + 9 − 3 + 1
15	27 − 9 − 3	35	27 + 9 − 1
16	27 − 9 − 3 + 1	36	27 + 9
17	27 − 9 − 1	37	27 + 9 + 1
18	27 − 9	38	27 + 9 + 3 − 1
19	27 − 9 + 1	39	27 + 9 + 3
20	27 − 9 + 3 − 1	40	27 + 9 + 3 + 1

$$\begin{array}{c} \bar{1} \\ 1020 \\ 1 \end{array}$$

The addition of a 1 and the subtraction of a 1 does not affect the value of the number. (We are really adding 1 × 3 and subtracting 1 × 3, but there is no real need for this extra step.) The addition of 2 and 1 means that we can carry a 1 to the adjacent column. The result may now be expressed as $11\bar{1}0$. This means 27 + 9 − 3, whose value is 33. Several solutions of this type are carried out:

20 pounds: 202_{three}

$$= \begin{array}{c} \bar{1}\ \bar{1} \\ 202 \\ 1\ 1 \end{array} = 1\bar{1}1\bar{1}$$

(This means 27 − 9 + 3 − 1 = 20.)

32 pounds: 1012_{three}

$$= \begin{array}{c} \bar{1} \\ 1012 \\ 1 \end{array} = 102\bar{1} = \begin{array}{c} \bar{1} \\ 102\bar{1} \\ 1 \end{array} = 11\bar{1}\bar{1}$$

(This means 27 + 9 − 3 − 1 = 32.)

23 pounds: 212_{three}

$$= \begin{array}{c} \bar{1}\ \bar{1} \\ 212 \\ 1\ 1 \end{array} = 1\bar{1}2\bar{1} = \begin{array}{c} \bar{1} \\ 1\bar{1}2\bar{1} \\ 1 \end{array} = 10\bar{1}\bar{1}$$

(This means 27 − 3 − 1 = 23.)

The first forty numbers in the decimal system have values in the ternary system, numerals 1, 0, $\bar{1}$, as shown in Table 3.

Table 3

1	1	21	$1\bar{1}10$
2	$1\bar{1}$	22	$1\bar{1}11$
3	10	23	$10\bar{1}\bar{1}$
4	11	24	$10\bar{1}0$
5	$1\bar{1}\bar{1}$	25	$10\bar{1}1$
6	$1\bar{1}0$	26	$100\bar{1}$
7	$1\bar{1}1$	27	1000
8	$10\bar{1}$	28	1001
9	100	29	$101\bar{1}$
10	101	30	1010
11	$11\bar{1}$	31	1011
12	110	32	$11\bar{1}\bar{1}$
13	111	33	$11\bar{1}0$
14	$1\bar{1}\bar{1}\bar{1}$	34	$11\bar{1}1$
15	$1\bar{1}\bar{1}0$	35	$110\bar{1}$
16	$1\bar{1}\bar{1}1$	36	1100
17	$1\bar{1}0\bar{1}$	37	1101
18	$1\bar{1}00$	38	$111\bar{1}$
19	$1\bar{1}01$	39	1110
20	$1\bar{1}1\bar{1}$	40	1111

Most puzzles of the type, "Find your age on each of the following cards and I'll tell you what it is," are based on the ternary system. So are a series of puzzles in which names are guessed. Following is a description of one of the better of the name puzzles.

First prepare four cards according to the directions which follow.

Girl's Name Puzzle

First card (Worth 27 points)

(Front) Brunette, Brown Eyes

Dinah	Florence	Helen
Dolores	Fredrika	Hilda
Dorothy	Gail	Ida
Edith	Geraldine	Ilene
Elizabeth	Gertrude	Ingeborg
Eloise	Grace	Ingrid
Ethel	Gwendolyn	Jacqueline
Fanny	Harriet	Jeannette
Flora	Hazel	Joy

(Back) Brunette, Blue Eyes

Nancy	Petunia	Ursula
Naomi	Rita	Vicki
Nina	Roberta	Viola
Nora	Sarah	Wanda
Ocelie	Suzanne	Wendy
Ophelia	Tallulah	Winifred
Pamela	Tina	Yvette
Patsy	Tess	Yvonne
Paula	Tracy	Zaza

Second card (Worth 9 points)

(Front) Blonde, Brown Eyes

Barbara	Joy	Nancy
Betsy	Helen	Naomi
Betty	Hilda	Nina
Bunny	Ida	Nora
Carmen	Ilene	Ocelie
Carol	Ingeborg	Ophelia
Charlotte	Ingrid	Pamela
Corinne	Jacqueline	Patsy
Diane	Jeannette	Paula

(Back) Blonde, Blue Eyes

Dinah	Lana	Ursula
Dolores	Lena	Vicki
Dorothy	Lisette	Viola
Edith	Louise	Wanda
Elizabeth	Mary	Wendy
Eloise	Margaret	Winifred
Ethel	Mildred	Yvette
Fanny	Marilyn	Yvonne
Flora	Nanette	Zaza

Third card (Worth 3 points)

(Front) Auburn, Brown Eyes

Alice	Gwendolyn	Nancy
Anne	Harriet	Naomi
Audrey	Hazel	Nina
Charlotte	Jacqueline	Petunia
Corinne	Jeannette	Rita
Diane	Joy	Roberta
Ethel	Lana	Ursula
Fanny	Lena	Vicki
Flora	Lisette	Viola

(Back) Auburn, Blue Eyes

Barbara	Helen	Pamela
Betsy	Hilda	Patsy
Betty	Ida	Paula
Dinah	Katherine	Tina
Dolores	Kay	Tess
Dorothy	Kyle	Tracy
Florence	Marilyn	Yvette
Fredrika	Mildred	Yvonne
Gail	Nanette	Zaza

Fourth card (Worth 1 point)

(Front) Red, Brown Eyes

Abigail	Grace	Nancy
Audrey	Hazel	Nora
Betty	Ida	Pamela
Carol	Ingrid	Petunia
Diane	Joy	Sarah
Dorothy	Katherine	Tina
Eloise	Lana	Ursula
Flora	Louise	Wanda
Gail	Mildred	Yvette

(Back) Red, Blue Eyes

Alice	Gwendolyn	Nina
Barbara	Helen	Ophelia
Bunny	Ilene	Paula
Charlotte	Jacqueline	Roberta
Dinah	Joyce	Tallulah
Edith	Kyle	Tracy
Ethel	Lisette	Viola
Florence	Margaret	Winifred
Geraldine	Nanette	Zaza

Following is a list of possible names, with numerical value.

Positive

1. Abigail	21. Fanny
2. Alice	22. Flora
3. Anne	23. Florence
4. Audrey	24. Fredrika
5. Barbara	25. Gail
6. Betsy	26. Geraldine
7. Betty	27. Gertrude
8. Bunny	28. Grace
9. Carmen	29. Gwendolyn
10. Carol	30. Harriet
11. Charlotte	31. Hazel
12. Corinne	32. Helen
13. Diane	33. Hilda
14. Dinah	34. Ida
15. Dolores	35. Ilene
16. Dorothy	36. Ingeborg
17. Edith	37. Ingrid
18. Elizabeth	38. Jacqueline
19. Eloise	39. Jeannette
20. Ethel	40. Joy

Negative

$^{-}1$. Joyce	$^{-}21$. Patsy
$^{-}2$. Katherine	$^{-}22$. Paula
$^{-}3$. Kay	$^{-}23$. Petunia
$^{-}4$. Kyle	$^{-}24$. Rita
$^{-}5$. Lana	$^{-}25$. Roberta
$^{-}6$. Lena	$^{-}26$. Sarah
$^{-}7$. Lisette	$^{-}27$. Suzanne
$^{-}8$. Louise	$^{-}28$. Tallulah
$^{-}9$. Mary	$^{-}29$. Tina
$^{-}10$. Margaret	$^{-}30$. Tess
$^{-}11$. Mildred	$^{-}31$. Tracy
$^{-}12$. Marilyn	$^{-}32$. Ursula
$^{-}13$. Nanette	$^{-}33$. Vicki
$^{-}14$. Nancy	$^{-}34$. Viola
$^{-}15$. Naomi	$^{-}35$. Wanda
$^{-}16$. Nina	$^{-}36$. Wendy
$^{-}17$. Nora	$^{-}37$. Winifred
$^{-}18$. Ocelie	$^{-}38$. Yvette
$^{-}19$. Ophelia	$^{-}39$. Yvonne
$^{-}20$. Pamela	$^{-}40$. Zaza

The puzzle is worked as follows: Someone is asked to pick a name from the cards and to identify the hair color and eyes each time the name is mentioned. The score on any card is considered positive when the eyes are brown. The score is considered negative when the eyes are blue. Thus: Rita—Brunette, Blue Eyes ($^{-}27$); Auburn, Brown Eyes (3). Add the scores algebraically. The sum is $^{-}24$. Look up $^{-}24$ in the list of names and identify "Rita."

Thus: Joy (27, 9, 3, 1) or 40
Yvonne ($^{-}27$, $^{-}9$, $^{-}3$) or $^{-}39$
Harriet (27, 3) or 30
Ethel (27, $^{-}9$, 3, $^{-}1$) or 20

The structure of the Girl's Name Puzzle is identical to that of the "butcher problem." Once the desired number has been converted to the Gardner positive-negative number, the name associated with that number is placed on the appropriate face of the proper card.

An examination of other number systems will reveal the unique nature of the ternary system. Suppose that the weights used in the "butcher problem" had been 1, 2^1, 2^2, 2^3, etc.—elements of the binary system. Then meat, in the amounts indicated below, could have been weighed as follows:

1. 1 or $4 - 2 - 1$, etc.
2. 2 or $8 - 4 - 2$, etc.
3. $1 + 2$ or $4 - 1$, etc.
4. 4 or $8 - 4$, etc.

Such weights could, of course, be used, but there would be nothing unique about such a system. Each weight could be produced in a variety of ways.

A base-four (quaternary) system presents its own problems. Let us attempt to weigh out the first twenty pounds; we may use only individual weights of 1, 4, 16, 64, etc.

1. 1	11. $16 - 4 - 1$
2. ?	12. $16 - 4$
3. $4 - 1$	13. $16 - 4 + 1$
4. 4	14. ?
5. $4 + 1$	15. $16 - 1$
6. ?	16. 16
7. ?	17. $16 + 1$
8. ?	18. ?
9. ?	19. $16 + 4 - 1$
10. ?	20. $16 + 4$

Certain "clusters" appear with a predictable pattern, but many "holes in the Swiss cheese" show up. This deficiency becomes aggravated when the number base increases.

It would appear that the ternary system alone allows us to achieve all integral weights in a unique manner.

4

Integers

The introduction of integers as early as third grade in recently published elementary school mathematics textbooks is an outgrowth of the natural encounter of elementary school children with many situations that require the use of integers. When primary-grade children learn to use thermometers, they note that thermometers can register temperature below zero as well as at or above zero. The countdowns for space launchings involve negative and positive numbers, and altitudes are recorded as above or below sea level. In football games, we make note of yardage gained and yardage lost, and we speak of budgets being in the red or in the black.

The study of integers in the primary and intermediate grades should be largely informal, and many such experiences can be provided through game activities. In the opening article, Demchik suggests a card game called "Integer Football" as one way of generating interest and enthusiasm for a practice exercise involving integer addition. Milne suggests a student-created game that could lead to the discovery of a method for subtraction of integers as a result of various moves on a game board.

In a game described by Mauthe, upward movements on a ladder correspond to positive integers, and downward movements correspond to negative integers. The composition of movements is identified as addition. Imagine the excitement of students as they reach the top of the ladder or as their opponents fall off the ladder. Frank incorporates much fun in her shuffleboard game while relating integers, number lines, and operations on integers. Children will enjoy this activity and gain practice in working with integers.

To provide more drill with the operations on integers, Milne has developed a spinner game; Cantlon, Homan, and Stone have devised a student-constructed game that uses Popsicle sticks. Both activities will provide much student involvement and painless practice.

In the closing pair of selections, we see how creative minds with an interest in mathematics can generate some challenging puzzles. Many upper elementary and middle school students will find "Grisly Grids" quite interesting.

Integer "football"

VIRGINIA C. DEMCHIK

Bowie State College, Bowie, Maryland

Generally, students in the upper elementary and junior high grades are interested in sports. The card game "Integer Football" uses that enthusiasm for sports as motivation for practice with directed numbers. The game is designed for use with an entire class or with a small group of students.

Materials needed

Integer football is played with a deck of 65 two-and-one-half-by-three-inch cards. The cards are assigned numerical values as follows:

One set is numbered 0 through $+20$.

Two sets are numbered -1 through -20.

There are four penalty cards: $+5$, $+15$, -5, -15.

Playing the game

Before play begins, the class is divided into two teams. The deck is shuffled and placed face down on the table. Each team starts on the "fifty yard line." A flip of a coin determines which team "kicks off."

Play begins as a member of the team that won the toss turns over the top card. The team adds the integer on the card to the fifty yards with which the team began. For example, if the top card is -16, the play is $50 + (-16) = 34$. This means that the team now needs 34 yards to score. On the other hand, if the drawn card is $+16$, the play is $50 + (+16) = 66$, and the team now needs 66 yards to score. Thus, it is to the team's advantage to draw a negative number. Similarly, a penalty card with a negative integer (-5 or -15) is to the team's advantage; a positive integer (+5 or +15) is to the team's disadvantage. The "ball goes to the other team" after each play. On the team's next play, the integer on the drawn card is added to the total from the previous play.

A team "makes a touchdown" and thereby scores six points whenever the team's total play becomes zero or less. When a team scores a touchdown, the scoring team plays for the extra point by drawing the next card. If the next card is negative, the play is good and the team scores the extra point. If the next card is positive, there is no extra point.

After every touchdown and the play for the extra point, play resumes again on the fifty yard line. This time the team that did not score begins the play. The game continues until time runs out. The team with the largest score at the end of the playing time wins the game.

The time limit for the game is optional; thirty minutes is suggested. At halftime, both teams go back to the fifty yard line and play is resumed.

Editorial Comment.—Most children gain extra enjoyment when a "football field" and "football" are used in playing this game. The football field can be easily constructed from posterboard and marked along the edges to indicate the "yard lines." A piece of sponge or a football drawn on posterboard may be used to mark the position of the ball.

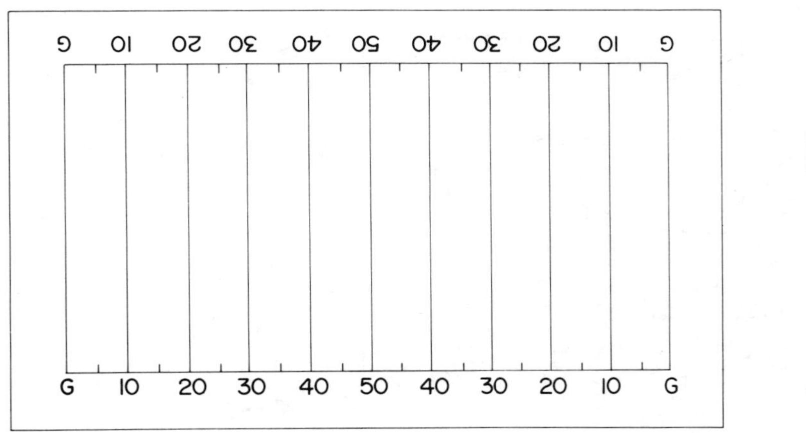

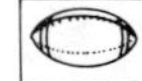

Subtraction of integers—discovered through a game

ESTHER MILNE

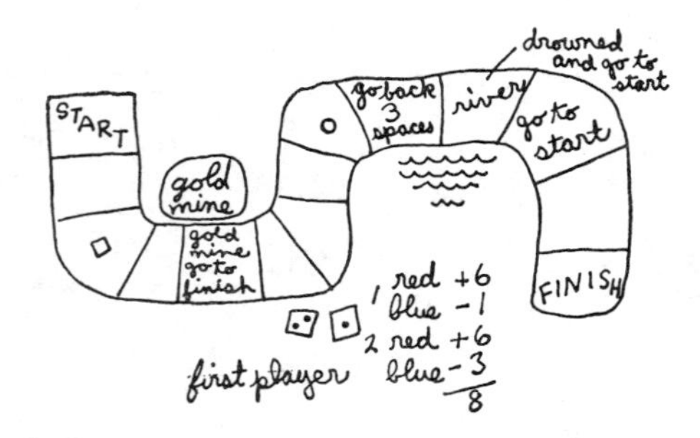

On a childs game for two peopl there were two dice. On was red and one was blue. When it was your turn you roll both dice. Then you advance the number of spaces as shown on the red die. Then do the opposite for the blue die.

This game description was a student's idea of a good problem requiring subtraction.

From this suggested game, a variation was devised to include negative integers. Thus it can be used to introduce subtraction of negative integers.

Materials needed are a game board, two or three markers (buttons or small toys), and several blank cubes to use in place of dice. The cubes are marked with names of integers.

Two cubes are used at a time. One represents addition: it has red numerals on it. The other one is for subtraction: it has blue numerals printed on it. It is suggested that each of the first cubes used be marked with 0, $^{+}1$, $^{+}2$, $^{+}3$, $^{+}4$, and $^{+}5$ in its respective color.

To start the game each player rolls the cubes. The person who rolls the higher value on the addition cube gets the first turn. In case of a tie, the player with the higher value on the other cube goes first.

It is agreed that subtraction means to do the opposite of addition. A $^{+}3$ on the addition cube means go forward three spaces. Then a $^{+}3$ on the subtraction cube means go backward three spaces.

Usually with the two cubes suggested, little progress is made by either player. (It won't take students long to realize that a $^{+}3$ on both cubes gets them exactly nowhere.) After several turns where little progress is made, a new cube should be introduced. A subtraction cube with $^{+}1$, $^{+}2$, $^{+}3$, $^{-}1$, $^{-}2$, and $^{-}3$ on the faces could speed up the game. (If this speed-up measure seems a bit inadequate after a fair trial, substitute a subtraction cube with 0, $^{-}1$, $^{-}2$, $^{-}3$, $^{-}4$, and $^{-}5$.) The person who reaches home first is the winner.

Suggestions for some other cubes:

Addition: $^{+}2$, $^{+}3$, $^{+}4$, $^{+}5$, $^{+}6$, $^{+}7$
Subtraction: 0, $^{+}1$, $^{+}2$, $^{+}3$, $^{+}4$, $^{+}5$ (Starter Cube)

Addition: $^{+}1$, $^{-}1$, $^{+}4$, $^{-}4$, $^{+}7$, $^{-}7$
Subtraction: Similar to, or same as, addition cube

Addition: $^{+}10$, $^{+}9$, $^{-}6$, $^{-}5$, $^{-}4$, $^{-}3$
Subtraction: $^{-}10$, $^{-}9$, $^{+}6$, $^{+}5$, $^{+}4$, $^{+}3$

Any combination of these or other cubes suggested by students should be tried.

After the students have played a few games, make a rule that the one who goes first also chooses the pair(s) of cubes with which to play the game. Of course, both players throw the same cubes.

This game, if introduced early enough (before any models for subtraction are offered), could be the vehicle by which students figure out a method for subtraction of integers. It is quite possible that they will intuitively arrive at the very "rule" so often presented to first-year algebra students in the past.

And what are the chances that this game could develop into a study in probability?

EDITOR'S NOTE.—There is a *good possibility* this might happen in a classroom in which the teacher creates a learning situation where someone can ask the question! It might be interesting to survey available children's games with reference to application of mathematical skills and principles. It may be that many more of them than we now realize can be used to introduce or to practice mathematical understandings.—CHARLOTTE W. JUNGE.

Climb the ladder

ALBERT H. MAUTHE

Albert Mauthe is chairman of the department of mathematics at A. D. Eisenhower High School at Norristown, Pennsylvania. He is on leave from his position this year to participate in the Experienced Teacher Fellowship Program in Mathematics for Prospective Supervisors at Teachers College, Columbia University.

Here is a new game. It is called "Climb the Ladder." The rules of the game are very easy to learn, and you will have lots of fun playing Climb the Ladder.

To play the game, you will have to make a ladder and a dial-spinner. (These will be shown later.)

As many players who wish to play the game together may do so. The object of the game is to go above the $^+12$ step on the ladder. Anyone who falls below the $^-12$ step on the ladder has fallen off the ladder, and he is out of the game. Thus a player can win the game in either of two ways:

1. if he is the first person to go above the $^+12$ step on the ladder, he wins the game; or

2. if he is the last person remaining on the ladder after all the other players have fallen off the ladder (by falling below the $^-12$ step), he wins the game.

Everyone starts by putting his marker at "Zero", the "0" step. We will call the numbers above 0 "positive numbers," and we will call the numbers below 0 "negative numbers." We will not call 0 either "positive" or "negative"; 0 will be called "zero."

The first person spins the spinner. If the spinner stops on a $^+1$, $^+2$, $^+3$, or $^+4$ space, the player moves his marker *up* the ladder the number of steps indicated (1, 2, 3, or 4). We will call the numbers $^+1$, $^+2$, $^+3$, and $^+4$ on the spinner-dial "positive numbers."

If the spinner stops on a $^-1$, $^-2$, $^-3$, or $^-4$ space, the player moves his marker *down* the ladder the number of steps indicated (1, 2, 3, or 4). We will call the numbers $^-1$, $^-2$, $^-3$, $^-4$ on the spinner-dial "negative numbers."

If the spinner stops on the 0 space, the player does not move up or down the ladder. We will call this number "zero."

Each player takes one turn at a time until he has fallen off the ladder or some player has been declared the winner.

Let us watch Betty, Bob, and Dave play the game. We will watch each player's progress up the ladder on the demonstration ladder. At the same time, we will keep a record of each player's steps at the right side of this page.

	BETTY	
Betty begins. She starts at Zero, 0.	Start	0
She spins a $^+3$. Thus, she goes 3 steps *up* the ladder.	Add	$^+3$
She is now on step $^+3$.	Step	$^+3$

Bob and Dave each take their first turns. Bob lands on $^-2$, Dave on $^-4$. The second turns for Betty, Bob, and Dave look like this:

	BETTY	
Betty was at $^+3$.	Start	$^+3$
She spins a $^+1$. Thus, she goes 1 step *up* the ladder.	Add	$^+1$
She is now on step $^+4$.	Step	$^+4$

	BOB	
Bob was at $^-2$.	Start	$^-2$
He spins a $^+3$. Thus, he goes 3 steps *up* the ladder.	Add	$^+3$
He is now on step $^+1$.	Step	$^+1$

	DAVE	
Dave was at $^{-}4$.	Start	$^{-}4$
He spins a $^{-}2$. Thus, he goes 2 steps *down* the ladder.	Add	$^{-}2$
He is now on step $^{-}6$.	Step	$^{-}6$

	BOB	
Bob spins a $^{-}2$ on his third turn. He was at $^{+}1$.	Start	$^{+}1$
Thus, he goes 2 steps *down* the ladder.	Add	$^{-}2$
He is now at $^{-}1$.	Step	$^{-}1$

	BOB	
Bob spins a 0 on his fourth turn. He was at $^{-}1$.	Start	$^{-}1$
Thus, he does not move either *up* or *down* the ladder.	Add	0
He is still at $^{-}1$.	Step	$^{-}1$

The game continues until one player passes $^{+}12$ or all but one player passes $^{-}12$.

Make your own game

Materials needed.—We'll want to play this game in class. But before we begin playing, we'll want to make our own game to keep. To make the game, each person in the room will make the game from these materials:

1. A square piece of heavy cardboard, about 15 centimeters on a side (15 centimeters is about 5¾ inches).
2. A piece of heavy paper or cardboard, approximately 8 centimeters by 28 centimeters (that is, about 3 inches by 11 inches).
3. A paper clip.
4. A thumbtack.
5. Several small pieces of cardboard or other objects to serve as markers.
6. A ruler, preferably one with centimeter markings.
7. A protractor.
8. Compasses (to draw a circle).

The dial-spinner.—On the heavy cardboard, draw a circle, with center in the middle of the cardboard, with a radius of 5 centimeters (about 2 inches).

Study the spinner-dial shown in Figure 1. Note that there are nine different sectors, four for the numbers $^{+}1$, $^{+}2$, $^{+}3$, and $^{+}4$, four for the numbers $^{-}1$, $^{-}2$, $^{-}3$, and $^{-}4$, and one for 0.

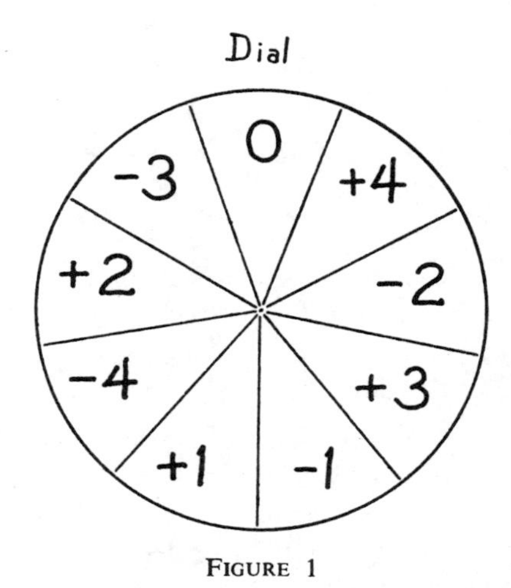

FIGURE 1

Do you remember that there are 360 degrees in one complete turn around the center of a circle? How many degrees must there be in each of the angles around the circle if we use 9 sectors that have equal angles?

Draw a radius from the center of the circle you have drawn to any point on the circle. Measuring with your protractor, draw another radius so that the two radii make an angle of 40 degrees at the center of the circle.

Continue making 40-degree angles until you have nine equally spaced radii, and nine sectors of the same size around the center of the circle. Label these sectors as shown in Figure 1.

Take a paper clip and bend the last loop out so that it looks like the one shown in Figure 2.

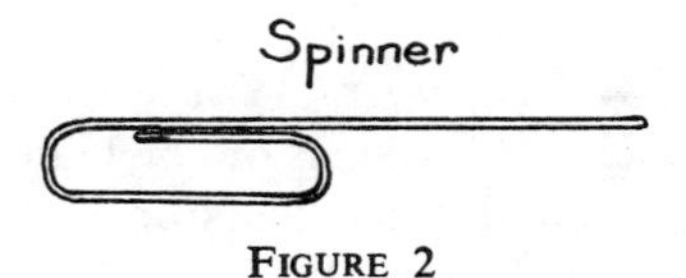

FIGURE 2

Place a thumbtack through the oval still remaining in the paper clip, and place the thumbtack through the center of the circle you have drawn. Your spinner-dial should look like the one shown in Figure 3.

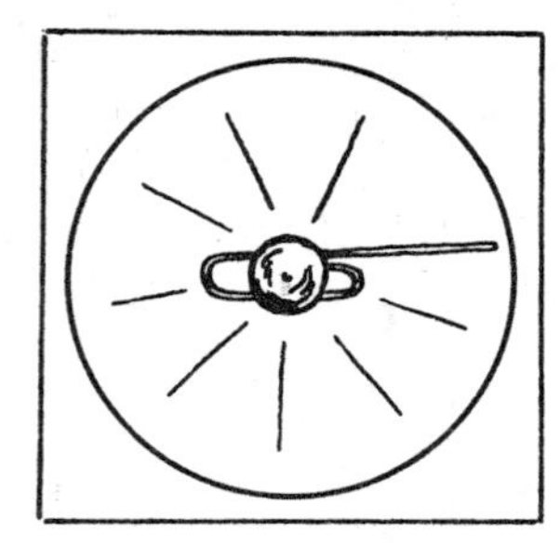

FIG. 3. —Thumbtack through oval of paper clip.

Make sure the paper clip spins freely around the dial.

The Ladder.—On a heavy piece of paper (or cardboard), draw a straight line segment 24 centimeters long (or 9 inches, if you do not have a ruler marked in centimeters). Make a mark at the bottom of the line segment. Then mark off each centimeter (use ⅜ inch, if you used a 9-inch line segment), until you get to the top of the line segment.

Label each step from $^{-}12$ to $^{+}12$, as was done with the ladder in Figure 4.

Let's play!

Although any number of players can play together, let's divide into groups of about three players each and try playing our new game. Each group will use one ladder; each player will use his own marker.

With each turn, be sure to move your marker up or down the ladder. And be sure to keep score, as we did for Betty, Bob, and Dave.

If your game finishes before the others, play another game. Keep score for this game also. If time runs out before anyone has won the game, the player highest on the ladder will be the winner.

Have fun!

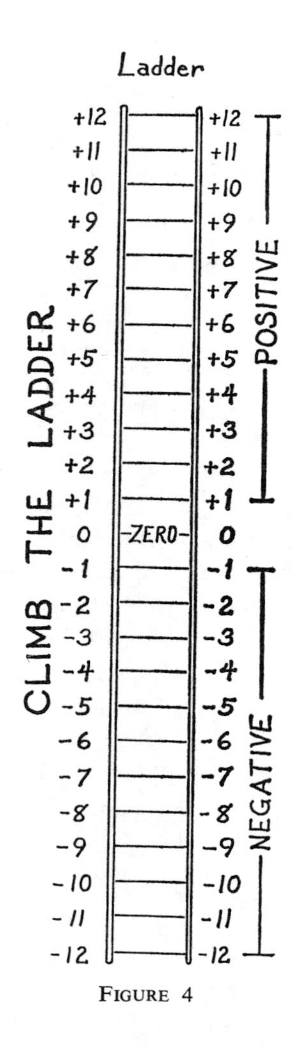

FIGURE 4

EDITORIAL COMMENT.—You may wish to have the student write a number sentence to describe each of his moves on the ladder. This will provide the connecting link between the manipulations and movements on the ladder and the symbolic representation of these movements. Many times the student fails to realize that these game experiences have any relationship to the symbolic language of mathematics.

Play shuffleboard with negative numbers

CHARLOTTE FRANK
Albert Einstein Intermediate School 131, Bronx, New York

"Shuffleboard, anyone?" Can you imagine how a student-filled math classroom would respond to that invitation? Compare that with the reaction to a teacher announcing, "Students, today we are going to learn the four fundamental operations of the set of integers." By the time the teacher had finished saying the word "fundamental" his pupils would have tuned out both him and his lesson. Negative Shuffleboard can be used to introduce the four basic operations of negative integers to the elementary school youngster in a fun setting.

For the few who have never played, regular Shuffleboard is a game in which players use long-handled cues to shove wooden disks into scoring beds of a diagram. Each player tries to push his own disk into one of the scoring areas or to knock his opponent's disk out of a scoring area and hopefully into a penalty area.

By adding positive and negative signs, the playing field (Fig. 1) can easily be transformed into Negative Shuffleboard (Fig. 2).

Other adjustments necessary to fit our new game to the facilities of a classsroom are the following:

1. Target diagrams are drawn on paper sheeting large enough to cover the student's table.

2. Red and black checker-size disks replace the larger disks normally used.

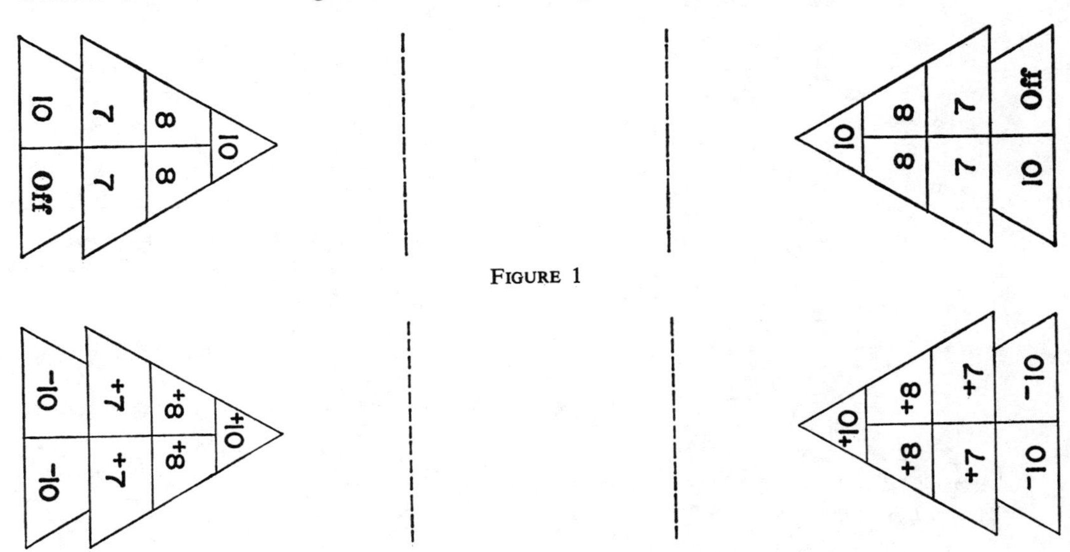

FIGURE 1

FIGURE 2

3. A pencil, pen, or ruler is used as a propelling instrument.

4. The ratio of the playing field to the disk is increased in order to allow for more scoring activity.

5. A number line is used to keep a running score.

When Anne and Harley played the game, their mathematical computations developed this way:

ROUND 1

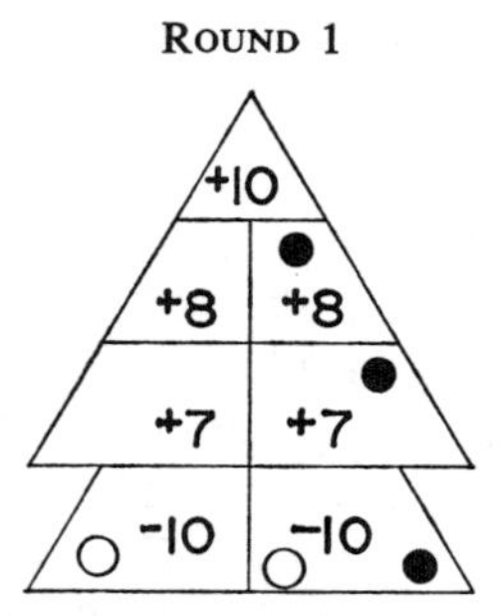

Anne (○ disks) $^{-}10 + {}^{-}10 = {}^{-}20$.
or $2 \times {}^{-}10 = {}^{-}20$.

Harley (● disks) $^{+}8 + {}^{+}7 + {}^{-}10 = {}^{+}5$.

Round 2

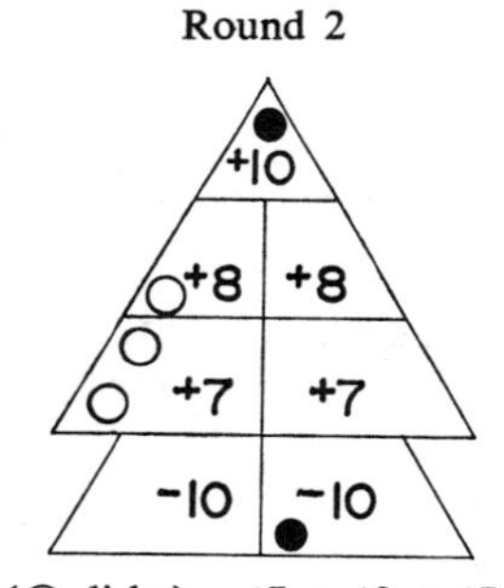

Anne (○ disks) $^{+}7 + {}^{+}8 + {}^{+}7 = {}^{+}22$.

Harley (● disks) $^{+}10 + {}^{-}10 = 0$.

Round 3

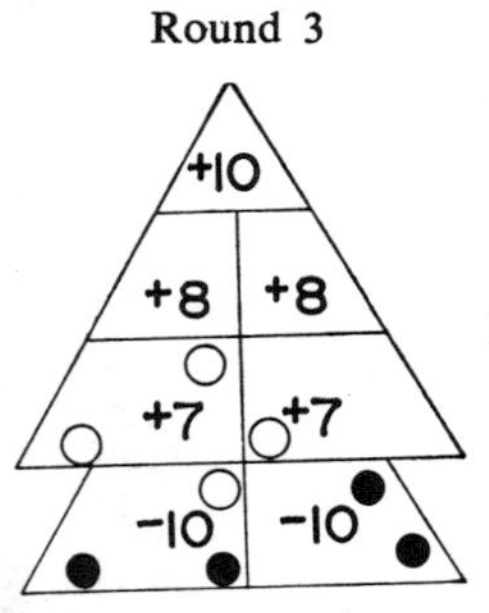

Anne (○ disks) $(3 \times {}^{+}7) + {}^{-}10 = 11$.

Harley (● disks) $4 \times {}^{-}10 = {}^{-}40$.

This game continues until a previously agreed upon winning score is reached.

Some of the computations at another table, where Matthew and Sidney were seated, were as follows:

Situation 1.—Matthew had $^{+}7$, $^{+}8$, $^{+}10$ and one disk out of limits, giving him a total of $^{+}25$. When Sidney's turn came he successfully knocked Matthew's disk out of the $^{+}8$ box and into the $^{-}10$ box. Matthew's $^{+}8$ was taken away; written mathematically, it was $-{}^{+}8$. This number sentence describes the last play:

$$^{+}8 - {}^{+}8 = 0.$$

Situation 2.—Sidney successfully knocked his own disk out of the $^{-}10$ penalty box. Translated into our language of numbers, Sidney had a $^{-}10$. When he took away the $^{-}10$, it gave him a 0 for the box. This number sentence is written:

$$^{-}10 - {}^{-}10 = 0.$$

Situation 3.—After Round 4 Matthew had 30 points more than Sidney had. With three disks still to shoot, Sidney told the group watching that he could win if Matthew were unlucky enough to end the round with a $^{-}30$. That would be written by the following number sentence:

$$^{-}30 \div 3 = {}^{-}10.$$

The target diagrams vary from table to table to meet the needs of the individual

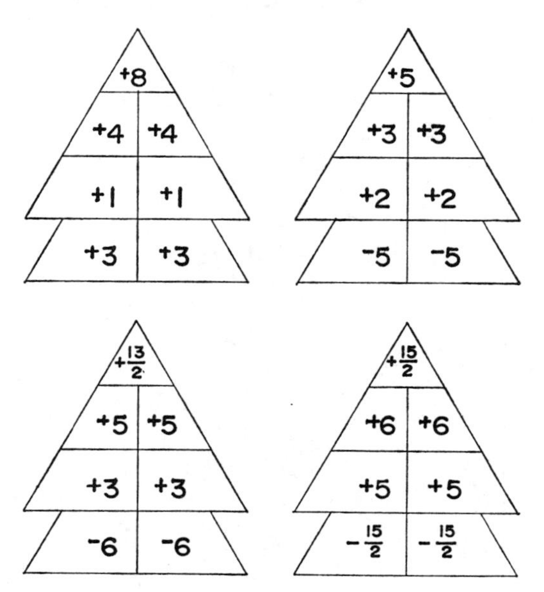

FIGURE 3

children. Some suggestions for a few possible changes to match the learning styles of your students are indicated in Figure 3.

EDITORIAL COMMENT.—Many variations of this activity are possible. A beanbag-game format could be used in a manner similar to that suggested for the shuffleboard activities. The game arrangement will actively involve the children and provide an interesting format for skill development. If the students can be encouraged to record their scores on a number line and write equations to represent the operations, they will gain additional insight into the operations with directed numbers.

Disguised practice for multiplication and addition of directed numbers

ESTHER MILNE

Esther Milne is a Mathematics Helping Teacher in the Tucson Public Schools, Tucson, Arizona.

Practicing computation can be drudgery. But practicing computation with integers is necessary. Games aren't drudgery. Let's combine games with practice. Then students can have fun as they improve the accuracy of their computation.

Try this game.

Integer Game

How many can play? Two, three, or a whole class divided into two teams.

What equipment is needed? Two spinners* and paper on which to write scores. (When a large group is playing as two teams, it is best to have replicas of the spinner dials on an overhead projector transparency.)

Mark some numerals + (positive) on the first spinner. Mark the rest of the numerals − (negative). On the second spinner label a couple of the numerals "power" and the rest positive and negative. (There might be a zero on a dial.)

How do you play? Spin the first spinner. The number on which the pointer stops is the first factor. Spin the second spinner. Where the pointer stops gives the second factor, or tells us the power to which the first factor is to be raised. (If a large group is playing, place a marker on each dial on the overhead projector to show where the pointer stopped. A coin could be used as a marker.) Record the product. Keep a running score.

Who wins? Decide before the game

* If you don't have your own favorite source for purchasing spinners, these are available from Vroman's, 367 S. Pasadena Ave., Pasadena, Calif. 91105.

starts how many turns will constitute a game. After the last turn the player or team whose score is farthest from zero is the winner. Score is kept by all players. Each one keeps his own score and that of his opponent(s).

If 10 turns is to be a game, the score sheet looks something like this:

ME

	Factors	Product	Running Score
1.	$^-5 \cdot {}^+4$	$^-20$	$^-20$
2.	$^+4 \cdot {}^+4$	$^+16$	$^-4$
3.	$(^+3)^3$	$^+27$	$^+23$
4.	$0 \cdot {}^-6$	0	$^+23$
5.	$0 \cdot {}^+1$	0	$^+23$
6.	$^-8 \cdot {}^-6$	$^+48$	$^+71$
7.	$(^+1)^2$	$^+1$	$^+72$
8.	$(^+3)^2$	$^+9$	$^+81$
9.	$0 \cdot {}^-5$	0	$^+81$
10.	$(^-7)^2$	$^+49$	$^+130$

THEE

	Factors	Product	Running Score
1.	$^-5 \cdot {}^-6$	$^+30$	$^+30$
2.	$^+3 \cdot {}^+4$	$^+12$	$^+42$
3.	$(^-5)^2$	$^+25$	$^+67$
4.	$0 \cdot {}^-6$	0	$^+67$
5.	$(1)^2$	$^+1$	$^+68$
6.	$^-6 \cdot {}^+1$	$^-6$	$^+62$
7.	$^+2 \cdot {}^-5$	$^-10$	$^+52$
8.	$(^-6)^3$	$^-216$	$^-164$
9.	$^-8 \cdot {}^+1$	$^-8$	$^-172$
10.	$(^+2)^3$	$^+8$	$^-164$

Thee is the winner!

EDITORIAL COMMENT.—Many variations of this game are possible. For example, a basic addition game may be played by having students add the two integers identified by spinning the two spinners. A subtraction game may be played by having a student spin the first spinner to identify the minuend and then spin the second spinner to identify the subtrahend. The student then finds the difference between the two integers.

If you are looking for a way to design spinners that work extremely well, borrow a pair of plastic lazy Susan turntables from your kitchen cabinets and cut posterboard disks to fit in the turntables. The disks can be marked and numbered according to your needs and a pointer can be placed on the desk top beside the turntable.

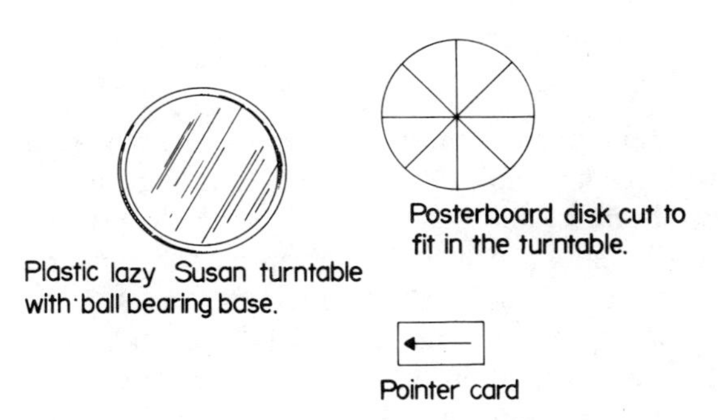

A student-constructed game for drill with integers

MERLE MAE CANTLON
Jefferson Junior High School, Caldwell, Idaho

DORIS HOMAN
Vallivue Junior-Senior High School, Caldwell, Idaho

BARBARA STONE
Central Junior High School, Nampa, Idaho

In teaching a unit on integers, teachers are invariably faced with the age-old problem of how to provide interesting and meaningful drill. Popsicle sticks can be used to provide a motivating, "hands on" laboratory exercise in which the students build their own game. The anticipation of using what they are studying in a fun situation seems to be an added motivational factor for students. The exercise shown in figure 1 is a suggestion for the first activity, in which the children make the materials and experiment with the "Popsicle-Stick Game."

For this activity, the students should be divided into groups of three or four. Without further instructions, each group is given its assignment sheet for the day (fig. 1).

The second activity involves the actual playing of the Popsicle-Stick Game with the numbered Popsicle sticks that the children have made. Instructions for playing the game follow:

1. The class is divided into teams of three or four students, and each team is given a set of numbered Popsicle sticks.

Building a Popsicle Stick Game

Names of team members:______________________________

Check the box at the right when the team has completed each step, please.

1. Get 15 Popsicle sticks and one felt marker. ☐
2. Mark $\underline{-1}$ on the first stick, keeping the sticks in vertical position. Mark $\underline{-2}$ on the second stick, and so on through $\underline{-5}$. ☐
3. Mark 0 on both sides of one stick. ☐
4. Mark $\underline{-1}$ on one side of a stick and $\underline{1}$ on the opposite side. Mark $\underline{-2}$ on one side of a stick and $\underline{2}$ on the opposite side, continuing on in this manner through $\underline{-9}$ and $\underline{9}$. ☐
5. Build with Popsicle sticks the following math phrases (expressions) that are equal to $\underline{-4}$.
   a. $\underline{-3 + -1} =$ ☐
   b. $\underline{-8 + 4} =$ ☐
   c. $(-4 \times 2) + 4 =$ ☐
6. With your sticks, write three math phrases that are equal to 48.
   a. ______________ = 48 ☐
   b. ______________ = 48 ☐
   c. ______________ = 48 ☐
7. Using at least one negative integer in each, write three math phrases equal to 3.
   a. ______________ = 3 ☐
   b. ______________ = 3 ☐
   c. ______________ = 3 ☐

Fig. 1

2. Students on each team take turns acting as captain, the honor rotating around the table with each new problem. A permanent scorekeeper should be chosen for each team.

3. The object of the game is to see which team, using one or more of the four operations on integers, can be the first to construct a mathematics problem with an answer equal to the integer that the teacher has written on the board. Numbers from -100 through 100 can be used for the integer answers, and for variation, these can be written on slips of paper and then drawn from a box by the captain of the last winning team.

4. When a team has constructed a problem with an answer equal to the integer on the board, all members of the team must raise their hands. The teacher notes the first team whose hands are all raised. When it is apparent that most groups have completed a problem, time is called.

5. The completed problems are written on the board (or on the overhead projector) by the respective team captains, and the accuracy of all problems is checked by the class. The team captains have the responsibility for writing the correct operations as well as the correct integers, and for indicating the order in which the operations are to be performed.

6. If the problem prepared by the first team whose hands were raised is correct, the team receives four points. Two points are awarded to each of the other teams that has a correct problem. Teams with incorrect problems receive a score of -1. Each team keeps its own score.

7. For variations, the teacher may place restrictions on the problems to be constructed. For example, some possible restrictions are:

a. There must be at least one negative integer in the problem.
b. At least two operations must be used in the problem.
c. The problem must be worked with a minimum of two sticks, or three sticks, and so on.

The Popsicle-Stick Game has several advantages. It promotes total student participation—groups are small. The honor of acting as captain is rotated within the groups. *All* members of the team must raise their hands in order for the team to be recognized as first. All students are involved in checking the problems that are put on the board, since numerous correct responses are possible. The scoring procedure provides additional opportunities for practice. "Hands on" materials are used, and little teacher direction is needed.

In summary, this experience provides students with useful and enjoyable practice for operations with integers through the use of manipulative materials in a laboratory setting. In the words of a Chinese philosopher: I hear and I forget; I see and I remember; I *do* and I understand.

EDITORIAL COMMENT.—The first activity shown in figure 1 requires the students to "write math phrases" in steps 5, 6, and 7, using Popsicle sticks; however, there is no provision for the operation signs. You may wish to provide each team of students with a number of paper squares on which they may write the desired operational symbols in order to complete their math phrases with the Popsicle sticks.

Game activities such as these can provide very interesting and stimulating practice experiences, but these experiences should be preceded by meaningful instruction to develop an understanding of the operations with integers.

Grisly grids

WILLIAM G. MEHL and DAVID W. MEHL

William Mehl is a teacher of mathematics at Wilson Junior High School in Pasadena, California.

David Mehl, his son, was a student in Mr. Mehl's summer class in exploratory algebra and assisted in the writing of the article.

The multiplication lattice grid is familiar to most teachers and many students throughout the country. A more challenging version of this grid type might include directed numbers. The writers have modified the traditional presentation and introduced a grid incorporating a new set of "rules." The grid can be used to stimulate and to entertain the most ambitious pre-algebra or algebra student.

Let's look at the grid's anatomy. Each small square shall be called a cell. Each cell consists of two congruent triangular regions. The upper region is designated *UR* and the lower region *LR* (Fig. 1). The total grid may be in the form of a square or other rectangle, thus enabling the teacher to regulate the number of factors and cells for any given presentation. All marginal entries at the top and right of the grid shall represent factors, while the entries at the left and bottom shall represent sums. Each cell contains a product. All entries consist of integers, and all marginal entries are integers x such that $^{-}10 < x < {}^{+}10$.

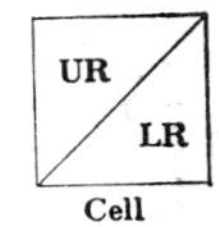

FIGURE 1

The following rules regulate the operation of the grids:

1. All tens are recorded in *UR*, while all ones are entered in *LR*.
2. A negative product consisting of two digits is always recorded such that the tens digit is negative and the ones digit is *not* negative.
3. A positive product consisting of two digits is always recorded such that neither digit may be negative.
4. The integers in the margins opposite any cell may have one digit only.
5. Marginal sums are found by adding obliquely from *right* to *left,* starting in the lower right corner of the rectangle.
6. The solution consists of any necessary marginal or tabular entries, and in some cases requires that the marginal sums be totaled and entered in a frame as directed by an arrow (Fig. 4).

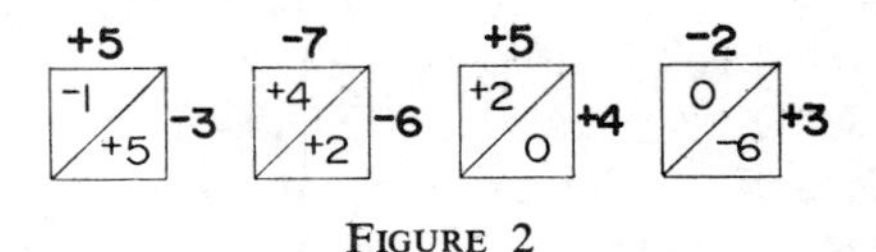

FIGURE 2

The teacher might begin simply by presenting a grid such as Figure 3. A variation of the first presentation might appear

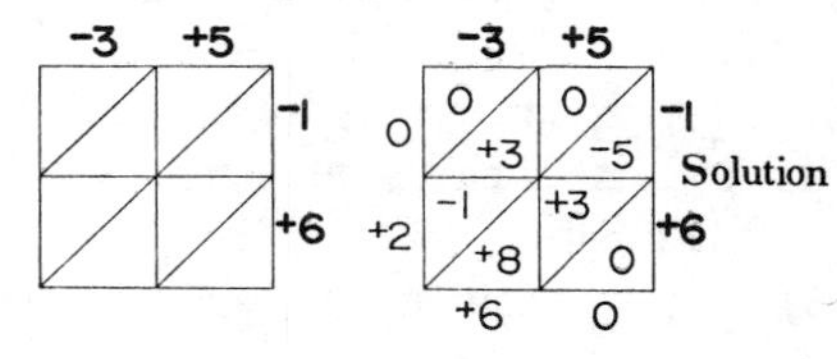

FIGURE 3

as shown in Figure 4. Note the required answer recorded in the frame at the head of the arrow.

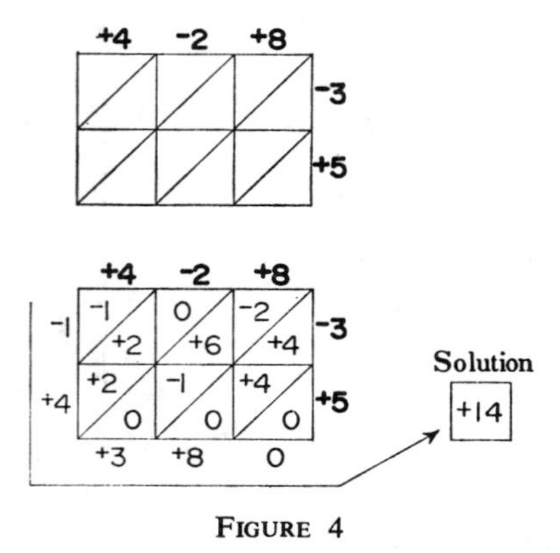

FIGURE 4

It is suggested that the teacher formulate several grids of the type previously shown to provide sufficient orientation for students and as a prerequisite to attempting the grid types that follow. The student must be thoroughly familiar with the multiplication, division, and addition of directed numbers, some minor rules for divisibility, and have a talent for observation and deductive reasoning. Of course, the teacher may elect to have students attempt some simpler solutions than those shown here, or confine the entire presentation to merely determining products.

Complete the grid (Fig. 5) and place your final answer in the frame at the head of the arrow.

As the students attempt the solution of this and similar grids, many observations may be noted, and the writers have found that successful students are most eager to share them. These observations include the following:

1. No product can exceed 81 or be less than $^{-}81$.
2. No factor can consist of two digits.
3. If *UR* is positive or negative, than *LR* is positive.
4. If *UR* is positive, the signs of both factors are alike.
5. If *UR* is negative, the signs of both factors are unlike.
6. If *UR* is 0, then *LR* is 0, negative

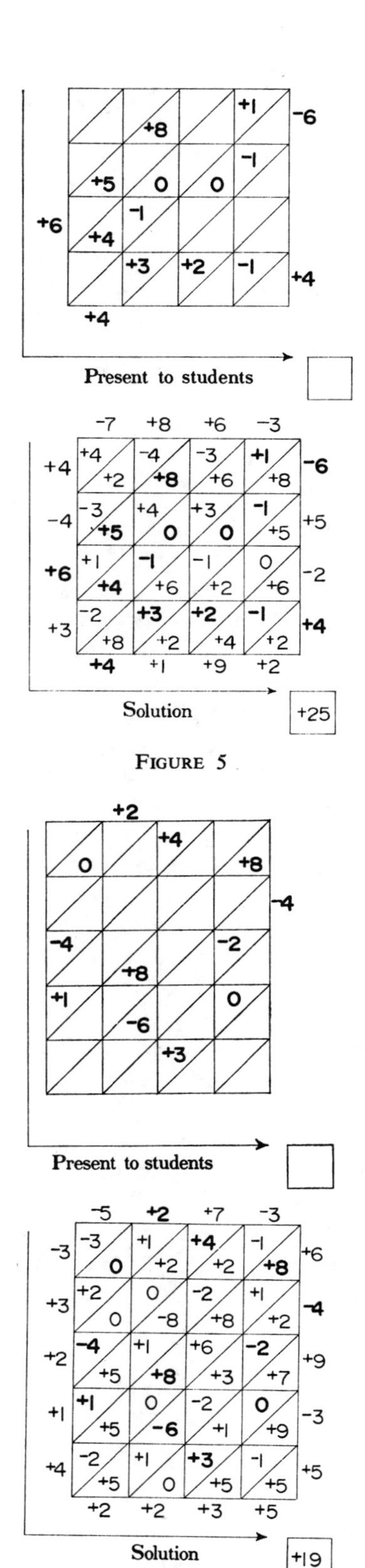

FIGURE 5

FIGURE 6

or positive.

7. If LR is negative, then UR is 0.
8. If LR is even, then either the upper or right-hand marginal entry for that cell is even.
9. If LR is 0 or 5 and $UR \neq 0$, then either the upper or right-hand marginal entry for that cell is 5 or $^{-}5$.
10. If "carrying" is necessary while adding obliquely, a positive sum "carries" a positive tens digit, while a negative sum "carries" a negative tens digit.

The reader may enjoy attempting the next grid (Fig. 6) while paying particular attention to the observations previously noted.

Perhaps some explanation or advice may be useful to the teacher who desires to formulate grids of his own. For best results, the writers recommend that a completed grid be formulated at the beginning. The teacher would then contribute to the difficulty of the solution by removing appropriate factors, products, sums, and partial products. Here it is advisable for the teacher to attempt his own solution in order to decide whether or not he may have abbreviated the original grid beyond the point of effecting a complete solution.

An alternate method of presenting and solving grids of our type follows.

Complete the grid in accordance with the mathematical instructions below:

1. A is not positive.
2. $d = 5$.
3. $^{-}3 < g < {}^{-}1$.
4. k is not negative.
5. $(A)(F) > 0$.
6. $B \neq 7$.
7. $7 < p < 9$.
8. $i > 0$.
9. $|A| + |B| = (C)(D)$.
10. $D - A = j$.
11. $q \nless 0$.

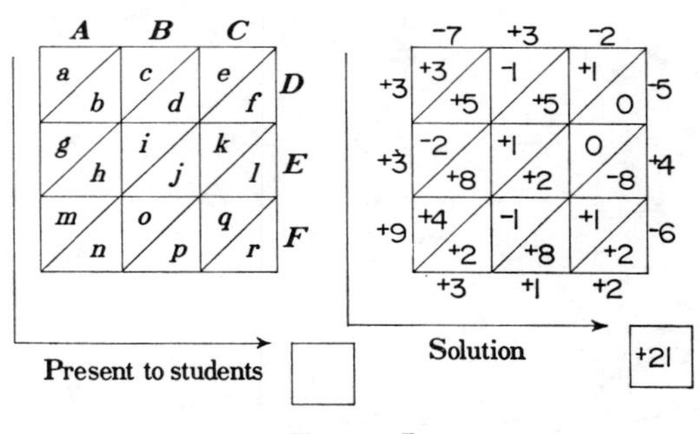

FIGURE 7

Editorial feedback

MARY HELEN BEAN
Bel Air Junior High School, Bel Air, Maryland

My eighth graders thoroughly enjoyed working with "Grisly Grids" (May 1969) after their study of integers. The grids proved to be a highly motivating type of drill.

The students found that one of the grids does not have a unique solution (fig. 5, p. 358). Some of the students examined it and found that the inclusion of one additional number would have resulted in a unique solution. (Note fig. 1.)

The circle can be replaced with ⁻8, ⁺8, ⁻7, or ⁺7, in order for the original grid to have but one solution. The four possible

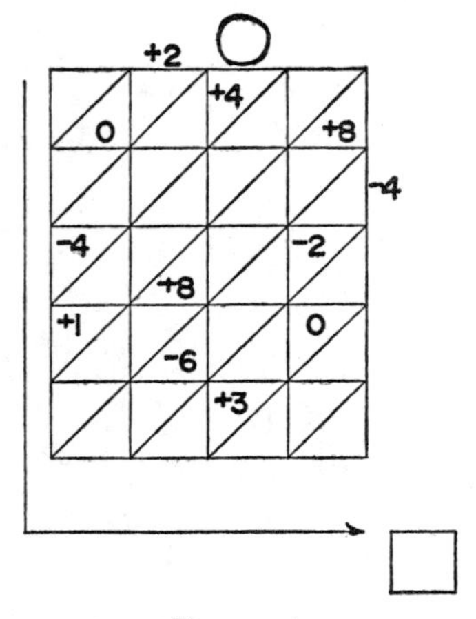

FIGURE 1

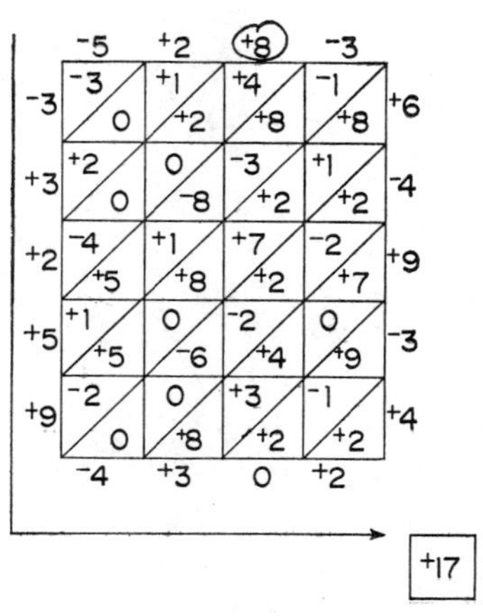

FIGURE 2

solutions obtained when the additional number clue is provided are shown in figures 2–5.

Our thanks to the authors William G. Mehl and David W. Mehl for making drill so enjoyable!

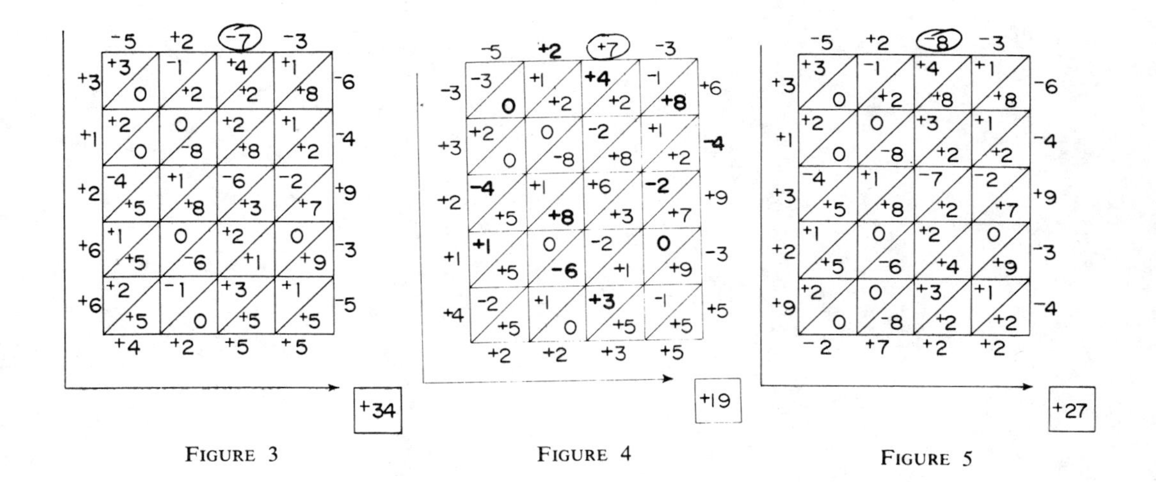

FIGURE 3

FIGURE 4

FIGURE 5

5

Rational Numbers

A considerable portion of the elementary school mathematics curriculum is devoted to the teaching of fractional numbers and their operations. Although whole numbers are extremely important in everyday living, it is the use of rational numbers with their fractional and decimal representations that has made progress possible in any field requiring measurement.

Many of the games selected for this section deal with the equivalence of names for rational numbers. These include equivalences of decimals and percents as well as the usual equivalences with fractions. To create instructional games for use with operations on fractions, many of the games suggested for whole-number operations may be adapted.

The first article, by Junge, gives directions for a most useful fraction activity: fraction strips. She suggests several games with these materials that are either based on equivalences or designed to show how fractional parts relate to make a whole.

If Carlisle's "Crazy Fractions" reminds you of Crazy Eights, it should. Students get much practice recognizing like denominators and fraction equivalences with this game.

"Fracto," by Molinoski, is a game of fraction, decimal, and percent equivalences. It is strongly reminiscent of the universal game called poker.

Lazerick's "Conversion Game" is a variation of Bingo using percent, decimal, and fraction equivalences. Be sure to let the winner of one game be the caller for the next game. Armstrong's "Fradécent" is a card game related to the same equivalences.

The activity discussed in "An Adventure in Division" helps children to understand the concept of repeating decimals and to recognize that rational numbers expressed in fraction form can also be expressed in repeating decimal notation. Lots of practice in long division as well as in graphing is provided in this activity.

The article "Introduction to Ratio and Proportion" will give you some ideas on presenting rate pairs to children and for showing how rate pairs differ from ordinary fractions. The function machine can be used effectively with this activity.

Rode's "Make-a-Whole" and Zytkowski's "Game with Fraction Numbers" are useful when working with operations on rational numbers. Make-a-Whole gives students an intuitive feeling for many ways to generate a unit region from fractional parts. "A Game with Fraction Numbers" provides much practice with computations of the four basic operations.

A game of fractions

CHARLOTTE W. JUNGE

This game may be played by two, three, or four children. Each one makes his own set of fractional parts as follows:

1. Cut from lightweight tagboard five rectangular regions 12″ × 4″.

2. Very carefully mark four of them into fractional parts, and label them as shown below. *Leave one region unmarked as a measure.*

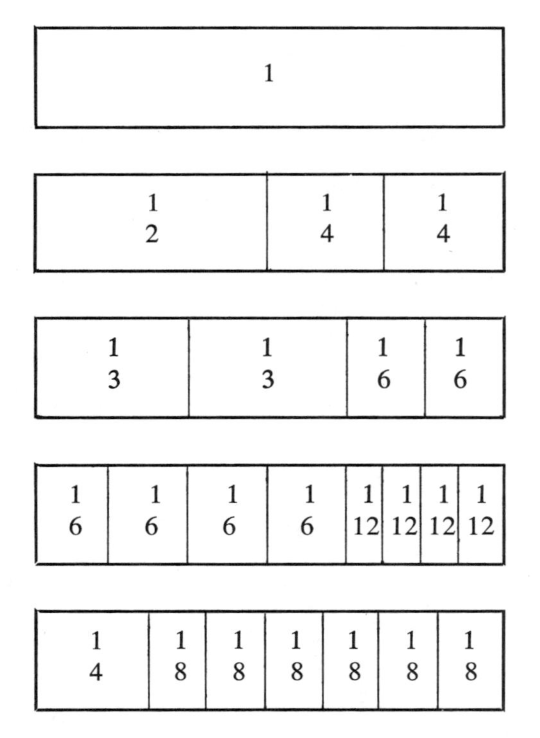

3. Cut along the lines drawn. Write your name or initial on the back of each of the fractional parts and place them in an envelope for ease of storing.

These parts may be used for several games. For example:

I. Each player empties all the fractional parts from his envelope into a common pile. Each player in turn draws a part until all have been drawn. Then each player begins assembling the parts he has drawn into rectangular regions the same shape and size as the uncut card. The first child to assemble three "wholes" wins the game.

II. Start as before. Set a limit of three to four minutes. The player who in this time has the greatest number of equivalent expressions for $\frac{1}{2}$ wins.

III. This game is something like "Authors." Each player in turn quickly draws two fractional parts from the common pile until he has taken ten. The remaining parts are put aside in a box and not used in the game. The player who first fits *all* of the parts he has drawn into a region or regions equivalent to one, wins. Besides the parts he has drawn, each player in turn may ask any other player for the parts he needs, giving in return an equivalent, thus: $\frac{3}{12}$ for $\frac{2}{8}$; $\frac{1}{4}$ for $\frac{2}{8}$; $\frac{2}{6}$ for $\frac{1}{3}$, etc.

A variation of this game may be played without reference to objective materials. Children often refer to this game as "Make One." One player starts the game, saying "$\frac{1}{3}$, $\frac{1}{6}$, make one." The next player adds another fraction which will "make one," thus: "$\frac{1}{3}$, $\frac{1}{6}$, $\frac{1}{2}$," "$\frac{1}{3}$, $\frac{1}{6}$, $\frac{3}{6}$," or "$\frac{1}{3}$, $\frac{1}{6}$, $\frac{4}{8}$." This game is played orally, without reference to written solutions, except where they are needed for clarification.

EDITORIAL COMMENT.—This game should provide some very interesting practice with a variety of fractional-number concepts. Additional variety could be provided by the construction of five circular regions, which are marked and cut into the same sets of fractional parts suggested for the rectangular regions. This would enable the students to view the fractional parts in another spectrum.

Crazy fractions: An equivalence game

EARNEST CARLISLE

Columbus College, Columbus, Georgia

The game "Crazy Fractions" is useful in helping fifth and sixth graders learn equivalent fractions and the renaming of whole numbers as fractions.

The game is played with a deck of seventy-two cards. There are cards for the following sets of fractions: halves, thirds, fourths, fifths, sixths, sevenths, eighths, ninths, tenths and fourteenths. There are also cards for the whole numbers one, two, three, and four.

The game works ideally with four players; however, any reasonable number of players may play. The game begins with the dealer giving each player seven cards, turning one card faceup (starting the discard pile), and placing the remaining cards in the center of the table facedown. Play starts with the player to the left of the dealer discarding a card of the same suit (same denominator) or of equivalent value to the top card of the discard pile, or a whole number card. Whole number cards act as wild cards because a player may rename a whole number card to any suit he chooses.

During his turn, a player may discard or draw a card from the deck. A player does not both draw and discard unless the drawn card is discardable. The object of the game is for a player to get rid of his cards.

A typical hand may begin as follows (refer to fig. 1): Dealer turns 0/4 faceup;

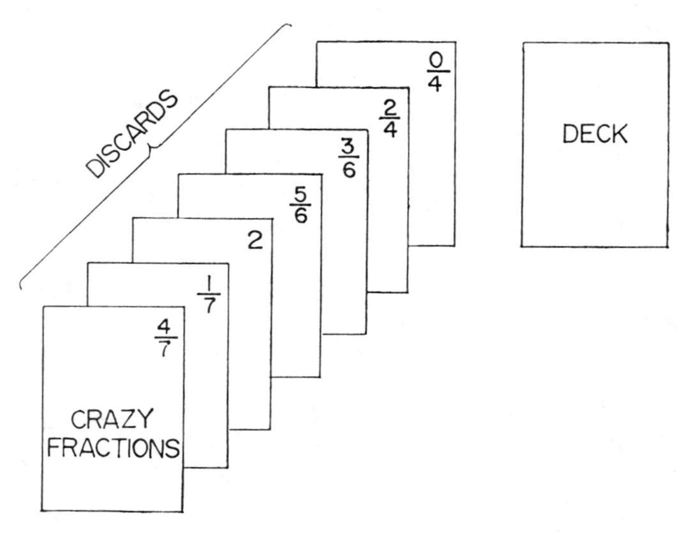

Fig. 1

player one discards 2/4; player two has no fourths but discards the equivalent fraction 3/6; player three discards 5/6; dealer has no sixths or equivalent fraction but draws 2 from the deck, which he plays and names sevenths as the suit; player one follows by discarding 1/7; player two discards 4/7; and the play continues until one player is rid of all of his cards.

When one player gets rid of all of his cards, or when no one can play, the hand is over and the points are totaled. In case of a blocked game, no one receives any points. Otherwise, a player receives twenty-five points for getting rid of his cards plus the total of the cards remaining in all other hands—one point for each fraction less than 1/2; two points for each fraction equal to or greater than 1/2; and three points for each whole number. Before play begins, players should agree on the number of points that will constitute the game. Fifty- and hundred-point games have worked well in the past.

EDITORIAL COMMENT.—The cards for homemade card games such as "Crazy Fractions" can be made more attractive and more durable if you cover the back side of the card with a colorful pattern of contact paper and then cover the front, or playing, side with clear contact paper. The cards may be cut from index cards, posterboard, or other lightweight cardboard.

Fracto

MARIE MOLINOSKI

Worth Junior High School, Worth, Illinois

Upper elementary and junior high mathematics teachers are always searching for ways of teaching or reviewing the relationships between decimals, percents, and fractions. *Fracto* is a card game designed to make the review a little more fun.

Materials needed

Fracto is played with a deck of fifty-two cards. The cards can be of any convenient size; index cards may be used. The make-up of a deck of *Fracto* cards can vary depending on the type of practice desired. Following are some suggested decimals, percents, and equivalent fractions:

One set consists of thirteen fractions. 1/5, 1/4, 3/8, 2/5, 1/2, 3/5, 5/8, 2/3, 3/4, 4/5, 5/6, 7/8, and 8/9.

One set consists of thirteen equivalent fractions with a denominator of 100.

$$\frac{20}{100}, \frac{25}{100}, \frac{37\ 1/2}{100}, \frac{40}{100}, \frac{50}{100}, \frac{60}{100}, \frac{62\ 1/2}{100}, \frac{80}{100}, \frac{66\ 2/3}{100}, \frac{75}{100}, \frac{83\ 1/3}{100}, \frac{87\ 1/2}{100}, \text{ and } \frac{88\ 8/9}{100}.$$

One set consists of thirteen equivalent decimals. .20, .25, .37 1/2, .40, .50, .60, .62 1/2, .66 2/3, .75, .80, .83 1/3, .87 1/2, and .88 8/9.

One set consists of thirteen equivalent percents. 20%, 25%, 37 1/2%, 40%, 50%, 60%, 62 1/2%, 66 2/3%, 75%, 80%, 83 1/3%, 87 1/2%, and 88 8/9%.

Number of players

The number of players can be from two to six. A class could be divided into groups of varied size.

Rules of the game

Shuffle the fifty-two cards and deal five cards to each player one at a time. Players sort their cards, trying to keep equivalent cards together. Starting with the player to the left of the dealer, each player is given a chance to discard from one to three cards and ask for new ones. A player may choose to keep his hand as it was dealt. After every player has had an opportunity to draw new cards, the player to the left of the dealer lays down his hand with the other players following in order. The player with the best hand wins. When the points have been recorded, a new hand is dealt and the game continues until a player reaches 300 points.

Point values of cards in hands.

Two of a kind or one pair	20 points
Two pairs	40 points
Three of a kind	50 points
Straight	60 points
Example: 5/8, 2/3, 3/4, 4/5, 5/6	
Four of a kind	70 points
Full house	80 points
Example: 1/2, .50, 50%, 1/4, .25	

The conversion game

BETH ELLEN LAZERICK
The Dalton School, New York, New York

Just how do you convince a sixth or seventh grader that he should be able quickly to recall fraction-decimal-percent equivalents? After students have used the concept of (for example)

$$\frac{1}{2} = .50 = 50\%$$

in word problems and computations, they frequently forget their facts; and to help them retain mastery of the various equivalencies I fashioned the "Conversion Game." This game also reinforces the concept that a "number" is represented by many "numerals."

Playing cards fashioned like bingo cards were made from brightly colored oaktag. I used cards measuring five inches by five inches, but three-by-three or four-by-four cards could be used with slight alterations. My key list contained 28 sets of basic facts and their equivalents. This number can vary according to the level of the class. The list included halves, thirds, fourths, fifths, sixths, eighths, tenths, elevenths, twentieths, and one-hundredths. Fractions in lowest terms, nonreduced fractions, decimals (repeating and nonrepeating), and percentages comprised the list. A portion of it is reproduced in figure 1.

The students had learned that .6 and $.\overline{6}$ ("point 6 repeating") are *not* the same and that $.\overline{6}$, $\frac{2}{3}$, $\frac{4}{6}$, and $66\frac{2}{3}$ percent are all numerals representing the same number.

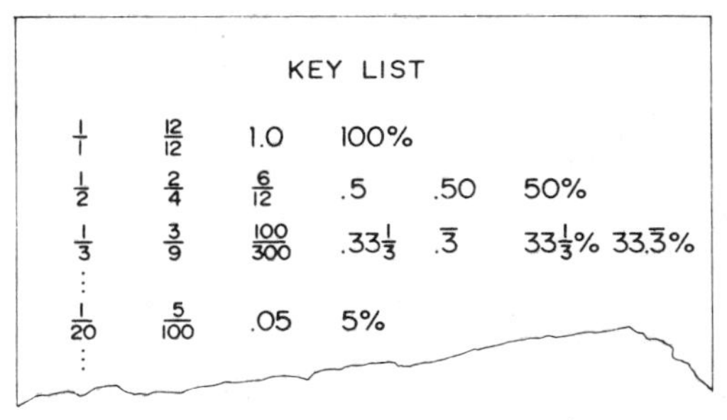

Fig. 1

These and other facts were placed randomly on the cards so that each card had an equivalent of 24 of the 28 facts and each card had about eight fractions, eight decimals, and eight percentages. Samples of the cards are shown in figures 2 and 3.

To play the game, the caller (teacher or student) simply chooses randomly from the key list any equivalent—or he can select from duplicates of the list that are placed in a box, as in bingo. He might select $\frac{1}{3}$ or $\frac{100}{300}$. He then calls the number and writes it on a chalkboard while the players search for any equivalent (i.e., $.\overline{3}$ or $33\frac{1}{3}$ percent) on their cards. When a player locates an equivalent on his card, he covers it with a simple marker. After many pulls from the key-numbers box or selections from the list, someone will complete a horizontal, vertical, or diagonal line and call out "Conversion!" The caller then checks the winning line by looking for equivalents of all the covered numerals as listed on the chalkboard.

The game is over if the student is correct. Boards can then be cleared and the game repeated two or three times without, in most cases, any loss of interest.

The advantage of putting an equivalent of nearly every key number on each card is that students then know they have to look carefully after each call. Better students simply find more answers on their cards, while slower students can always identify halves, thirds, and other simple equivalents.

Since most students enjoy games, the Conversion Game proves to be a fun way to review important facts long after the "real" mathematics of these concepts have been stored away!

The Conversion Game				
$8\frac{1}{3}\%$	$\frac{1}{4}$	90%	$\frac{6}{10}$	.07
$\frac{1}{6}$	100%	$\frac{6}{40}$	.5	$.83\frac{1}{3}$
1%	37.5%	FREE	5%	20%
$\frac{80}{90}$	$\frac{3}{33}$	$87\frac{1}{2}\%$	.3	$.\overline{6}$
$\frac{7}{10}$	$\frac{100}{300}$	$\frac{2}{16}$	80%	$\frac{2}{20}$

Fig. 2

The Conversion Game				
.375	$\frac{7}{100}$	60%	15%	$33\frac{1}{3}\%$
$66\frac{2}{3}\%$	.1	$\frac{1}{12}$	.4	$\frac{5}{6}$
$\frac{14}{16}$	$\frac{2}{200}$	FREE	70%	$\frac{2}{8}$
$16\frac{2}{3}\%$	$\frac{50}{100}$	.05	$\frac{4}{5}$	1.0
.90	$\frac{1}{11}$	.75	30%	$\frac{3}{24}$

Fig. 3

"Fradécent"—a game using equivalent fractions, decimals, and percents

CHARLES ARMSTRONG

Claypit Hill School
Wayland, Massachusetts

Fradécent is a card game that is useful in helping sixth graders learn equivalent fractions, decimals, and percents as well as aiding them in adding and subtracting fractions. The word *fradécent* contains parts of the words fraction, decimal, and percent.

The game is played with a deck of eighty-one cards. The following fractions are used: 1/2, 2/4, 4/8, 5/10, 1/4, 2/8, 4/16, 3/4, 6/8, 12/16, 1/5, 2/10, 3/15, 2/5, 4/10, 6/15, 4/5, 8/10, 12/15, 1/8, 3/8, 5/8, 7/8, 1/10, 3/10, 7/10, and 9/10. Cards with the corresponding equivalent decimals and percents are used with these fraction cards.

Play begins with the dealer giving each player seven cards and placing the remaining cards in the center of the table face down. The player to the left of the dealer begins by drawing the top card from the deck. Each player in turn tries to make a scoring set. There are two possible scoring sets: (1) a set of three cards consisting of an equivalent fraction, a decimal, and a percent (one of each), and (2) a set of three cards consisting of three equivalent fractions. When a player is able to make a scoring set, he places the cards face up in front of himself. Whether or not he is able to make a scoring set, each player must discard (place a card face up beside the deck of cards in the center of the table) to complete his turn. A player may not go out without discarding. Play continues clockwise.

Additional scoring is possible by adding on to existing scoring sets. This may be

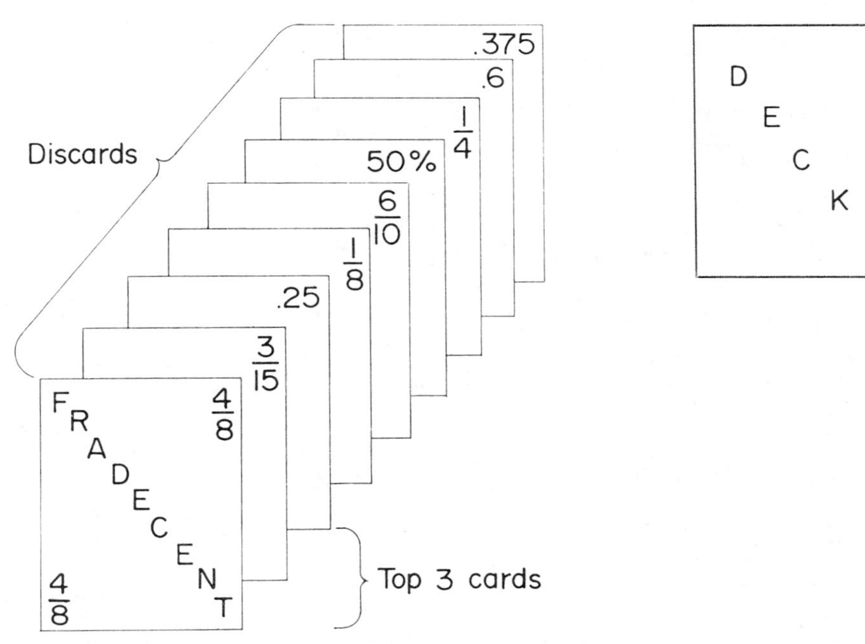

Fig. 1. Drawing from the discards

done by any player during his turn. He may build on his own sets or on the sets of his opponents. A player builds on his opponents' sets by placing the scoring card in front of himself and declaring the set on which he is building.

During his turn, a player may draw a card from the deck in the center of the table or any number of cards from the pile of discards. If he draws from the discards, he must use the last card he takes in a scoring set during that turn and keep all the other discards in his hand. In figure 1, for example, if the player takes the top three cards, he must use the .25 card in a scoring set and add the 3/15 and 4/8 cards to his hand.

When a player gets rid of all his cards, the hand is over and the points are added. The face values of the cards are counted for points. The total of the cards remaining in a player's hand is deducted from the total of his scoring cards. The winner of the game is the player who gets five or ten points first. Before play starts, players should agree on whether they will play a five-point game or a ten-point game.

EDITORIAL COMMENT.—This game should provide some meaningful practice for students, and once the rules are learned, they should be able to play independently. As a possible variation of the game, you might suggest that the goal is to build a set of four cards equal to one (for example: 1/8, 3/15, 30%, .375).

An adventure in division

LOIS STEPHENS *Camarillo, California*

Lois Stephens is principal of McKinna School in Oxnard, California. She recently completed her doctoral dissertation and is now making a study of mathematics in the primary grades.

The adventure upon which you are about to embark requires some computation, so before reading any further, get a pencil and some scratch paper.

Ready? Here we go!

Until a few years ago I had viewed division as a process to be used only when needed to solve a practical problem. What changed my attitude was an investigation of the patterns to be found in the repetends of the decimal equivalents of common fractions having prime numbers as denominators. Try these operations with me:

For a start, find the equivalents of the sevenths, carrying each division problem to the number of places necessary to repeat the pattern twice.

Finished?

You should have these answers:

$\frac{1}{7} = 0.142857142857\ldots$ $\frac{4}{7} = 0.571428571428\ldots$

$\frac{2}{7} = 0.285714285714\ldots$ $\frac{5}{7} = 0.714285714285\ldots$

$\frac{3}{7} = 0.428571428571\ldots$ $\frac{6}{7} = 0.857142857142\ldots$

Although at a glance these appear to be different patterns, it is easily shown that the same sequence of six digits appears in each of the decimal equivalents, and there is only one pattern for the sevenths, this one made up of six digits.

Moving on to the next prime number, eleven, find the decimal values of the ten proper fractions, then check your answers with those given below.

$\frac{1}{11} = 0.0909\ldots$ $\frac{6}{11} = 0.5454\ldots$

$\frac{2}{11} = 0.1818\ldots$ $\frac{7}{11} = 0.6363\ldots$

$\frac{3}{11} = 0.2727\ldots$ $\frac{8}{11} = 0.7272\ldots$

$\frac{4}{11} = 0.3636\ldots$ $\frac{9}{11} = 0.8181\ldots$

$\frac{5}{11} = 0.4545\ldots$ $\frac{10}{11} = 0.9090\ldots$

Again, at a glance there appear to be ten patterns, but closer inspection reveals that one eleventh and ten elevenths follow the same pattern, two and nine elevenths the same, three and eight, and so on, for a total of five different patterns of two digits each.

Let's try one more easy series—the thirteenths. You know now what to do, so perform the operations and check your results with the answers given here:

$\frac{1}{13} = 0.076923076923\ldots$ $\frac{7}{13} = 0.538461538461\ldots$

$\frac{2}{13} = 0.153846153846\ldots$ $\frac{8}{13} = 0.615384615384\ldots$

$\frac{3}{13} = 0.230769230769\ldots$ $\frac{9}{13} = 0.692307692307\ldots$

$\frac{4}{13} = 0.307692307692\ldots$ $\frac{10}{13} = 0.769230769230\ldots$

$\frac{5}{13} = 0.384615384615\ldots$ $\frac{11}{13} = 0.846153846153\ldots$

$\frac{6}{13} = 0.461538461538\ldots$ $\frac{12}{13} = 0.923076923076\ldots$

This time, did you find the two patterns of six digits each?

Now try putting all the results you have

obtained so far into a chart with four columns. Head the first column "N" for the prime number which is the denominator. The second column will be "D" for the number of digits in each pattern. Heading the third column will be "P" for the number of different patterns. You can determine the heading for the fourth column by discovering the relationship among the entries made in each row. Your chart should look like the one in Figure 1.

N	D	P	?
7	6	1	
11	2	5	
13	6	2	

FIGURE 1

If you are having difficulty in heading the last column, go on to 17, 19, and other prime numbers to obtain more entries for your chart. (More about this later in the article.)

Now let's go down another path on this adventurous journey. Looking back to the decimal values of the elevenths, designate the different patterns by letters, thus:

$\frac{1}{11}$ — A $\quad \frac{5}{11}$ — E $\quad \frac{9}{11}$ — B

$\frac{2}{11}$ — B $\quad \frac{6}{11}$ — E $\quad \frac{10}{11}$ — A

$\frac{3}{11}$ — C $\quad \frac{7}{11}$ — D

$\frac{4}{11}$ — D $\quad \frac{8}{11}$ — C

These patterns can then be graphed in the manner shown in Figure 2.

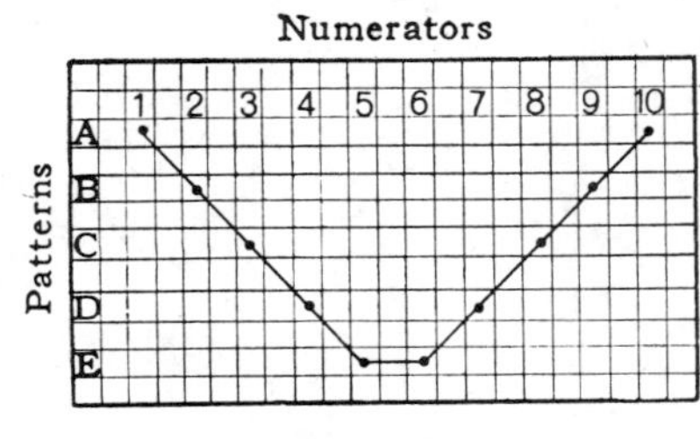

FIGURE 2

And using the same kind of designation, the thirteenths produced the graph shown in Figure 3.

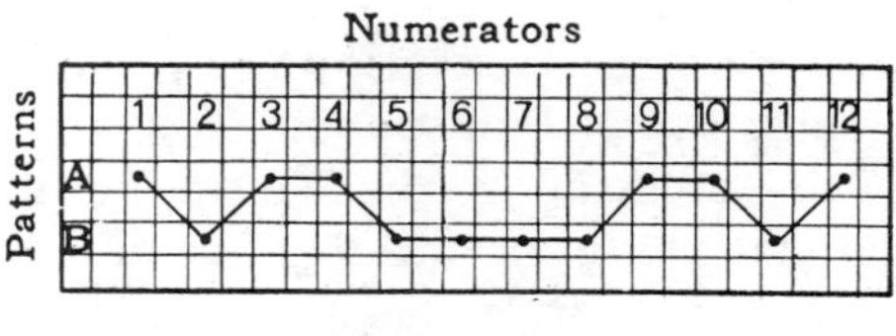

FIGURE 3

Going back to the chart you made, did you try the product of D and P and find it to be $N - 1$? Test this as a possible heading for the fourth column by calculating the decimal equivalents of the seventy-thirds. Before you begin calculations, predict the number of digits and the number of patterns by making a factor tree for 72, $N - 1$ in this case. (See Fig. 4.)

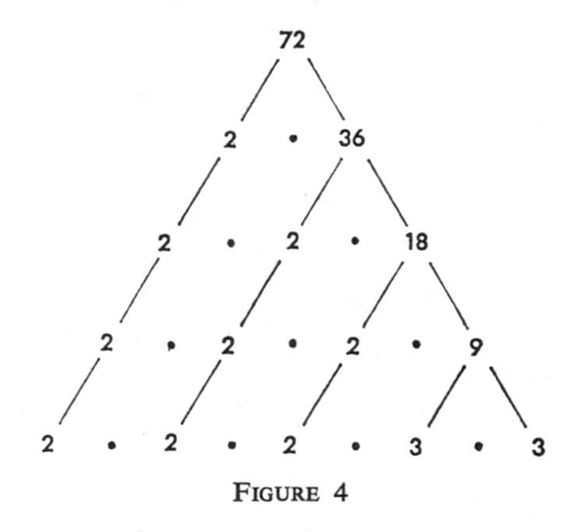

FIGURE 4

This gives many possibilities:

$$\begin{aligned} N &= 2, & D &= 36; \\ N &= 4, & D &= 18; \\ N &= 8, & D &= 9; \\ N &= 12, & D &= 6; \\ N &= 24, & D &= 3; \end{aligned}$$

and the opposites of all these!

Write your prediction at the top of your paper, and start dividing.

Finished? Now wasn't that fun? And was your prediction correct?

You may ask, "What is the value of this so-called adventure?" Not only have I had many pleasant surprises as I have explored these repetends myself, but I have found many junior high school students who would spend hours on problems such as 1/97 = ?, even though they had previously avoided all practice in division. (This problem is not recommended unless you are ready for a lot of practice in division, which of course also means practice in multiplication and subtraction!)

One successful method of using this computational game was to allow one student to work at the chalkboard while the other students looked on. In this way, without any teacher direction, the rest of the class became self-appointed "checkers," and all were participating in camouflaged drill as a sideline while the regular classroom assignment was being pursued.

If you have followed me down this path so far, do not think you have reached the end, for I predict you will find yourself wondering such things as, "How do you suppose 997 behaves as a denominator?" . . . . And do try 101 next!

Have fun!

EDITORIAL COMMENT.—If you are looking for something a little different after the work in division suggested by Stephens, have your students do the following puzzle by coloring in each region that is labeled with a symbol for one-half. This puzzle would be especially good just before Halloween.

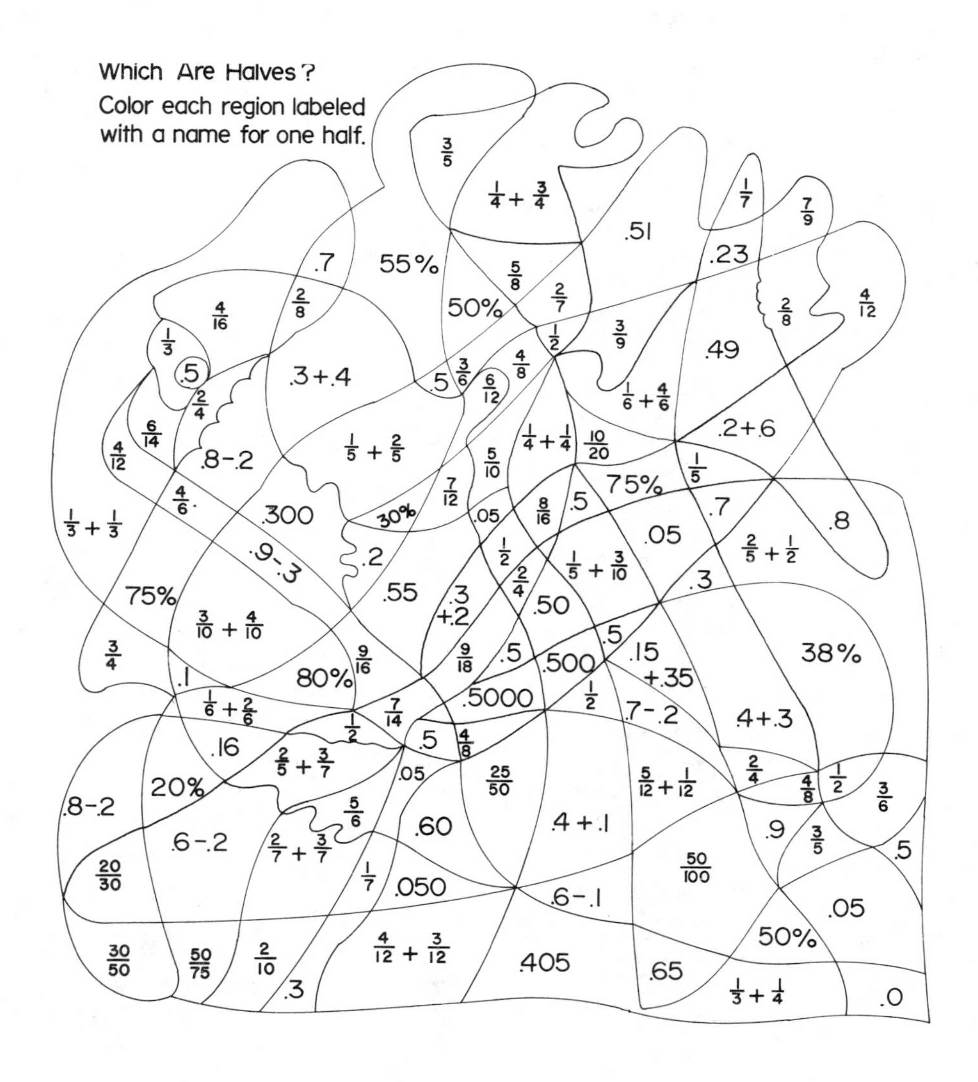

Introduction to ratio and proportion

ROLAND L. BROUSSEAU
North Attleboro High School, North Attleboro, Massachusetts

THOMAS A. BROWN
Merrill Junior High School, Des Moines, Iowa

PETER J. JOHNSON
Jim Bridger Junior High School, North Las Vegas, Nevada

The objective of this lesson is to introduce fifth- and sixth-grade students to ratio and proportion through a discovery situation. This approach would also be applicable for junior and senior high general mathematics students. This technique should allow students to find patterns that will enable them to verbalize algorithms for the solution of proportions. The following is from an actual demonstration with fifth graders.

As a starting point, in developing a feeling for ratio, the children took objects from a container. In order to keep the number property small, large objects were used. A comparison of the number of objects drawn by each of two children was made. A table was constructed to keep a record of the number of objects drawn. (See Example 1.) Each time the children came to the container, they were told to draw the same number of objects they had drawn on their first turn. The objects drawn were displayed where they could be seen by the entire group.

EXAMPLE 1

Girl	*Boy*
3	4
6	8
9	12
.	.
.	.
.	.
15	20

The girl in the example cited drew three objects, while the boy drew four. Thereafter, and with each successive drawing, the total number of objects was recorded in the table. (If by chance the same number of objects are drawn by both children, these numbers should be recorded, but another drawing is made to get different numbers. After three or four drawings, children should be able to supply new values without actually drawing objects from the container. If the children have trouble supplying the values, they should return to the container for additional drawings.)

After the first table was completed, a second table was constructed without having the children draw objects from the container. At this point the word "rate" was introduced. The children were asked to think of the table as a record of the *rate* at which these objects were selected from the container (Example 2). The girl drew three to the boy's five—expressed as a rate of 3 to 5.

EXAMPLE 2

Girl	*Boy*	*(G, B)*
3	5	(3, 5)
6	10	(6, 10)
12	20	(12, 20)
15	25	(15, 25)
.	.	. .
.	.	. .
.	.	. .

Children were then asked to fill in the table. Once again it should be stressed that the table can be verified by having the children actually draw the number of objects specified.

At this stage, the notation of ordered pairs was introduced. The tables were constructed vertically to make this transition easy for the children. The values listed in the second table were written using ordered pairs which were referred to as "rate pairs."

Once the concept of rate pairs had been established, the next step was to show equivalence between rate pairs. At this point, a child was asked to explain to his classmates how the objects were drawn from the container and how the tables were constructed. Children were asked to volunteer other rate pairs that showed the same rate of drawing. (See Example 3.)

EXAMPLE 3

$$(3, 5) = (9, 15)$$
$$(3, 5) = (6, 10)$$
$$\vdots \qquad \vdots$$

It is necessary for the children to see that equivalence exists between other rate pairs from the same table. The children were asked to find a rate pair that was equivalent to (21, 35), but they were told not to use (3, 5). A second rate pair was introduced and children supplied rate pairs.

The term "equivalent rate pairs" was used to express the same rate. The pair (3, 5) expresses the same rate as (6, 10) even though different numbers are used.

As further exercises on finding equivalent rate pairs, examples can be devised which call for replacements supplied by the students.

EXAMPLE 4

$$(2, \square) = (6, 9)$$
$$(12, 16) = (\square, 4)$$

The symbol of replacement should be used in all positions.

A game based on equivalent pairs was devised to help the children see the relationship between the product of the means and extremes. The game revolves around an expression like this: $(3, \triangle) = (\square, 4)$. The rules of the game are as follows: The girls pick a pair of numbers that would yield equivalent rate pairs. They then give one value and let the boys find the other. In a similar fashion the boys could pick a pair of replacement values.

A table like the one below could be constructed to record replacement values.

$\triangle$	12	1	3	4	6	2
$\square$	1	12	4	3	2	6

Other good examples for this game would be $(\square, 6) = (5, \triangle)$ and $(8, \square) = (\triangle, 2)$.

Hopefully, the children will soon realize that the product of the replacements listed in the table is the same as the product of the numbers in the examples. This might set the stage for a verbalization of the rule.

It is possible to see that replacement could also include fractions. A last example might be one such as this: $(4, \frac{1}{2}) = (\triangle, 2)$. If the children have discovered the rule of the game, examples such as the one above should not cause any frustration, but merely make the game more interesting.

In conclusion, the brief development of proportion as outlined above would, with extension, lend itself to problems involving percent, scale drawing, conversions, similar figures, area, and volume.

Make a whole—a game using simple fractions

JOANN RODE
University of West Florida, Pensacola, Florida

The game called "Make a Whole" helps develop the concept of fractional numbers by using concrete examples; it gives reinforcing experience with the use of fractions and their equivalents. It is suggested for use at third-grade level and above, depending on the achievement of the students.

The material required for the game is pictured in figure 1 and described more completely in a later section. The sections are made in four different colors, equally distributed; that is, of 8 sections representing $\frac{1}{2}$, 2 could be red, 2 blue, 2 green, and 2 yellow. The area for the "Whole" is also prepared in four different colors. The faces of the die are marked with a star and the following fractions: $\frac{1}{2}$, $\frac{1}{3}$, $\frac{1}{4}$, $\frac{1}{6}$, and $\frac{1}{8}$.

Rules of play

The object of the game is to use the fractional sections to construct a multicolored disk, the "Whole," in the black mat frame. The Whole may be constructed with any of the differently colored sections so long as they fit together to be equivalent to a whole.

Two to eight children may play the game in individual competition (in which case the color of the area for the Whole makes no difference), or they may play on color teams as determined by the color of that area. ⟶

EDITORIAL COMMENT.—The game may be extended to include shapes other than circles. The game may also be played on a number line with the draw pile being pieces of paper with fraction names. A student draws a piece from the pile and marks on his number line a segment representing that fraction.

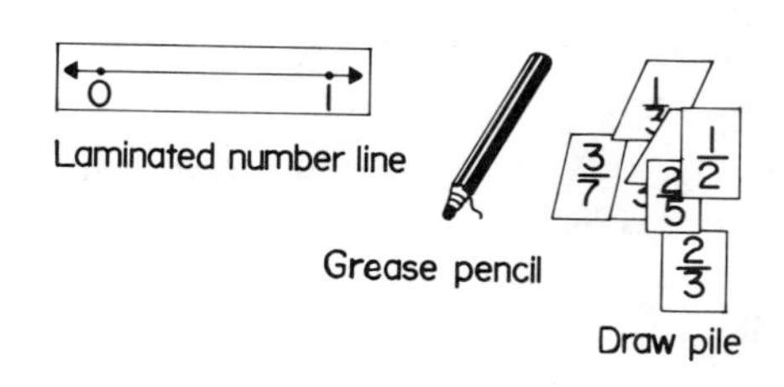

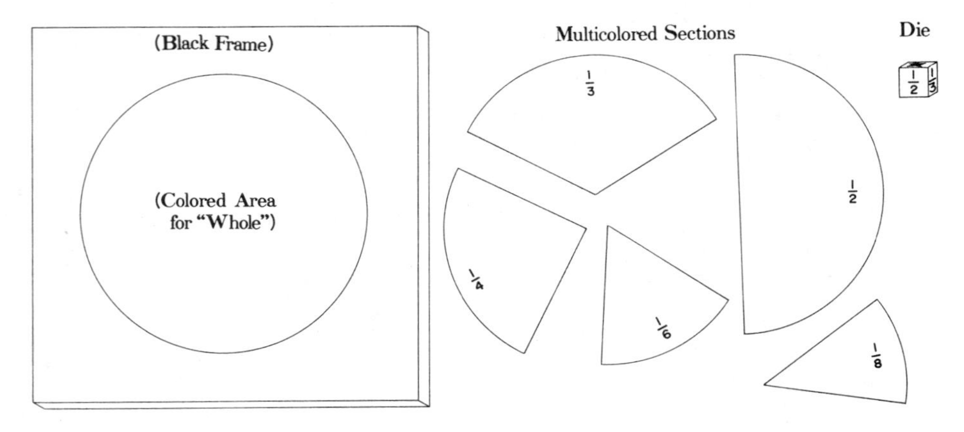

Fig. 1

The fractional sections are placed in a box, from which they are to be selected by the children as they roll the die, in clockwise order.

To determine who goes first, each child rolls the die seen in figure 1. The child rolling the star or the fraction with the greatest value goes first.

At each turn the child must decide whether he can use one of the sections represented by the fraction showing on top of the die.

If the player cannot use the section, he passes.

If he decides to take the section and can use it, he places the section in his mat. (If the section taken completes the Whole for that player, then the game is over and that player has won. If it does not, the roll of the die passes to the next player.)

If he decides to take the section and cannot use it, he loses his next turn.

If he passes up a section he could have used, he loses his next turn.

If the star is rolled, the player may choose any section he can use.

The winner is the first to construct a multicolored Whole.

The game may be played in various other ways, according to rules designed by the teacher. For example, a child might be asked to construct five Wholes with no more than one of a kind of fractional section (½, ⅓, etc.) to go into each. For advanced students, the numeral identification may be omitted from the sections.

The materials may also be used by the teacher to show equivalent fractions in classroom demonstrations and by the individual student in working with equivalences at his desk.

Construction of game materials

A complete list of pieces follows, the color of the mat disks and sections being equally divided among whatever four colors are used.

Mats, 8
½ sections, 8
⅓ sections, 12
¼ sections, 16
⅙ sections, 24
⅛ sections, 32
Die, 1

The material used for the black mat frame and the fractional sections is inexpensive tagboard (poster board) in different colors. Because the Wholes constructed by the students are to be multicolored and because the fractional sections come in equal numbers of the different colors, the colors give no clues to value.

The mats were made first by cutting 8 eight-inch squares from black tagboard. Disks six inches in diameter were cut from the center of these squares, and the frames were then glued to other eight-inch squares, 2 of each of the four colors chosen.

Five disks in each of the four colors were then constructed, with a diameter of six inches. A compass was used for accuracy. The disks were then divided, with the aid of a protractor, into sections corresponding to $\frac{1}{2}$, $\frac{1}{3}$, $\frac{1}{4}$, $\frac{1}{6}$, and $\frac{1}{8}$. The central-angle measurements for the five sections are 180°, 120°, 90°, 60°, and 45°, respectively. Depending on the level of the children, the sections may or may not be marked with the fractions to which they correspond, as seen in figure 1.

The die was made from a cube cut at a lumber yard so that each edge had a measurement of three-fourths of an inch. The five fractions were painted, one to a face on each of five faces of the cube. A star was placed on the sixth face. The cube was sprayed with a clear varnish to prevent smudging.

A game with fraction numbers

RICHARD THOMAS ZYTKOWSKI
University Heights, Ohio

The game described below is a unique way to provide practice in the addition, subtraction, multiplication, and division of fractions. The game can also be used to teach ratios as well as the sequence of numbers.

Use a 2-by-2 ft. piece of oaktag or cardboard to make a playing board. Divide it into spaces as shown below, then write the fractions as figure 1 shows.

A	Start →	$\frac{1}{2}$	$\frac{2}{4}$	$\frac{3}{6}$	$\frac{5}{10}$	$\frac{6}{12}$
B	$\frac{2}{3}$	$\frac{4}{6}$	$\frac{6}{9}$	$\frac{8}{12}$	$\frac{10}{15}$	$\frac{12}{18}$
C	$\frac{3}{4}$	$\frac{6}{8}$	$\frac{9}{12}$	$\frac{12}{16}$	$\frac{15}{20}$	$\frac{18}{24}$
D	$\frac{4}{5}$	$\frac{8}{10}$	$\frac{12}{15}$	$\frac{16}{20}$	$\frac{20}{25}$	$\frac{24}{30}$
E	$\frac{5}{6}$	$\frac{10}{12}$	$\frac{15}{18}$	$\frac{20}{24}$	$\frac{25}{30}$	$\frac{30}{36}$
F	★ Winner	$\frac{6}{7}$	$\frac{12}{14}$	$\frac{18}{21}$	$\frac{24}{28}$	$\frac{30}{35}$

FIGURE 1

On some small pieces of tagboard $3\frac{1}{2}$ in. by 8 in. write the following symbols and numerals: $-\frac{1}{2}$; $+\frac{1}{3}$; $\times\frac{1}{4}$; $\div\frac{1}{5}$; $+\frac{1}{6}$; $\times\frac{1}{7}$; $-\frac{1}{8}$; $\times\frac{1}{9}$; $+1\frac{1}{2}$; $\times2\frac{2}{3}$; $\div3\frac{3}{4}$; $+4\frac{4}{5}$; $\times5\frac{5}{6}$; $-6\frac{6}{7}$; $+7\frac{7}{8}$; and $\times8\frac{8}{9}$.

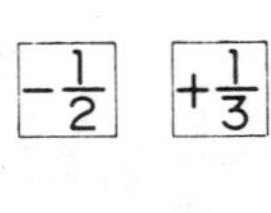

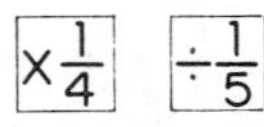

FIGURE 2

In figure 2, examples are shown of how the front side of the small pieces of tagboard would be labeled.

On the back of each small piece of tagboard write the answers for the problem that is on the front. There will only be six possible answers, since there are only six rows: *A, B, C, D, E, F;* the rest of the fractions in the row are equivalent to the first fraction in that row. Figure 3 shows how the back side of $-\frac{1}{2}$, $+\frac{1}{3}$, $\times\frac{1}{4}$, $\div\frac{1}{5}$ would look respectively.

Two to six students can easily play this game, or the class may be divided into groups.

$-\frac{1}{2}$	$+\frac{1}{3}$	$\times\frac{1}{4}$	$\div\frac{1}{5}$
A 0	A $\frac{5}{6}$	A $\frac{1}{8}$	A $2\frac{1}{2}$
B $\frac{1}{6}$	B 1	B $\frac{1}{6}$	B $3\frac{1}{3}$
C $\frac{1}{4}$	C $1\frac{1}{12}$	C $\frac{3}{16}$	C $3\frac{3}{4}$
D $\frac{3}{10}$	D $1\frac{2}{15}$	D $\frac{1}{5}$	D 5
E $\frac{1}{3}$	E $1\frac{1}{6}$	E $\frac{5}{24}$	E $4\frac{1}{6}$
F $\frac{5}{14}$	F $1\frac{4}{21}$	F $\frac{3}{14}$	F $4\frac{2}{7}$

FIGURE 3

The rules of the game are as follows:

Each player will begin with the first numeral on the large tagboard, which is the fraction $1/2$.

He will, without looking, pick a small piece of tagboard with a symbol and fraction on it.

He will then perform the mathematical operation in order to arrive at the correct answer.

If the player gets the answer correct, he moves one space. If he is wrong, he remains where he is. (The answer can be verified by checking the back of the card used by the player in performing the mathematical operation.)

The first student to reach "winner" is the "mathematical champ" or "whiz kid."

All answers must be reduced to their lowest terms. Improper fractions must be changed to mixed numbers.

Each player is given only one chance per card, then it is the next player's turn.

The fraction on the small piece of tagboard or cardboard should always be considered to represent the second numeral in the problem; thus, it will either be the second addend, the minuend, the multiplier, or the divisor.

Example:

To begin the game, player A, without looking, picks a card. The card states $\times\ 1/4$. Since all players begin with the first square on the large oaktag board ($1/2$), the player must multiply $1/2 \times 1/4$.

Player A's computations:

$$\tfrac{1}{2} \times \tfrac{1}{4} = \tfrac{1}{8}.$$

Player B checks the back of the card being used to see if the answer is correct. The answer is correct, so player A moves to the next space, $2/4$. It is now player B's turn. Player B misses his problem. Therefore, he must remain at the same place until it is his turn again.

If there are no more than two players, it will be player A's turn again. Only this time instead of being at $1/2$, he has now moved to $2/4$. He draws another card; the card states $-\ 1/2$. Player A must subtract $-\ 1/2$ from $2/4$.

Player A's computations:

$$\tfrac{2}{4} - \tfrac{1}{2} = \tfrac{2}{4} - \tfrac{2}{4} = 0.$$

Player B checks his answer by looking at the back side of $-\ 1/2$ to find the answer. He sees that the answer is correct, so player A moves to the next space, $3/6$. It is now player B's turn again. Player B should always get his turn last, since player A was first.

If the game is used by the teacher in row competition a few times, the pupils should be led to see the relationship between numerals. The student should be asked what numeral would follow the sequence in each row; then this can be expanded to lead the pupils to see the relationship between the numerator and the denominator of each fraction.

6

Number Theory and Patterns

Most contemporary elementary and middle school programs contain topics from number theory—factors, primes, multiples, divisibility, patterns, sequences, and so on. There are two major reasons for including these topics in the curriculum:

1. Number theory topics are convenient tools used in conjunction with fractional number computations.
2. There is an intrinsic fascination in the relatedness of number properties—this natural intrigue can be used to provide many opportunities for practice with whole-number computations.

Most of the articles in this section reflect the sentiments expressed by Kapp and Hamada in the lead article, "Fun Can Be Mathematics"—"Every mathematics classroom can be a laboratory where students experiment with numerical ideas. . . . The games suggested . . . can stimulate the students' mathematical thinking by the use of number sequences and patterns."

The first article gives several suggestions for turning sequence investigations into games. There is an especially good game for developing the concept of greatest common factor. This game is easily adapted to the concept of lowest common multiple.

Allen's description of "Bang, Buzz, Buzz-Bang, and Prime" is guaranteed to give a vigorous presentation of multiples and primes. Be prepared for a high level of excitement when you play this game.

The ever-popular Bingo game is the basis for Holdan's game of "Prime." This is a game of disguised practice in the recognition of prime and composite numbers. Most students enjoy playing Bingo games, especially when the reward for winning a game is being the next caller.

If you are interested in a game requiring the development of a strategy to win rather than just chance, then try Harkin and Martin's "Factor Game." The game rules are presented in the form of a computer printout—however, you don't need a computer to play the game! If you do have programming capabilities, then by all means make the computer the opponent.

You will find additional strategy-based games in Trotter's "Five 'Non-trivial' Number Games." The secrets behind several very clever and useful games are exposed. One of the secrets involves the Fibonacci sequence.

At first glance, Sawada's article on "Magic Squares" may seem somewhat complex for elementary or middle school students. However, the author's purpose is not to encourage the teaching of abstract concepts but rather to "suggest the strategy of taking interesting ideas, such as magic squares, and extending them in ways that lead to the discovery of other mathematical structures, all the while providing interesting practice of fundamentals as originally intended." In addition, the article gives some very good illustrations of activity cards, using magic squares as the subject.

The final two articles provide a potpourri of tricks and games designed as "Interest Getters" and "Just for Fun." Such diversions provide the perfect fillers for those in-between times.

Fun can be mathematics

AUDREY KOPP and ROBERT HAMADA
Edison Model Mathematics Demonstration Center,
Los Angeles City Schools, California

Every mathematics classroom can be a laboratory where students experiment with numerical ideas. Two-way communication between teacher and class by means of games can foster an atmosphere of eager participation in mathematical activities. The games suggested below have characteristics that can stimulate the student's mathematical thinking by the use of number ideas and number sequences and patterns. Some of the exercises call for use of paper and pencil by students and either the chalkboard or the overhead projector for the teacher to show collection of data. Often each child may be asked for an oral response, thus allowing all to participate, as well as permitting the teacher to check if each student understands the rules of the game.

1. An easy way to establish the notion of patterns and sequences is to first ask the class to count by five. Tell the students to listen for a secret as each child in order gives an answer. Then again count by five but use 1, 2, 3, or 4 as the initial number. For example, count: 3, 8, 13, 18, 23. Again suggest that students listen for the secret. Try a third time. The teacher can judge if students know the secret by the promptness of their responses. If they have not caught on by the third round, the teacher may repeat each answer, enabling students to hear the terminal digits. By this time, usually everyone knows the secret and you may have a perfect response on the fourth try.

The game can be used at the end of a class period in order to set the groundwork for the lessons to come on patterns and sequence. If only a few students discover the secret one day, try again the next.

2. Number sequences may be introduced by the teacher presenting the first numbers on the chalkboard:

$$1, 4, 7, \text{—}, \text{—}, \text{—}, \cdots$$

The class discovers more numbers in the sequence

$$1, 4, 7, 10, 13, 16, 19, \cdots$$

Then the teacher asks the class to verbalize the rule governing the sequence. If practical, the class should be asked to write the rule in some form depending on the ability of the class: Add 3 to the previous number: $1 + (3 \cdot 0)$, $1 + (3 \cdot 1)$, $1 + (3 \cdot 2)$, $1 + [3 \cdot (n - 1)]$ where n is the nth term in the sequence. Here is another example:

$$1, 4, 9, \text{—}, \text{—}, \text{—}, \cdots$$

After the class suggests the correct numbers in sequence:

$$1, 4, 9, \underline{16}, \underline{25}, \underline{36}, \cdots$$

it is time to formulate rules again. Some classes may come up with the rule of squares; others may suggest adding the odd numbers in sequence. Teachers may find sequences or make up some that are commensurate with class levels. The patterns should not be difficult at first so that students can easily be acquainted with the idea of formulating a rule. After a few sequences the students can participate by offering sequences themselves. One student gives a series of numbers which the teacher records on the board. The student calls on other members of the class to complete the sequence and can either supply the rule himself or ask another student to do so. Rules and sequences may be recorded by all students.

Occasionally not enough data is supplied to predict just one pattern.

1, 2, 4, —, —, —, ···

As many possibilities as can be found should be examined along with their rules.

1, 2, 4, $\underline{8}$, $\underline{16}$, $\underline{32}$, ···

1, 2, 4, $\underline{7}$, $\underline{11}$, $\underline{16}$, ···

3. Student Participation Game

Each child in the room is assigned a number in sequence. Students may count off in order to make sure each knows his own number and those of the others around them. The teacher then asks students to stand according to a particular rule.

The relationship between ordinal and cardinal numbers may be illustrated by having every third student stand, or asking each student after the tenth student to stand. The game extends to practice with factors and primes. After the teacher makes a statement, the class members respond by standing, making it easy to check answers and locate missing ones. Sample directions:

The even numbers
Multiples of three
Factors of twelve
Numbers with only two factors
The prime numbers

Eventually, students will be able to make up appropriate directions themselves. Students often express ideas as they play this game. After a series of statements regarding factors, Carl asked, "Why do I have to stand all the time?" Dave replied immediately, "Because 1 is a factor of every number." When Ernest was 1, he announced that he would stand for every question that had to do with factors. He learned that he was wrong when the teacher asked all the prime factors to stand. Questions are easily found for each direction. After the even numbers are standing, for instance, one may ask 8 why she is standing. If she doesn't know, somebody will probably state that it's because 8 contains a factor of 2. The question of why 13 should stand in the prime numbers was answered by Mary as "13 has only 2 different factors." Jerome, number 17, was tired of sitting so he suggested "the factors of 34 stand." Here is another example of a series of statements.

Factors of 12:
Students assigned to numbers 1, 2, 3, 4, 6, and 12 should stand.

Factors of 18:
Students assigned to numbers 1, 2, 3, 6, 9, and 18 should stand.

Common factors of 12 and 18:
Only those students who stood up both times previously should stand.

Greatest common factor of 12 and 18:
The student with the largest number remains standing.

Another drill on fundamentals would have the teacher make a statement such as:

4 plus 3, multiplied by 2, minus 2, divided by 6.

Obviously, the student assigned to the number 2 should stand up.

Instead of having pupils stand, each may be given a 3 × 5 card with his number written on both sides in bold color. Cards may be held up for answers. After a while the numbers may be given out randomly, which may increase the difficulty of the game.

4. Ordered pairs lend themselves to games wherein students select and verbalize rules. Visual reproduction of a function machine is often helpful. Basic parts are Input, Output, and Function Rule. Many

rules can be used, depending again on the level and needs of the class. Here are two sample sets of number pairs:

(2, 4) (3, 6) (8, 16) (12, —) (—, 100)

(4, 9) (5, 11) (6, 13) (9, 19) (10, —) (—, 51)

Students enjoy working out their own rules and trying them on the other members of the class. Thus the notion of function can be introduced at even a very elementary level.

EDITORIAL COMMENT.—The student-participation game described above may be adapted for use with lowest common multiples (LCM). To find the LCM of 6 and 4, give directions such as the following:

"Multiples of 6 stand:" 6, 12, 18, 24, 30, . . .

"Now sit down."

"Multiples of 4 stand:" 4, 8, 12, 16, 20, 24, 28, . . .

"Only those of you who stood up both times remain standing:" The student with the lowest number remains standing. This will be the LCM of 6 and 4.

Bang, buzz, buzz-bang, and prime

ERNEST E. ALLEN

Southern Colorado State College, Pueblo, Colorado

Games often provide an excellent technique to stimulate motivation and maintain interest. This article discusses four versions of a simple counting game. Though the game is easily played by primary youngsters, it is also applicable to older children and adults.

"Bang" is by far the easiest version. Here are the rules:

1. Seat a small number of students, say less than ten, in a circle.
2. Have a student begin the counting in a clockwise manner.

3. When it is a student's turn to say "5" or any multiple of five, such as 10, 15, 20, . . . , he says "bang" *instead.*
4. At this point the counting direction reverses, beginning in a counter-clockwise manner, so that the person who said "4" now says "6," etc.
5. The game continues in this manner. When a student—
   a) utters the wrong number name
   b) forgets to say "bang"
   c) doesn't say anything (as when counting reverses direction),

   the game starts over with this person saying "1."

"Buzz" is played much like Bang but with the following exception:

The student says "buzz" on multiples of 7 and numbers which have names that include the numeral 7, such as 17, 27, 37, ...

"Buzz-Bang" is a combination of the above two games. It requires a better knowledge of our number/numeral system and more concentration. The following diagram indicates the correct moves up to thirty.

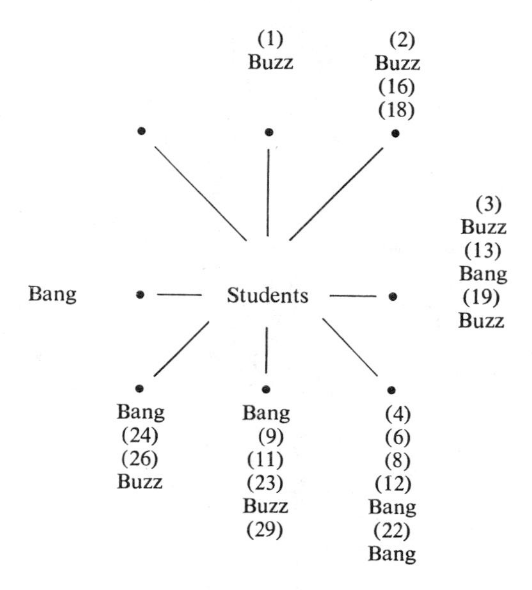

"Prime" is played exactly like Buzz except that the student says "prime" instead of the particular name for the prime number.

Prime: a drill in the recognition of prime and composite numbers

GREGORY HOLDAN *Indiana University of Pennsylvania*

Drill has always been of fundamental importance in mathematics for mastering and reinforcing the basic concepts. To discourage the development of unfavorable attitudes towards mathematics, a drill exercise should not become a boring, rote exercise. Instead, drill should be made meaningful for the student so that its contributions become functional in later learning experiences. Whenever drill is disguised in some other not-so-obvious form, such as a game, it tends to be more fruitful.

"Prime," as I have devised it, is basically a disguised drill in the recognition of prime and composite numbers less than 100. The game sheets, rule slips, and rules for playing "Prime" are very similar to those for "Bingo." As you read the rules for constructing and playing "Prime," you will probably find that the game is quite versatile; it can be readily adapted, in terms of subject matter and the methods used in constructing and playing it, to practically any situation.

Construction of the game sheets

On a ditto master, construct a five-inch square; subdivide it into twenty-five one-inch squares. Above the five columns of squares you have constructed, print the letters P-R-I-M-E, one letter per column. From this master ditto, duplicate the number of game sheets you will need.

Select sets of twenty-five numbers between 1 and 99 from a table of random numbers. Place one number in each square on each game sheet. (When I constructed the game sheets, I used a black ditto master and, on the copies, wrote in the numbers with a red fine-tipped felt pen.)

Your game sheets should look something like that shown in Figure 1.

P	R	I	M	E
47	13	54	77	38
72	38	16	11	5
28	46	9	76	23
27	62	61	99	41
30	97	54	20	39

FIGURE 1

Constructing the rules

On slips of paper, write or type an appropriate selection of the following statements for *each* of the letters P-R-I-M-E:

(In items 21 and 22, you may select the number to take the place of the blank.)

1. A prime number
2. The largest prime number
3. The smallest prime number
4. A composite number
5. The largest composite number
6. The smallest composite number
7. A twin prime
8. A composite number between a pair of twin primes
9. A single-digit prime number
10. A two-digit prime number
11. A two-digit composite number
12. A two-digit number if *either* of its digits is a prime number
13. A two-digit number, the sum of whose digits is a composite number
14. A two-digit number, the sum of whose digits is a prime number
15. A two-digit number with a prime number in the ones place
16. A two-digit number with a composite number in the ones place
17. A two-digit number with a prime number in the tens place
18. A two-digit number with a composite number in the tens place
19. A two-digit composite number with both of its digits prime numbers
20. A prime number less than______
21. A prime number greater than______
22. An even prime number
23. An odd prime number

Your rule slips should look something like this:

M: a two-digit prime number

Playing the game

Pass out the game sheets to the students. Spread out the rule slips *face down* in front of you. Randomly choose a rule and read it aloud to the students. If a student has a number on his game sheet, under the appropriate letter P, R, I, M, or

E, which has the property defined by the rule read, he is to put an X in that square. (If you require that only a pencil can be used for marking the game sheets, and that the X's are to be marked lightly, the game sheets can easily be erased and reused.)

For checking purposes, have the students write the number of the rule in the lower right-hand corner of the square in which he places an X. For instance, the first rule read would be considered number one; the second rule read would be considered number two, and so on. As each rule is read, place it to the side. Keep these rules in the order that they are read!

Make sure that the students understand that only one square can be crossed out per rule, and that after a square is crossed out, the number in it cannot be considered for any following rules.

Continue selecting rules and reading them aloud to the students until someone has crossed out all the blocks in a single column, row, or diagonal.

Possible winning game sheets might look like one of those pictured in Figure 2. The student signifies that he feels he is a winner by saying aloud the word "Prime."

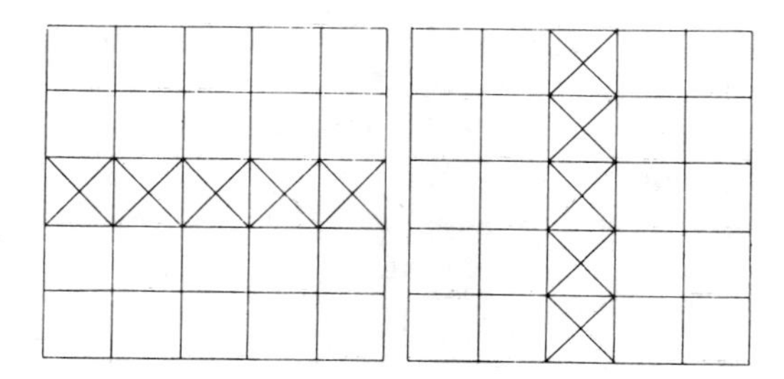

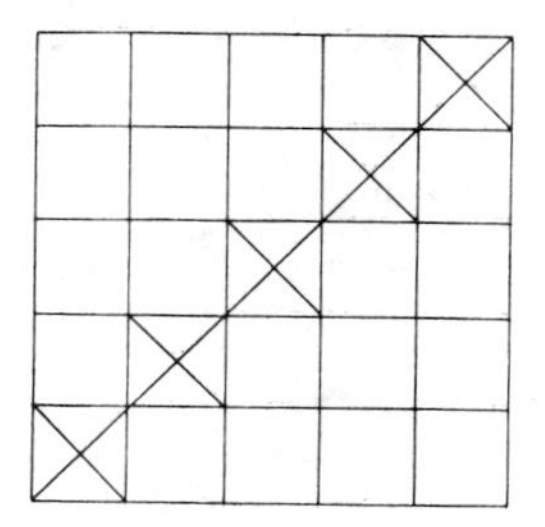

FIGURE 2

Checking

Have the winner read aloud the numbers in the row, column, or diagonal, along with the rule that permitted him to cross out each of those numbers. Record these rules and numbers on the chalkboard and then have the class help you to check the validity of the winner's card.

Anyone for a game of "Prime"?

EDITORIAL COMMENT.—To help students discover which whole numbers are prime and which are composite, you may develop a game using the pegboard. Show children that prime numbers have peg arrays that are single rows or single columns. Composite numbers have other kinds of arrays.

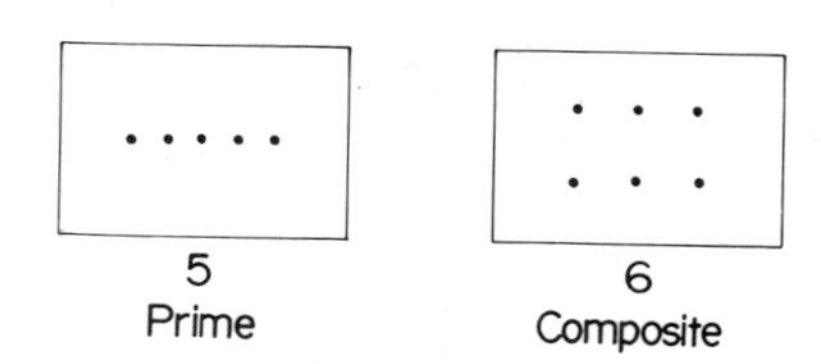

Give the students a pegboard and a supply of pegs. Call out a number. The first team (or student) to determine whether the number is prime or composite using an array scores a point. Play the game to a fixed number of points.

The factor game

J. B. HARKIN and
D. S. MARTIN

Both authors are on the faculty of New York State University College at Brockport. J. B. Harkin is a professor of mathematics with responsibility for preservice and in-service mathematics programs for elementary school teachers. D. S. Martin is an instructor of mathematics with an interest in computer-assisted instruction in the mathematics curriculum of the elementary school.

In a systems approach to teaching, particularly in the individualization of activities, manipulatives and other media hardware play an essential role. An activity package in the delivery system is a sequence of multigroup activities developed for a content unit. A variety of media hardware (a term that embraces everything from a geoboard, a projector, or a video set to a piece of chalk) can be exploited in the activity package. This includes a computer, when it is available to a school.

The "factor game" has appealed to preservice and in-service elementary teachers because of the diversity of skills and concepts embodied in the game. It involves the student with concepts of prime and composite as well as applications of the fundamental theorem of arithmetic. These experiences are basic preparation for addition of rationals.

In the framework of an activity package designed to treat algorithms for the addition of rationals, we included the factor game as an individualized computer instruction activity. The following computer print out displays the student and machine dialogue that reveals how the factor game is played as well as the strategies of the game.

WELCOME TO THE GAME OF FACTORING. THE RULES ARE EASY.
WE START WITH THE FOLLOWING TABLE:

2	3	4	5	6	7
8	9	10	11	12	13
14	15	16	17	18	19
20	21	22	23	24	25
26	27	28	29	30	31
32	33	34	35	36	37

THE IDEA IS SIMPLE. WE TAKE TURNS PICKING NUMBERS FROM THE BOARD. AFTER ONE OF US PICKS A NUMBER, THE OTHER MAY THEN CLAIM ALL THE DIVISORS OF THE NUMBER THAT ARE NOT ALREADY USED. EACH NUMBER IN THE TABLE MAY BE USED ONLY ONCE, WHETHER IT IS PICKED OR CLAIMED. THE SCORE WILL BE THE SUM OF ALL THE NUMBERS PICKED OR CLAIMED BY A PLAYER. HIGHER SCORE WINS.

IF YOU WANT TO SEE AN UP-TO-DATE CHART AT ANY TIME, JUST PUSH THE RETURN BUTTON.

YOU WILL START THE FIRST GAME AND WE WILL ALTERNATE AFTER THAT. YOU MAY WANT TO RIP OFF THE PAPER CONTAINING THE BOARD, SO THAT YOU CAN X OUT THE NUMBERS ALREADY USED.

START BY TYPING ANY NUMBER ON THE CHART. REMEMBER, YOU MUST PUSH THE RETURN BUTTON AFTER YOU TYPE.

```
?37
YOUR SCORE IS NOW          37
MY CHOICE IS       31      AND MY SCORE IS NOW       31
IT IS NOW YOUR TURN. YOU CAN EITHER CLAIM A DIVISOR
OF MY NUMBER (IF THERE IS ONE NOT YET USED), OR YOU
MAY PICK A NUMBER OF YOUR OWN.
  IF YOU WANT TO CLAIM, TYPE 1.
  IF YOU WANT TO PICK, TYPE THE NUMBER.
  IF YOU WANT TO SEE THE CHART, JUST PUSH THE RETURN.
YOUR CHOICE IS         ?29
YOUR SCORE IS NOW          66
MY CHOICE IS       23      AND MY SCORE IS NOW       54
IT IS YOUR TURN. TYPE EITHER 1 (TO CLAIM), A
NUMBER, OR JUST PUSH THE RETURN BUTTON. YOUR CHOICE
IS       ?35
YOUR SCORE IS NOW          101
I CLAIM THE NUMBER (S)         5     ,      7
MY SCORE IS NOW       66
MY CHOICE IS       25      AND MY SCORE IS NOW       91
IT IS YOUR TURN. TYPE EITHER 1 (TO CLAIM), A
NUMBER, OR JUST PUSH THE RETURN BUTTON. YOUR CHOICE
IS       ?1
WHAT IS A NUMBER YOU WISH TO CLAIM  ?5
I AM SORRY BUT WE HAVE ALREADY USED THAT NUMBER.
IF YOU WANT TO CLAIM ANY OTHER NUMBERS, TYPE ONE OF
THEM IN. OTHERWISE, TYPE 1. YOUR CHOICE IS  ?1
YOUR NEXT CHOICE IS  ?19
YOUR SCORE IS NOW       120
MY CHOICE IS       33      AND MY SCORE IS NOW       124
IT IS YOUR TURN. TYPE EITHER A 1 (TO CLAIM), A
NUMBER, OR JUST PUSH THE RETURN BUTTON, YOUR CHOICE
IS  ?1
WHAT IS A NUMBER YOU WISH TO CLAIM     ?3
YOUR SCORE IS NOW       123
IF YOU WANT TO CLAIM ANY OTHER NUMBERS, TYPE ONE OF
THEM IN. OTHERWISE, TYPE 1. YOUR CHOICE IS  ?11
YOUR SCORE IS NOW       134
IF YOU WANT TO CLAIM ANY OTHER NUMBERS, TYPE ONE OF
THEM IN. OTHERWISE, TYPE 1. YOUR CHOICE IS  ?7
I AM SORRY BUT WE HAVE ALREADY USED THAT NUMBER.
IF YOU WANT TO CLAIM ANY OTHER NUMBERS, TYPE ONE OF
THEM IN. OTHERWISE, TYPE 1. YOUR CHOICE IS  ?0
            2        x        4        x        6        x
            8        9        10       x        12       13
            14       15       16       17       18       x
            20       21       22       x        24       x
            26       27       28       x        30       x
            32       x        34       x        36       x
YOUR NEXT CHOICE IS  ?9
NO,  9  IS NOT A FACTOR OF  33  33  /  9 = 3.66667
IF YOU WANT TO CLAIM ANY OTHER NUMBERS, TYPE ONE OF
THEM IN. OTHERWISE, TYPE 1. YOUR CHOICE IS  ?1
YOUR NEXT CHOICE IS  ?34
YOUR SCORE IS NOW       168
I CLAIM THE NUMBER(S)       2    ,      17
MY SCORE IS NOW       143
```

Program strategies

The numbers 2 through 37 were chosen (1) to avoid 1, (2) to fit into a convenient array (here, six by six), and (3) to give a nice spread of both primes and composities that are low enough for easy calculation by the student.

The first strategy is to choose (in order) 37, 31, 29, and 23. The choice of 19 next is not optimal as, for example, the choice of 35 will result in a forfeit of 5 and 7 and the net gain is thus 35 − 12 = 23, which is greater than 19.

This quantity—choice minus forfeit—can be calculated very easily by the machine. This net gain is a measure of the value of the choice. Without attempting to justify our decision by game theoretic considerations, we will say that a choice is optimal if (1) no other choice has a larger net gain and (2) among all choices with the same net gain, it is the largest in numerical value (for reasons of convenience, the program actually minimizes the negative of the net gain rather than maximizing the net gain).

If the machine were to choose the optimal choice every time, the benefit to the student would be limited and, perhaps, psychologically destructive. In this program, the machine will pick the optimal choice one-half of the time. The rest of the time, the machine will pick randomly (occasionally picking an optimal choice).

After two losses, the machine will pick the random choice only one-third of the time. Experience has shown that on the first try, most students can play two or three games in thirty minutes. More experienced players will probably encounter the better machine strategy.

EDITORIAL COMMENT.—(1) Adapt the card game Fish for use with prime factorizations. In this game, a "run" consists of a card for a whole number and cards for its prime factorization (12, 2, 2, 3). When you prepare the playing deck, include cards for the numbers 2 through 24 and a liberal number of cards for the prime factors 2, 3, and 5. Include a few cards for the prime factors 7 and 11. See the description of Fish in Hunt's article in chapter 9.

(2) You can develop dice games to work the "multiple of" and "factor of" relationships. Start with two blank cubes of different colors. On one cube write the numerals 2, 3, 4, 5, 6, and 7. On the other cube write the numerals 10, 12, 14, 16, 18, and 20. Players roll the two cubes. Points are scored if the number named on one cube is "a multiple of" or "a factor of" the number named on the other cube.

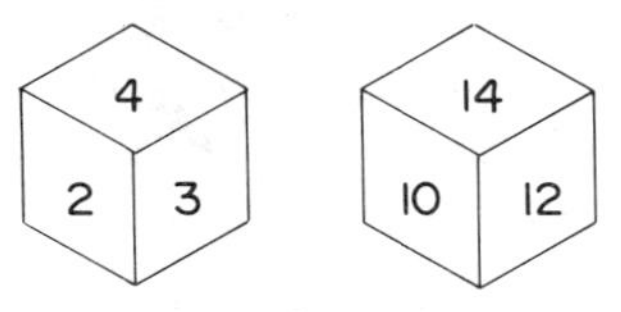

No Score
14 is not a multiple of 4
4 is not a factor of 14

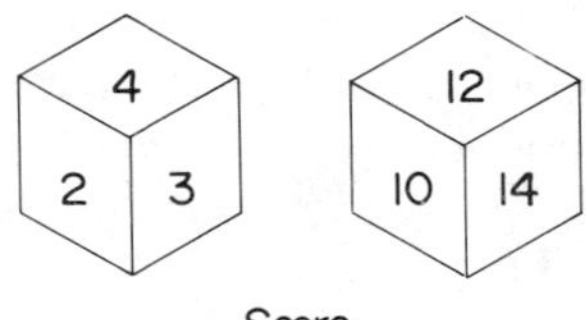

Score
12 is a multiple of 4
4 is a factor of 12.

Five "nontrivial" number games

TERREL TROTTER, JR.

Formerly a teacher of high school mathematics at the Norwich High School in Norwich, Kansas, Terrel Trotter is a participant in the 1971–72 Academic Year Institute for mathematics teachers at the University of Illinois.

Number games can provide elementary students with a good opportunity to test their observation skills for problem solving, and at the same time they will be practicing such drill skills as adding and subtracting. Here are five trivial number games guaranteed to provide hours of fascination and lots of hidden drill. The word *trivial* is used here with two meanings in mind. The games are trivial in the usual mathematical sense—namely, once the secret is known to both players, the game's winner is predetermined, based on the starting number and who begins play. The games are *non*trivial from the point of view that they are valuable teaching aids for motivating student thinking.

The rules for the games will be presented first, followed by discussions of the respective winning strategies. You are encouraged to try the games before reading the solutions section. This way you may have the pleasure of discovering the secrets for yourself before challenging your students to attempt the same.

Rules

Game 1. The first player selects any integer from 1 to 10. Then the two players alternately add any integer from 1 to 10 to the sum left by the opponent. Play continues until one player can make an addition giving a grand total of 100. That player is thereby declared the winner.

Game 2. The rules are the same as those for game 1, *except* that now the winner is the player who forces his opponent to make the total 100 or more.

Game 3. The name of this game is "Aliquot." It was devised by David L. Silverman and appeared in the problem section of the *Journal of Recreational Mathematics*. Here is Mr. Silverman's own description from that journal (1970[a]):

> Two players start with a positive integer and alternately subtract any aliquot part (factor) with the exception of the number itself from the number left by the opponent. Winner is the last player able to perform such a subtraction.
>
> By way of example, if the original number is 12, first player may subtract either 1, 2, 3, 4, or 6 (but not 12). If he subtracts 2, leaving 10, second player may subtract 1, 2, or 5.
>
> The objective is to leave your opponent without a move. This can only be done by leaving him a 1, since 1 is the only positive integer with no aliquot part other than itself.

Game 4. This game, called "Proper Aliquot," was also devised by Mr. Silverman. In the same issue of the *Journal of Recreational Mathematics,* he gives the rules thus:

> The rules are the same as those of Aliquot with the exception that only *proper* divisors may be subtracted. Consider 1 an improper divisor.

Game 5. The first player selects any reasonable large number. Then the two players alternately subtract any number they choose from the number left by the opponent, provided the chosen number meets one requirement: the number subtracted must not exceed twice the value of the number subtracted by the opponent on the previous play. For example, if player A subtracts 4, then player B can select any integer from 1 to 8. The game is won when one player can "take it all," that is, leave 0.

Solutions

Game 1. After playing this game a time or two, most students realize that whoever makes the total 89 can force the win. A player does this by adding the difference between 11 and whatever his opponent adds next. That is, if A adds 3 to 89, B adds 11 − 3, or 8, making 100 and winning.

Similar reasoning leads to the conclusion that 78 is the next desirable total to obtain, since 78 = 89 − 11. A continuation of the same reasoning yields the winning sequence 67, 56, 45, 34, 23, 12, and 1. (An easy way to remember these numbers is to recognize that the tens digit is one less than the units digit.) Therefore, by starting with 1, the first player can force the win, regardless of what the other player does. At that point the game becomes trivial. (Of course, if a player doesn't know the secret, his opponent can conceal it from him longer by simply "entering" the sequence at some later point.)

Game 2. Inasmuch as 99 is the largest total a player can make without leaving or exceeding 100, it and the sequence derived by subtracting 11s are the secret totals. (Naturally, they are the multiples of 11.) Therefore, by allowing his opponent to select the first number, a player can once again force the win.

Game 3. The proper way to analyze this game is also to begin at the *end.* Only now the *end* is the integer 1. First, examine all the possible subtractions for some integers near the end of the game, say 2 through 10:

$$\begin{array}{r} 2 \\ -1 \\ \hline 1 \end{array} \quad \begin{array}{r} 3 \\ -1 \\ \hline 2 \end{array} \quad \begin{array}{rr} 4 & 4 \\ -2 & -1 \\ \hline 2 & 3 \end{array} \quad \begin{array}{r} 5 \\ -1 \\ \hline 4 \end{array} \quad \begin{array}{rrr} 6 & 6 & 6 \\ -3 & -2 & -1 \\ \hline 3 & 4 & 5 \end{array}$$

$$\begin{array}{rrr} 8 & 8 & 8 \\ -4 & -2 & -1 \\ \hline 4 & 6 & 7 \end{array} \quad \begin{array}{rr} 9 & 9 \\ -3 & -1 \\ \hline 6 & 8 \end{array} \quad \begin{array}{rrr} 10 & 10 & 10 \\ -5 & -2 & -1 \\ \hline 5 & 8 & 9 \end{array}$$

If a *winning number* is defined to be one from which a player can force a win, even with the best play from his opponent, a *loser* will be a number from which a player can only leave a winning number to the opponent, no matter what factor is subtracted. Thus, 2 is a winner—the ultimate winner, of course, since it leads directly to the game's objective. On the other hand, 3 is a loser because it leaves 2 to the opponent. Though 4 has two possible subtractions, it still can be classified as a winner because the best play (the factor 1) leaves a loser.

Further investigation along this line yields the conclusion that all even numbers are winners and all odd numbers are losers. This suggests a simple rule for winning: If player A faces an even number, he subtracts one of its *odd* factors so that he leaves an odd number for his opponent to play from. The logic behind this is that the only factors of an odd number are also odd, and odd − odd = even. Therefore, the opponent must return some other even number, and the cycle is established. Eventually he must leave the 2, and then player A wins.

Game 4. The exclusion of 1 from the set of factors makes some interesting changes in the strategy. Now all the primes are the objectives of the game and therefore are immediate losers. Testing the other numbers as was done for game 3 leads to the conclusion that all odds are still losers, but not all evens are winners. For example, 8 is the first even loser—after 2, that is, which is also prime. Identification of the losing even numbers becomes less difficult when the next even loser (32) is found. Factoring these three even losers yields 2^1, 2^3, and 2^5. These are odd powers of 2, that is, *2^n where n is odd.*

In summary then, the winning procedure is to subtract such factors that leave either an odd number or an odd power of 2.

Game 5. The secret behind this game is a great deal more sophisticated and elusive than the secrets behind the other four. It relies on some high-powered concepts from higher mathematics. But, the secret is relatively easy to learn and use.

A preliminary observation about the game should be noted. A player should never subtract a value that is one-third or more of the number. If he does, the opponent can always take all the remainder. For example, if a player faced 17 and subtracted 6, leaving 11, his opponent could "take it all" because the opponent can go as high as 12.

The Fibonacci numbers are an essential part of the secret. Simply stated, the Fibonacci numbers are integers from the sequence 1, 2, 3, 5, 8, 13, 21, . . . , where succeeding members are found by adding the two preceding members. A theorem from number theory, the Zeckendorf theorem, says, in effect, that all non-Fibonacci numbers can be represented uniquely as the sum of two or more nonconsecutive Fibonacci numbers. (For a proof of this theorem, see Hoggatt [1969, p. 74].) Here are a few examples to illustrate this concept:

a) $17 = 13 + 3 + 1;$

b) $29 = 21 + 8;$

c) $40 = 33 + 5 + 2.$

It turns out that when playing from a non-Fibonacci number, a player can force the win by subtracting the smallest Fibonacci number appearing in the representation. This succeeds as a result of the easily proven fact that for any two nonconsecutive Fibonacci numbers, the smaller one is always less than one-half the larger. The significance of this is that it always allows the player to give his opponent a Fibonacci number at some future stage of the game. From there, the opponent must then return a non-Fibonacci number to the player. It will have a new representation with a smallest Fibonacci addend, and the cycle is repeated.

An illustration will serve to clear up the preceding discussion. Playing from 17, which equals 13 + 3 + 1, player A will take the 1. Now B can only subtract 1 or 2, neither of which will take all the 3 in the representation. Whatever B subtracts, A will then take whatever is left of the 3. This presents B with the Fibonacci number 13. The greatest possible number B can now subtract is 4, depending on how the two previous subtractions were made. In any case, A will have a new non-Fibonacci number to play from, and he can return to the winning procedure described above.

As mentioned in the beginning, these games are indeed trivial from a pure, mathematical standpoint. Obviously, if both players know the secret, there would be no point in playing a game. The first to play would hand the other a losing number, and for all practical purposes the game would be over right away. If only one player knew a game's secret, he could always win too, much to the consternation of his patsy. But, if both players are unaware of the secrets, the games provide a lot of practice in basic skills. The importance is that *the drill is only a means to an end, not an end to itself.*

In order to enjoy mathematics, students need to experience some success along the way. These games often allow the poorer student to achieve some wins over the better student, especially at first, before the strategies are either discovered by the students themselves (it is hoped) or revealed to them. The more talented student should be encouraged to seek the strategies as exercises in true problem solving. It is recommended that the secrets not be revealed too soon, if at all. Telling the secrets of the games destroys the fun and denies an individual the satisfaction of finding them for himself. Either way, the games are far from trivial as learning experiences.

References

Hoggatt, Verner E., Jr. *Fibonacci and Lucas Numbers.* Boston: Houghton Mifflin Co., 1969.

Silverman, David L. (a). "Problems and Conjectures." *Journal of Recreational Mathematics* 2 (October 1970): 227.

——— (b). *Your Move.* New York: McGraw-Hill Book Co., 1971.

Magic squares: extensions into mathematics

DAIYO SAWADA

An assistant professor in the Department of Elementary Education at the University of Alberta in Edmonton, Daiyo Sawada's main responsibility is the preparation of teachers for the teaching of mathematics.

Many a potentially dull practice session on addition and subtraction has been transformed into a lively encounter with mathematics through the introduction of magic squares. Figure 1 shows three of the more familiar ways in which magic squares are used. Normally work with magic squares in the elementary school does not go beyond the types of examples shown there. Further exploration of magic squares is usually limited to finding simple procedures for constructing 3×3 squares, 4×4 squares, and $n \times n$ squares where n is odd. The procedures are frequently presented in a rather rote fashion since the emphasis is on practice for addition, not on the properties of magic squares per se.

As an introduction to more novel ways of using magic squares, consider the sequence of activity cards shown in figures 2, 3, and 4. After reading the cards in order, the reader can probably sense the direction of the activities. When students have completed the three activities, the discussion that follows might consider other ways of adding magic squares, including ways that the students themselves suggest. Different ways of adding magic squares could be compared and, indeed, a fourth activity card could investigate the closure property of the operation of addition with magic squares. Subsequent cards could ask about the associative and commutative properties of addition with magic squares, the identity magic square for addition, and the existence of additive inverses.

Any consideration of these properties with operations with magic squares assumes that the students have had prior experience with the *system* of whole numbers. Some students may think of comparing magic squares with integers as a system. The emphasis would not be on proof but rather on the use of basic properties to characterize some new mathematical entities. In short, the students would be participating in the construction and exploration of mathematical systems—experiences that they seldom get in the usual elementary mathematics program. Furthermore, the strategy of asking questions concerning closure, associativity, commutativity, existence of inverses, and so on whenever interesting and new ideas are encountered is a highly mathematical behavior. A spirit of inquiry can be encouraged in students without insisting on formalization.

All of the suggestions so far have been rather general. The rest of this article focuses on details that teachers should find helpful if they are interested in exploring the idea of magic squares as a system.

Magic squares as a mathematical system

For teachers who believe that it is worthwhile to provide some informal experiences with mathematical systems for

Testing Magic Squares

Magic squares have row, column, and diagonal sums that are all equal. Test to see if the array below is a magic square.

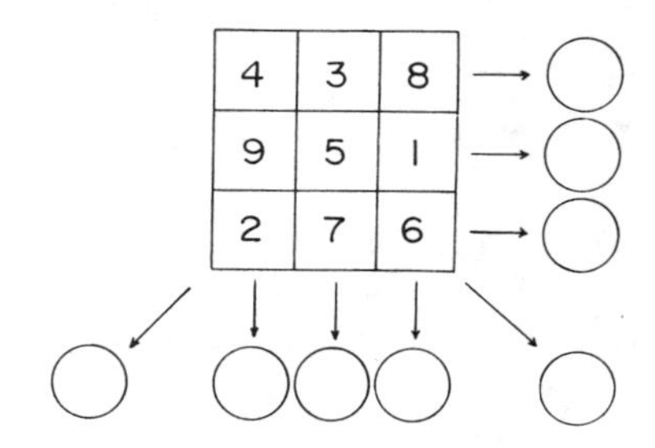

Completing Magic Squares

Complete the following magic square.

12	89	
	45	
		78

Constructing Magic Squares

Construct a magic square that has a sum of 27.

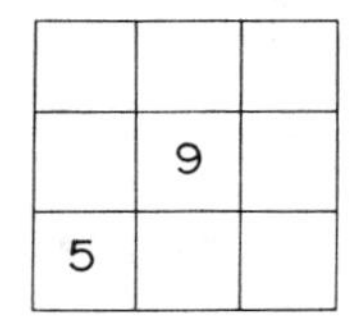

Fig. 1

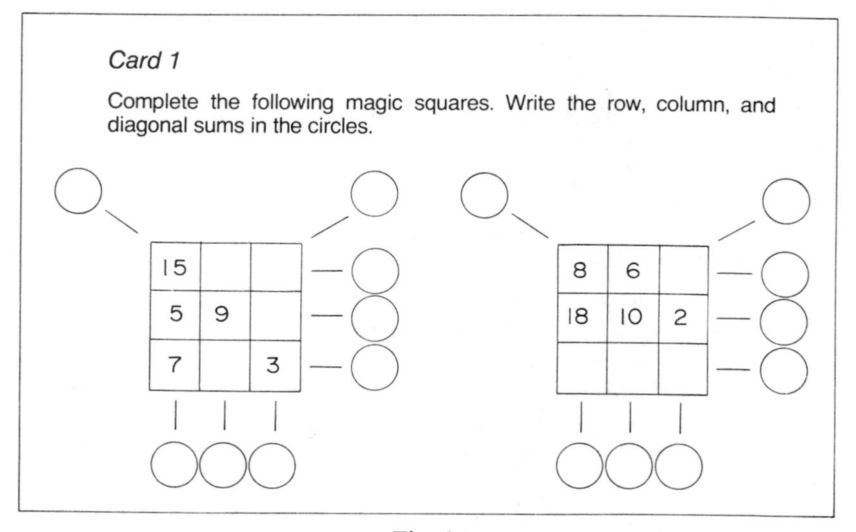

Fig. 2

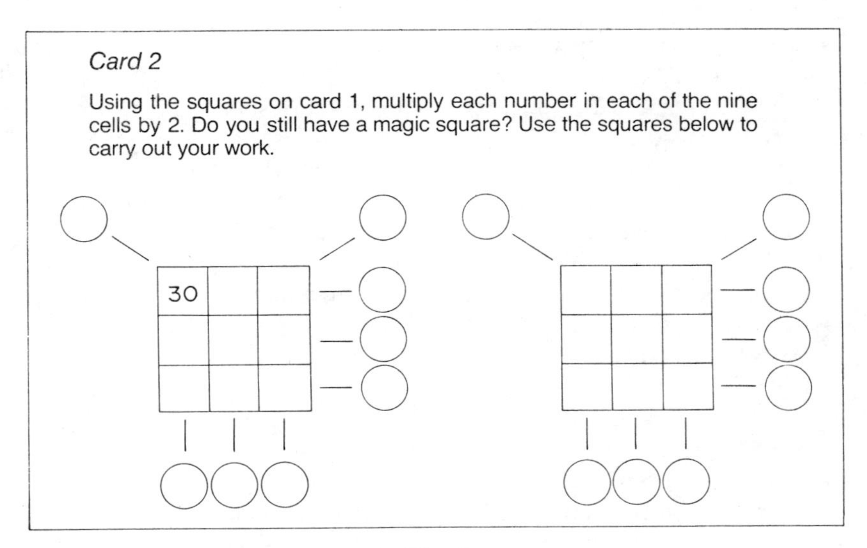

Card 2

Using the squares on card 1, multiply each number in each of the nine cells by 2. Do you still have a magic square? Use the squares below to carry out your work.

Fig. 3

Card 3

Suppose we wanted to add the squares given on card 1. One way of adding is started below. Can you finish the adding?

Did you get another magic square? Test it.

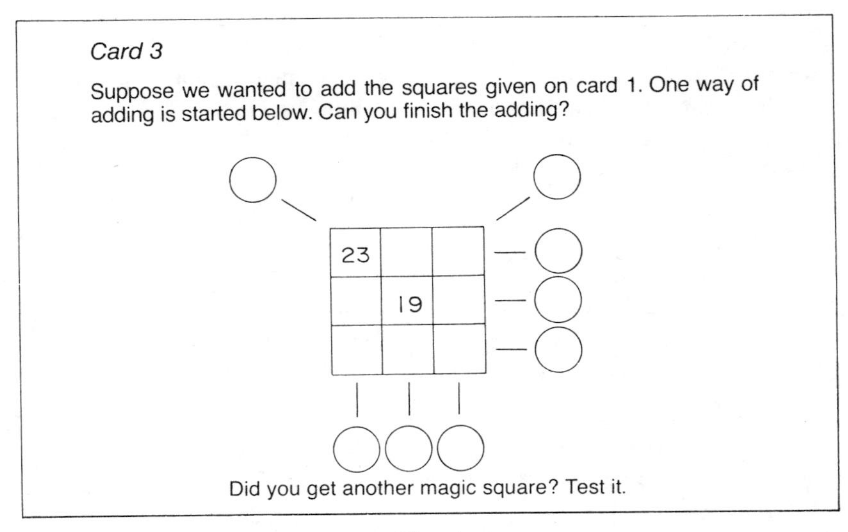

Fig. 4

their students, it is useful to know what kinds of mathematical systems can be easily constructed using magic squares as the elements and addition as the operation. As the reader has probably already guessed, the set of 3×3 magic squares can be used to form a group. That is, if we let M designate the set of 3×3 magic squares with integer entries in the cells and let $\oplus$ designate magic square addition as suggested on card 3 (see fig. 4), then for any magic squares $X, Y, Z \in M$ the properties shown in figure 5 hold. (Students can test these properties on some magic squares.) These four properties (closure, associativity, identity, inverse) are the axioms of a group. In

(i) *Closure.* $X \oplus Y$ is a magic square.
(ii) *Associativity.* $(X \oplus Y) \oplus Z = X \oplus (Y \oplus Z)$
(iii) *Identity.* The identity square is

0	0	0
0	0	0
0	0	0

(iv) *Inverse.* The inverse of a given magic square is generated by multiplying each cell entry by -1.

Fig. 5

other words, the set M for the operation of addition $\oplus$ is a group. Since it is also true that $X \oplus Y = Y \oplus X$—that is, the property of commutativity holds—the set M for the operation of addition is a commutative group.

As indicated earlier, it would not be a good idea to expect all students to construct proofs of these properties of addition of 3×3 magic squares. The proofs, however, are relatively simple, if you begin with the assumption that the properties hold for integers, and they are rather typical of elementary proofs involved in number systems—like proving that addition of rational numbers is commutative, for example. Interested students may find it very worthwhile to engage in some "proving" experiences.

The proof for $X \oplus Y = Y \oplus X$, namely, that the operation of addition for 3×3 magic squares is commutative, is shown in figure 6. The other proofs are similar. (For readers who have had experiences with matrices, it is probably evident that the proofs correspond quite closely to proofs related to the properties of matrix addition.)

Perhaps the property that is most interesting is the property of closure. Many students have some difficulty grasping what is meant by closure as related to the set of natural numbers under addition or multiplication. To them it is too obvious that "when you add two natural numbers you get another natural number"—why prove what is so obviously true? When students are confronted with the analogous problem with magic squares, however, it is not so obvious that adding two magic squares will produce another magic square. Hence the property of closure begins to take on meaning; it begins to make a difference. Indeed many students will want to test it out on several magic squares before accepting the hypothesis that magic square addition is closed.

To be proved: $X \oplus Y = Y \oplus X$

Proof:

Let $X = \begin{bmatrix} a & b & c \\ d & e & f \\ g & h & i \end{bmatrix}$ and $Y = \begin{bmatrix} j & k & l \\ m & n & o \\ p & q & r \end{bmatrix}$

where $a, b, c, \cdots, r$ are integers.

Then $X \oplus Y = \begin{bmatrix} a & b & c \\ d & e & f \\ g & h & i \end{bmatrix} \oplus \begin{bmatrix} j & k & l \\ m & n & o \\ p & q & r \end{bmatrix}$

$= \begin{bmatrix} a+j & b+k & c+l \\ d+m & e+n & f+o \\ g+p & h+q & i+r \end{bmatrix}$ By definition of $\oplus$.

$= \begin{bmatrix} j+a & k+b & l+c \\ m+d & n+e & o+f \\ p+g & q+h & r+i \end{bmatrix}$ By commutativity of addition with integers

$= \begin{bmatrix} j & k & l \\ m & n & o \\ p & q & r \end{bmatrix} \oplus \begin{bmatrix} a & b & c \\ d & e & f \\ g & h & i \end{bmatrix}$ By definition of $\oplus$

$= Y \oplus X$

Fig. 6

As suggested on activity card 2, a multiplication operation can also be defined. For example,

$$2\begin{bmatrix}6 & 1 & 8\\ 7 & 5 & 3\\ 2 & 9 & 4\end{bmatrix} = \begin{bmatrix}2\times 6 & 2\times 1 & 2\times 8\\ 2\times 7 & 2\times 5 & 2\times 3\\ 2\times 2 & 2\times 9 & 2\times 4\end{bmatrix}$$

$$= \begin{bmatrix}12 & 2 & 16\\ 14 & 10 & 6\\ 4 & 18 & 8\end{bmatrix},$$

where the multiplication of the number 2 and the magic square is indicated merely by writing them adjacent to each other. With this new operation we can get an additional property, $a(X \oplus Y) = aX \oplus aY$, where a is an integer. For example,

$$2\left(\begin{bmatrix}6 & 7 & 2\\ 1 & 5 & 9\\ 8 & 3 & 4\end{bmatrix} \oplus \begin{bmatrix}4 & 3 & 8\\ 9 & 5 & 1\\ 2 & 7 & 6\end{bmatrix}\right)$$

$$= 2\begin{bmatrix}6 & 7 & 2\\ 1 & 5 & 9\\ 8 & 3 & 4\end{bmatrix} \oplus 2\begin{bmatrix}4 & 3 & 8\\ 9 & 5 & 1\\ 2 & 7 & 6\end{bmatrix}.$$

It is easy to show that the property holds for this example, and other examples can be tried. Some students may even try a proof. Students may also recognize this as an example of the distributive property.

"1089 Overture"

As a closing example, consider an enrichment activity based on the interesting number pattern in figure 7. As a catalyst for discussion, column (b) is generated *after* the products are written in column (a). The teacher could say, "Let's pretend we have a mirror beside our column of products. What would the mirror images look like?" The images would then be written as shown in column (b). Students could then be encouraged to look for interesting relationships like the following:

1. Column (b) is column (a) written "upside down."

2. In column (a)—

- the units digits run from 1 through 9. Similar patterns hold for the tens, hundreds, and thousands columns.
- $1089 \rightarrow 10 + 89 = 99$
$2178 \rightarrow 21 + 78 = 99$
and so on.
- $1089 \rightarrow 1 + 0 + 8 + 9 = 18$
$2178 \rightarrow 2 + 1 + 7 + 8 = 18$
and so on.

3. Patterns observed for column (a) could be tested for column (b). Other relationships may be discovered. But what has this to do with magic squares?

			(a)	(b)
1	× 1089	=	1089	9801
2			2178	8712
3			3267	7623
4			4356	6534
5			5445	5445
6			6534	4356
7			7623	3267
8			8712	2178
9			9801	1089

Fig. 7

Two squares are shown in figure 8. Square A is the standard 3 × 3 magic square. Square B can be generated from the numbers in column (a) of figure 7 by looking at the thousands digit and matching it with the numbers in square A. (Is square B a magic square? Students could find out by adding.)

8	1	6
3	5	7
4	9	2

8712	1089	6534
3267	5445	7623
4356	9801	2178

Fig. 8

At some point, however, it should be noted that the solution to the open sentence in figure 9 is relevant. That is, what is the result if square A is multiplied by

$$1089\begin{bmatrix}8 & 1 & 6\\ 3 & 5 & 7\\ 4 & 9 & 2\end{bmatrix} = ?$$

Fig. 9

1089? If the students have already convinced themselves that multiplication of a 3 × 3 magic square by a whole number or integer is a closed operation, then the answer to this question must be yes, without calculating. Or the students can confirm the result by actual calculation.

Several other questions could be asked about square B:

1. Considering only the units digits in each cell, do you get a magic square?
2. Similarly, do you get a magic square if you consider just the tens digits? the hundreds digits? the thousands digits?
3. Knowing that the row sum of square A is 15, can you figure out the row sum of square B without adding? (Hint: Use multiplication.)

Summary

It is hoped that enough detail has been given so that many teachers will be able to use magic squares in new ways. Activities such as those described can lead to some profound mathematics (groups, modules, theory of multiples) at the same time they provide needed computational practice with whole numbers and integers. Or if students are working with fractions, magic squares could be multiplied by fractions. The latter could lead to the exploration of the well known structure of a vector space.

Although modules, vector spaces, multiples, and groups are important mathematical concepts in and of themselves, the purpose in writing this article was *not* to implore teachers of upper elementary grades, or even junior high, to teach such concepts. Rather, the purpose was to suggest the strategy of taking interesting ideas, such as magic squares, and extending them in ways that lead to the discovery of other mathematical structures, all the while providing interesting practice of fundamentals as originally intended. It is this process of extension that is important, not the particular outcome. Such extensions require the asking of the right questions. As students discover what sort of questions lead to interesting payoffs, they will be well on the road to converting interesting practice into interesting mathematics.

References

Heath, R. V. *Mathemagic*. New York: Dover Publications, Inc., 1933.

Keedy, M. L. *A Modern Introduction to Basic Mathematics*. Reading, Mass.: Addison-Wesley, 1963.

EDITORIAL COMMENT.—In the March 1974 issue of the *Arithmetic Teacher*, Anne Mae Cox suggests using magic spider webs in "Magic While They Are Young."

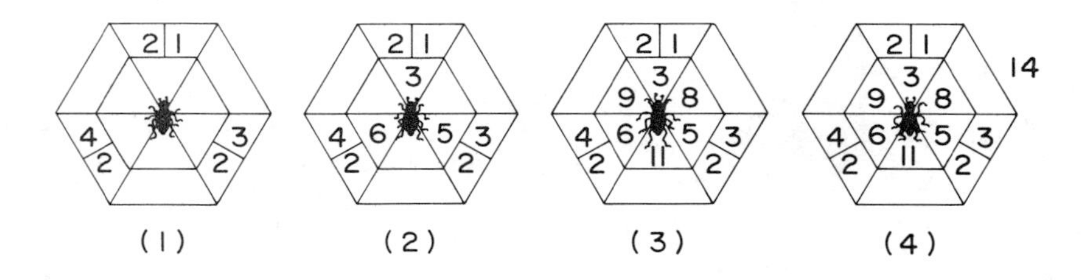

1. Place pairs of numbers at three locations in the outer ring.
2. Write the sum of these numbers in the adjacent triangle.
3. In the remaining triangles, write the sum of the numbers in the two adjacent triangles.
4. Now find the sums of the numbers in opposite triangles. Surprise!

"Interest getters"

KARL G. ZAHN *W. P. Roseman Campus Elementary School*
Wisconsin State University, Whitewater, Wisconsin

It is generally recognized that it is important for a teacher to maintain pupil interest in arithmetic. The use of a variety of learning activities is an excellent technique for retaining pupil interest. One technique is to use one class period per week for presenting "interest getters." Friday is usually a good day to present these, for the teacher can review the past week's activities and then present some stimulating items to create and maintain pupil interest.

Continued interest in arithmetic is also maintained by patient, sympathetic guidance given to the student by the teacher.

The following are examples of some activities that I have had success in using with elementary and junior high school students:

EXAMPLE 1

$1 \times 1 = 1$
$11 \times 11 = 121$
$111 \times 111 = 12{,}321$
$1{,}111 \times 1{,}111 = ?$
$11{,}111 \times 11{,}111 = ?$
$111{,}111 \times 111{,}111 = ?$
$1{,}111{,}111 \times 1{,}111{,}111 = ?$
$11{,}111{,}111 \times 11{,}111{,}111 = ?$
$111{,}111{,}111 \times 111{,}111{,}111 = ?$

EXAMPLE 2

$1\ 2\ 3\ 4\ 5\ 6\ 7\ 9 \times 9 = 111{,}111{,}111$
" $\times 18 = ?$
" $\times 27 = ?$
" $\times 36 = ?$
" $\times 45 = ?$
" $\times 54 = ?$
" $\times 63 = ?$
" $\times 72 = ?$
" $\times 81 = ?$
" $\times 90 + 1 = ?$

EXAMPLE 3

$9^2 = ?$
$99^2 = ?$
$999^2 = ?$
$9{,}999^2 = ?$

EXAMPLE 4

$222{,}222{,}222 \times 9 = 1{,}999{,}999{,}998$
$333{,}333{,}333 \times 9 = 2{,}999{,}999{,}997$
$444{,}444{,}444 \times 9 = ?$
$555{,}555{,}555 \times 9 = ?$
$666{,}666{,}666 \times 9 = ?$
$777{,}777{,}777 \times 9 = ?$
$888{,}888{,}888 \times 9 = ?$
$999{,}999{,}999 \times 9 = ?$

One purpose for using Examples 1, 2, 3, and 4 is to develop interest in arithmetic for the student. Another purpose might be to practice computational skills. Third, but not least, it is hoped that these exercises will help students develop insight and see relationships in numbers. In each exercise, I would ask students to work out a few of the problems and then make some "educated guesses" concerning the pattern that is evolving.

Dividing 10 by 81 has an interesting result. Ask your students to do this and carry it out for three or four places. Then ask them if they can predict what the eighth, ninth, and tenth digits will be.

Students always enjoy stories that are humorous or challenging. Here are two favorites that I have told to children (and adults) for a number of years:

Pat and Mike

Pat and Mike went into the coal business. Pat said, "Seven tons of coal at $16 per ton is $49." Mike said, "I had better multiply that and check to see if you are correct."

7 times 6 is 42. *Put down the 42.*	16 ×7
7 times 1 is 7. *Put it under the 42.*	42 +7
Add.	49
Result is 49.	

Pat said, "I had better check your multiplying by dividing 7 into 49 and see if we get 16. Seven doesn't go into 4, but it goes into 9 once.

1 times 7 is 7.	16
Subtract 7 from 9 and bring down the 4.	7) 49
7 divides into 42 six times.	7 / 42
It's correct!	

Mike said, "Let's check one more time by adding."

6 + 6 = 12 + 6 = 18 + 6 = 24 + 6 = 30 + 6 = 36 + 6 = 42

(then coming down the 1's at the top) 43, 44, 45, 46, 47, 48, 49.

16
16
16
16
16
16
16
—
49

It checks! (Now try this with 7 and 13.)

The Arab and His Three Sons

There is a story related about an Arab who died, leaving 17 horses to be divided among his three sons. According to the Arab's will, the oldest son was to receive one-half the horses, the second son was to receive one-third, and the youngest son's share was to be one-ninth. The sons could not agree upon the division of the horses and quarreled violently. Their dispute was overheard one day by an ancient wise man who was riding by on an old gray mare. The wise man considered the problem at length and at last presented his old mare to the sons, thereby giving them 18 horses to divide. The sons joyfully made the division; the eldest son took one-half of 18 or 9 horses, the second son one-third of 18 or 6 horses, and the third son one-ninth of 18 or 2 horses for a total of 17 horses. So the wise man mounted the remaining horse and rode away, leaving the sons happy and contented.

An interesting game that can be played in any classroom is Nim. I have usually introduced this game by putting marks on the chalkboard.

Row 1	1
Row 2	111
Row 3	11111
Row 4	1111111

Any number of marks or rows may be used; however, I have always used the 1, 3, 5, 7 combination with children at the beginning for the sake of simplicity. Two players take turns in making their moves. Each player, during his turn, may cross off all or only some of the marks from any one row, but he must cross off at least one mark. The play continues until the player who succeeds in crossing off the last mark wins the game.

To illustrate, let us follow the moves in a game:

Step 1

Row 1	1	Player No. 1 crosses out all of Row 4.
Row 2	111	
Row 3	11111	
Row 4	ǂǂǂǂǂǂǂ	

Step 2

1
111
111ǂǂ

Player No. 2 crosses off 2 in Row 3.

Step 3

ǂ
111
111

Player No. 1 crosses off 1 in Row 1.

Step 4

111
1ǂǂ

Player No. 2 crosses off 2 in Row 3.

Step 5

1ǂǂ
1

Player No. 1 crosses off 2 in Row 2.

Now player No. 2 can only cross off 1, so No. 1 is the winner.

This game can also be played with matches, pebbles, or toothpicks. It is a good "rainy day" game children can play at their desks or at the chalkboard.

Nim is frequently used as a gambling game, but a player who knows the secret can almost always win. After some experience, children can usually tell what the game-ending winning distributions should be. Can you? If someone becomes suspicious of your winning frequently, change the game so that the last mark crossed out is the loser rather than the winner.

EDITORIAL COMMENT.—Here are some other curiosities to consider:

A "number pyramid." Place 1's at each end. Find the numbers in the pyramid blocks by taking the sum of the two numbers in the blocks above.

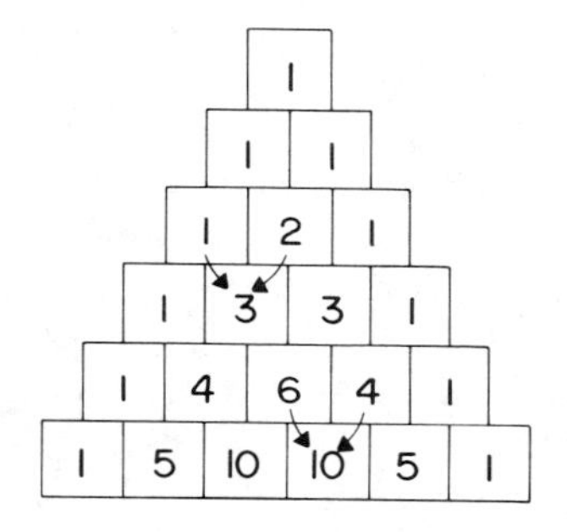

Do you recognize this as Pascal's triangle?

Predictable answers!

$$1 \times 8 + 1 = 9$$
$$12 \times 8 + 2 = 98$$
$$123 \times 8 + 3 = 987$$
$$1234 \times 8 + 4 = 9876$$
$$12345 \times 8 + 5 = 98765$$
$$123456 \times 8 + 6 = ?$$

$$0 \times 9 + 1 = 1$$
$$1 \times 9 + 2 = 11$$
$$12 \times 9 + 3 = 111$$
$$123 \times 9 + 4 = 1111$$
$$1234 \times 9 + 5 = 11111$$
$$12345 \times 9 + 6 = ?$$

Just for fun

J. D. CALDWELL *Public Schools, Windsor, Ontario*

Some of our readers might like to try the following diversion with their classes. As many times as I have seen it used, it never fails to stimulate pupil interest and further exploration. We might also note that this particular lesson is a good example of the inductive-deductive approach.

The teacher asks a pupil to give the last five digits of his telephone number. These are recorded on the chalkboard. Let us suppose the recorded number is 34,487. The teacher suggests that they will add four more 5-digit numbers to this one, and he is prepared to predict the answer. He says the sum will be 234,485 and places this in the appropriate position. He then asks a pupil to supply the second 5-digit number, which is recorded. The teacher supplies the third number (if he wants it to work), a pupil supplies the fourth number, and the teacher the fifth. At this point the addition is carried out orally, and the predicted answer is seen to be correct.

Successive examples are done. As the teacher writes down his prediction and the numbers which he is supplying, he says them aloud and encourages the children to say them with him, as soon as they have discovered how it is done. You will be surprised at how quickly some of the children can join the chorus and how pleased they will be with their own perception. The examples can go on until at least half the class has seen through the "trick," at which time a pupil might be asked to explain the procedure to everyone.

Follow-up on "Why does this work?" produces much worthwhile thought.

Examples

A few examples of the chalkboard work are given below, and you are left to figure out the trick for yourself. The numbers are given in the following order (but remember that the prediction is the second step):

Starting number
Number supplied by pupil
Number supplied by teacher
Number supplied by pupil
Number supplied by teacher
Prediction

34,487	99,896	66,831	48,320
26,356	18,723	73,491	61,133
73,643	81,276	26,508	38,866
48,915	43,711	91,735	29,114
51,084	56,288	8,264*	70,885
234,485	299,894	266,829	248,318

* With an apology from the teacher that in this case he can use only a 4-digit number.

7

Geometry and Measurement

Interest in geometry and related measurement activities has increased in the elementary school mathematics program during the past several years. On the one hand, geometric concepts are seen as important in helping the child give structure to his environment. On the other hand, there is a movement in the United States to adopt the metric system of measurement. Both forces have combined to make us more aware of the need for sound instruction in geometrical ideas in the elementary school years.

One measurement concept that is taught early in the mathematics program is telling time. Some children come to school already possessing this skill. In the first article, by Porlier, we find not only a game that helps children learn this skill but also a technique for maintaining the interest of children who already know how to tell time.

The second article, by Trueblood and Szabo, gives hints on games to use for introducing the metric system. The checklist they provide for helping teachers develop games is especially good, and it can also be applied to many other games.

The next pair of articles is related to the concept of area measurement. In addition to relationships developed in Berman's "Geo-Gin," we find the germ of an idea for many different card games. Games such as Old Maid,Fish, and rummy can be converted into activities for review and practice with geometric terms and symbols. The tangram square puzzle has already become familiar to many teachers. The puzzle pieces take the form of many of the basic geometric shapes included in the elementary curriculum. Dickoff's presentation uses many intuitive geometric constructions to create the seven pieces through paper folding. The paper-folding activity reminds us that the oriental art of origami is also a very appealing activity for children.

Grogan's article, "A Game with Shapes," and the article on mirror cards, by Walter, describe games designed to help students see the relationships involved in the geometric transformations known as reflections, rotations, and translations. Both games are easy and exciting for children to play; both are loaded with significant geometric ideas.

Popsicle sticks and tongue depressors have found many uses in the mathematics program. They make excellent counters and bundle easily into groups of ten for work on place-value concepts. The children in Lund's class probably have found the way to have the most fun with Popsicle sticks—making polygons out of them! And in addition to the fun of making them, they fly, too! This sounds like a good outdoor activity.

Reminiscent of the tangram activity is Hall's "Pythagorean Puzzle." The puzzle activity confirms the basic formulation of the Pythagorean theorem and is also used to explore the relationship between areas of various figures—triangles, squares, and parallelograms.

The last games in this section provide a means to develop and practice skills associated with graphing with Cartesian coordinates. Timmon's tic-tac-toe game is a very good example of using a game to develop a significant mathematical idea. One might even expect some students to perceive the numerical relationships existing between coordinates of points falling along a line. The article by Overholser shows how to incorporate area concepts into work with coordinates. On the lighter side, and especially suitable for the earlier elementary grades, are Deatsman's "Holiday Plot Dots" and Bell's "Cartesian Coordinates and Battleship," a game that most children (and adults) attack with fervor.

Don't miss the train

CORINNA PORLIER

Goldie King School, East Gary, Indiana

In our fast-moving society of schedules, everyone finds a need for telling time, but whose responsibility is it to teach this critical topic? Is it the teacher's? Is it the parent's? Or, is the child left to learn on his own?

Evidently, the textbook authors feel that the child already knows how to read a clock before coming to school. A check of several arithmetic textbooks will show that an average of only two to four pages per grade level (for the first, second, and third grades) are devoted to reading clocks. There is usually no space at all given to telling time after the third grade.

Many parents do teach their children to tell time at a very young age and sometimes reward them with a watch or clock of their very own when the children can successfully read them. But what about the child who is left to learn on his own? He may struggle through for a few years telling time by the locations of the "big hand" and the "little hand," or he may just avoid having to deal with exact times.

In the first grade, where the textbook deals only with the hour and half-hour, the child who already knows that is bored. On the other hand, the child who does not know this much is not given enough classroom time or experiences to learn to read these times on the clock. The same thing happens in the second and third grades. By the time a child is in the fifth or sixth grade, he may be thoroughly confused and too embarrassed to ask for help from the teacher or his parents.

This problem has bothered me for many years. After much thought, I have come up with a solution.

I have developed a game that not only provides a fun way to teach time to children, but also holds the interest of children who already know how to tell time.

In playing this game, the older child can learn more about telling time without having to be classified by his classmates as a dummy.

Space Marker
(1 red 1 green
1 blue 1 yellow)

BONUS

ADVANCE
CLOCK
25
MINUTES

Bonus cards (20)

TIME

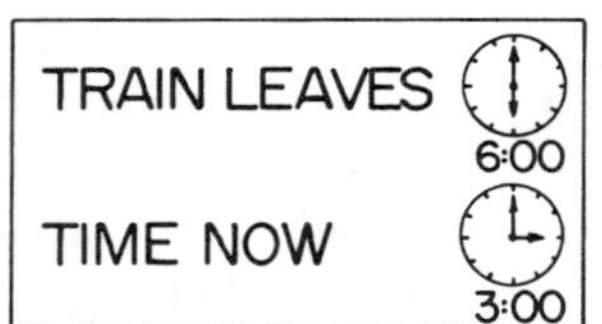

Time cards (20)

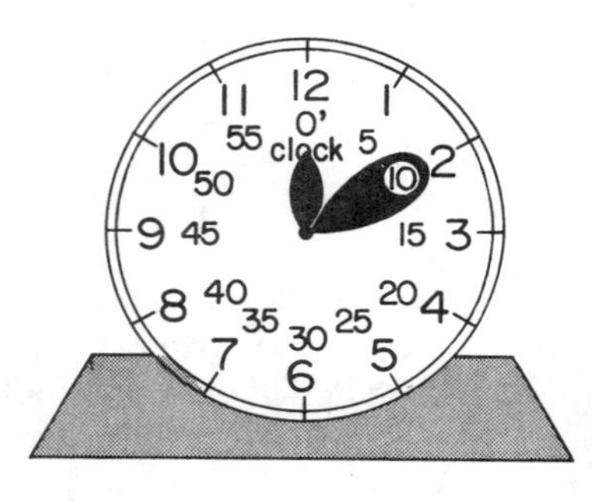

Players' clocks
(1 red 1 green
1 blue 1 yellow)

Dice

EXAMPLE
GAME BOARD

Don't Miss The Train

(Rules for playing the clock game.)

How the game is played:

The game is played by 2, 3, or 4 players. Each player moves his marker around the railroad track according to the throw of the dice. A player may advance his clock when he lands on a space that tells him to do so or when he draws a BONUS card. There are a few occasions when a player must set his clock back.

The master clock on the playing board is set at the time the train is to leave. The train's departure time is determined by the TIME card. The player whose clock first shows the departure time of the train is the one who "catches the train" and is the winner. All other players miss the train.

The object of the game: Don't miss the train.

Equipment:

1 playing board
4 clocks of different colors
4 colored space markers to match the clocks
2 dice
1 pack of BONUS cards (20)
1 pack of TIME cards (20)
1 set of directions

Playing board. The playing board shows a station with four spur lines, a circular track that is marked off as a master clock, and spaces at the bottom of the board for the BONUS and TIME cards. The master clock has its own set of movable hands. In some of the spaces between the railroad ties, there are special instructions for players; other spaces are blank.

Clocks. There are four clocks, one yellow, one blue, one green, and one red. Each player has his own clock on which he keeps his time as he advances. These clocks have special faces and hands that make telling time easier for the players.

Space markers. There are four colored space markers that match the colors of the players' clocks. The markers are used to pace off the spaces on the round track.

BONUS *cards.* The BONUS cards give players extra chances to advance their clocks. A player must land on a space marked BONUS in order to draw a card from the top of the BONUS pile. Bonuses range from five minutes to one hour.

TIME *cards.* The TIME cards state the time that the train leaves and the present, or starting, time. The top card of the stack of TIME cards is turned up at the beginning of the game; only one TIME card is used in each game. The first player sets the master clock with the time shown on the TIME card that he turns up, and each player sets his own clock at the present time shown on the card.

Playing the game:

1. The players throw the dice to see who starts first. The player with the largest total plays first, and play thereafter passes to the left.

2. At the beginning of the game, the markers are placed on the spur tracks in the station. The cards are shuffled and placed in their proper places. The first player turns up the top TIME card and sets the master clock at the time indicated on the TIME card. Each player sets his own clock at the "time now" indicated on the TIME card. When the game begins, the clocks of all of the players are set at the same time.

3. The beginning player must roll either a one or a six on at least one of the dice in order to bring his marker out of the station. If he does so, he places his marker on the space marked *start*. He then totals his throw of the dice and advances his marker clockwise around the tracks for that many spaces. If he lands on a space that has special instructions, he follows the directions that are given. If a player does not throw a one or a six on his first throw, he has to remain in the station and the

play moves to the player on his left.

4. If a player throws doubles, he gets another turn. If he throws doubles a second time, he gets an added turn. If he throws doubles a third time, he does not take the count, and his turn of play is ended.

5. A player is entitled to only the top BONUS card when his marker lands on a space marked BONUS. After he reads the BONUS card and advances his clock as directed, the player places the card face down on the bottom of the BONUS stack.

6. If a player's marker lands on a space where there is already a marker, the first marker must be moved back to *start*. The penalized player does not have to move his clock back, only his marker is moved back.

7. A player advances his clock fifteen minutes every time his marker moves around the entire circle and passes the space with the star.

8. When a player is sent back to *start*, he does not get the additional advancement of 15 minutes on his clock. The advancement is given only when a player has completed the circle.

9. When a player's clock reaches or passes the time shown on the master clock, that player wins the game—he has caught the train. All others have missed the train.

EDITORIAL COMMENT.—To provide practice in reading a clock, try CLOCK bingo. Two versions are suitable. In one version, the CLOCK card contains the names of times. The caller shows a large clockface. Players read the time from the clock and find the notation on the card.

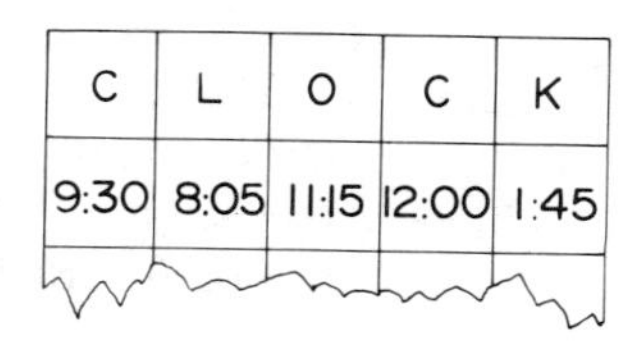

In the second version, the TIME card contains pictures of clockfaces set to specific times. The caller names a time. The players look for a clockface showing that time.

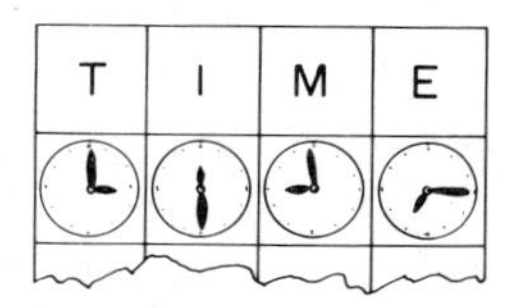

With some groups it is wise to use cards with fewer cells—perhaps a 3 × 3 arrangement. This makes the game go faster.

Procedures for designing your own metric games for pupil involvement

CECIL R. TRUEBLOOD and
MICHAEL SZABO

Currently an associate professor in mathematics education at Pennsylvania State University, Cecil Trueblood is particularly interested in the teaching of mathematics in the elementary school. Michael Szabo, also at Pennsylvania State, is an associate professor of science education. His educational interests center on instructional development, individualized instruction, and complex problem-solving.

Although much has been written on the values of mathematical games in the elementary grades and many game books have been published, little has been written that would help classroom teachers design, produce, and evaluate games for use in their classroom. The focus of this article is to present a set of seven criteria that were developed in a summer workshop for inservice elementary teachers who decided that they wanted to be able to produce metric games and related activities that would fit into their "metrication" program.

The teachers in the workshop began by asking a practical question: Why should I be interested in producing my own metric games? They concluded that the game format provided them with specific activities for pupils who did not respond to the more typical patterns of instruction. They felt that in the game format they could provide activities of a higher cognitive level for pupils who had difficulty responding to material requiring advanced reading skills.

The teachers then asked a second question: Does the literature on the use of mathematics games contain any evidence that would encourage busy classroom teachers to use planning time to develop their own games? The available professional opinion supported the following conclusions:

1. Games can be used with modest success with verbally unskilled and emotionally disturbed students, and students for whom English is a second language.

2. Games have helped some teachers deal with students who present discipline problems because they are bored with the regular classroom routine.

3. Games seem to fit well into classrooms where the laboratory or learning-center approach is used. This seems related to the feature that games can be operated independent of direct teacher control thus freeing the teacher to observe and provide individual pupils with assistance on the same or related content.

Plan for development

If for any of the reasons just cited you are interested in designing and evaluating several of your own metric games, how should you begin? Simply use the following checklist as a step-by-step guide to help you generate the materials needed to create your game. Use the exemplar that follows the checklist as a source for more detailed suggestions. Each item in the checklist has been keyed to the exemplar to facilitate cross referencing.

CHECKLIST GUIDE

_____ Write down what you want your students to learn from playing your game. (Establish specific outcomes)

_____ Develop the materials required to play the game. (Make simple materials)

_____ Develop the rules and procedures needed to tell each player how to participate in the game. (Write simple rules and procedures)

_____ Decide how you want students to obtain knowledge of results. (Provide immediate feedback)

_____ Create some way for chance to enter into the playing of the game. (Build in some suspense)

_____ Pick out the features that can be easily changed to vary the focus or rules of the game. (Create the materials to allow variation)

_____ Find out what the students think of the game and decide whether they learned what you intended them to learn. (Evaluate the game)

The exemplar

Establish specific outcomes

By carefully choosing objectives that involve both mathematics and science processes—such as observing, measuring, and classifying—the teachers created a game that involves players in the integrated activities. This approach reinforces the philosophy that science and mathematics can be taught together when the activities are mutually beneficial. That is, in many instances integrated activities can be used to conserve instructional time and to promote the transfer of process skills from one subject area to the other. The exemplar's objectives are labeled to show their relationship to science and mathematical processes.

1. Given a set of common objects, the students estimate the objects' weight correct to the nearest kilogram. (Observation and estimation)

2. Given an equal-arm balance, the students weigh and record the weights of common objects correct to the nearest centigram. (Measurement)

3. Given an object's estimated and observed weight correct to the nearest centigram, the student computes the amount over or under his estimate. (Computation and number relationships)

Make simple materials

The following materials were constructed or assembled to help students attain the objectives previously stated in an interesting and challenging manner.

1. Sets of 3-by-5 cards with tasks given on the front and correct answers and points to be scored on the back. (See fig. 1.)

2. A cardboard track (see fig. 2) made from oak tag. Shuffle the *E*'s (estimate cards), *O*'s (observed cards), and the *D*'s (difference cards) and place them on the gameboard in the places indicated.

3. An equal-arm balance that can weigh objects up to 7 kilograms.

4. A pair of dice and one different colored button per player.

5. A set of common objects that weigh less than 7 kilograms and more than 1 kilogram.

6. Student record card. (See fig. 3.)

Write simple rules and procedures

The rules and procedures are crucial to making a game self-instructional. In the following set of directions notice how a student leader and an answer card deck serve to ease the answer processing needed to keep the game moving smoothly from one player to another. It is essential to keep the rules simple and straightforward so that play moves quickly from one student to the other.

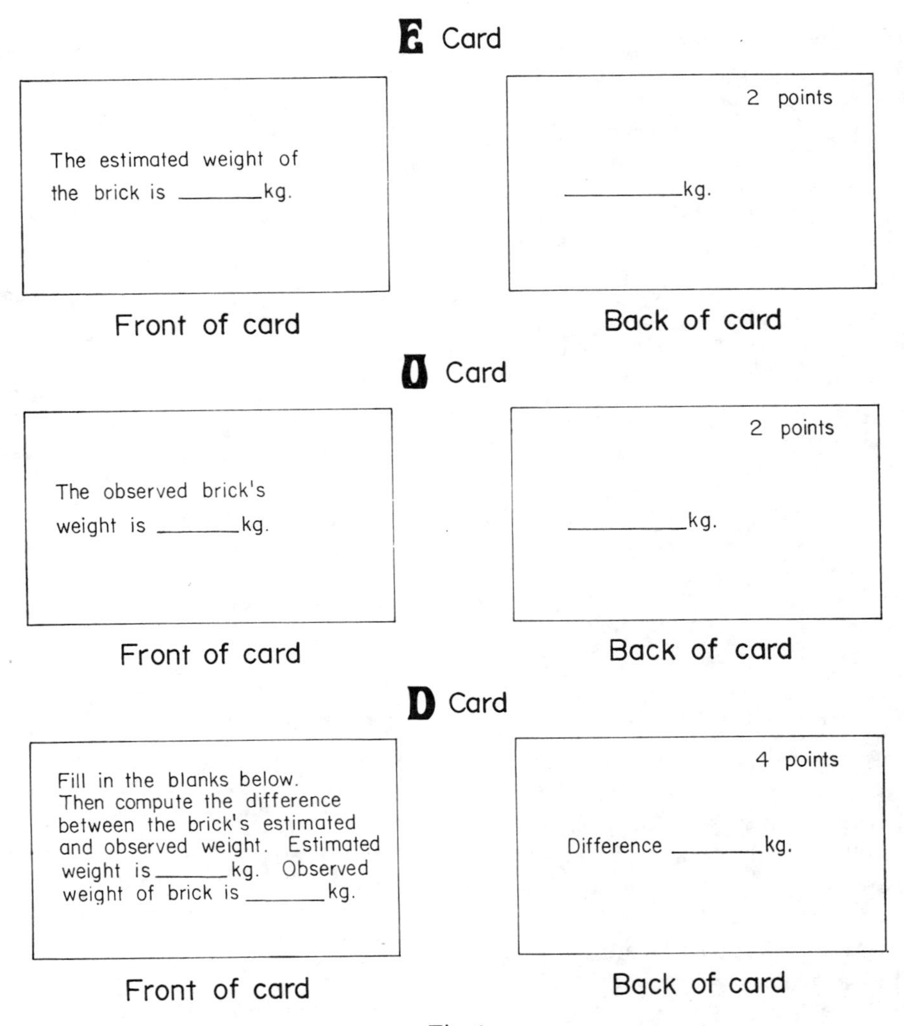

Fig. 1

1. Number of players, two to six.

2. The student leader or teacher aide begins by rolling the dice.

The highest roll goes first. All players start with their buttons in the "Start Here" block. The first player rolls one die and moves his button the number of spaces indicated on the die. If he lands on a space containing an *E*, *O*, or *D* he must choose the top card in the appropriate deck located in the center of the playing board or track and perform the task indicated. (In the example shown in figure 1 this would be Card E_3.) The player then records the card number, his answer, and the points awarded by the student leader on his record card. The student leader checks each player's answer and awards the appropriate number of points by reading the back side on the task card. He then places that card on the bottom of the appropriate deck and play moves to the right of the first player. The player who reaches "Home" square with the highest number of points is the winner. At the end of the play each player turns in his score card to the student leader who gives them to the teacher.

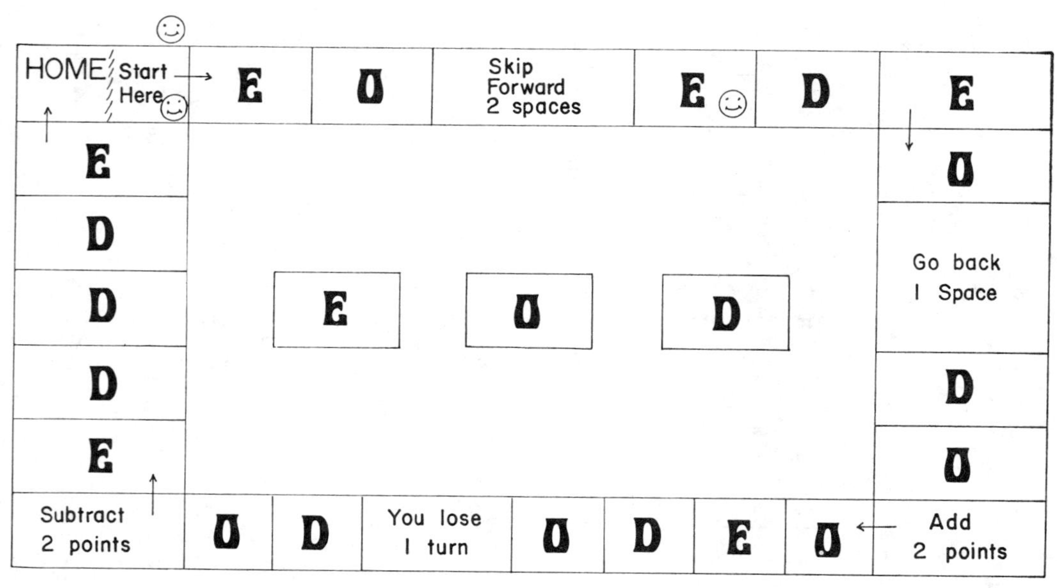

Fig. 2

Provide immediate feedback

By placing the answer on the back of the task card and appointing a student leader, the teacher who developed this game built into the game an important characteristic, immediate knowledge of the results of each player's performance. In most cases this feedback feature can be built into a game—by using the back of task cards, by creating an answer deck, or by using a student leader whose level of performance would permit him to judge the adequacy of other students' performance in a reliable manner. Feedback is one of the key features of an instructional game because it has motivational as well as instructional impact.

Have students record diagnostic information. The student record card is an important feature of the game. The cards help the teacher to judge when the difficulty of the task card should be altered and which players should play together in a game, and to designate student leaders for succeeding games. The card also provides the player with a record that shows his scores and motivates him to improve.

This evaluative feature can be built into most games by using an individual record card, by having the student leader pile cards yielding right answers in one pile and cards with wrong answers in another pile, or by having the student leader record the results of each play on a class record sheet.

Build in some suspense

Experience has shown that games enjoyed by students contain some element of risk or chance. In this particular game a player gets a task card based upon the roll of the die. He also has the possibility of being skipped forward or skipped back spaces, or of losing his turn. Skipping back builds in the possibility of getting additional opportunities to score points; this feature helps low-scoring students catch up. Skipping forward cuts the number of opportunities a high-scoring player has to accumulate points. The possibility of adding or subtracting points also helps create some suspense. These suspense-creating features help make the game what the students call "a fun game."

Student's name ______________________ Date__________

Card number	Answer given	Number of points
E_3	2 kg.	2
E_2	1 kg.	1

Fig. 3

Create the materials to allow variation

A game that has the potential for variation with minor modifications of the rules or materials has at least two advantages. First, it allows a new game to be created without a large time investment on the part of the teacher. Second, it keeps the game from becoming stale because the students know all the answers. For instance, the exemplar game can be quickly changed by making new task cards that require that students estimate and measure the area of common surfaces found in the classroom such as a desk or table tops. By combining the two decks mixed practice could be provided.

Evaluate the game

Try the game and variations with a small group of students and observe their actions. Use the first-round record cards as a pretest. Keep the succeeding record cards for each student in correct order. By comparing the last-round record cards with the first-round record cards for a specific student, you can keep track of the progress a particular student is making. Filing the cards by student names will provide a longitudinal record of a student's progress for a given skill as well as diagnostic information for future instruction.

Finally, decide whether the students enjoy the game. The best way is to use a self-report form containing several single questions like the following, which can be answered in an interview or in writing:

1. Would you recommend the game to someone else in the class? __Yes __No

2. Which face indicates how you felt when you were playing the game?

3. What part of the game did you like best?

4. How would you improve the game?

Concluding remarks

The procedure just illustrated can be generalized to other topics in science and mathematics. The following list provides some suggested topics.

1. Classifying objects measured in metric units by weight and shape
2. Measuring volume and weight with metric instruments
3. Measuring length and area with metric instruments
4. Classifying objects measured in metric units by size and shape
5. Comparing the weight of a liquid to its volume
6. Comparing the weight of a liquid with the weight of an equal volume of water
7. Predicting what will happen to a block on an inclined plane
8. Comparing the weights of different metals of equal volume

Why don't you try and create some games for each of these topics? Then share the results with your colleagues. Additional examples developed by the authors are available in "Metric Games and Bulletin Boards" in *The Instructor Handbook Series* No. 319 (Dansville, New York, 1973).

Geo-gin

JANIS A. BERMAN

Student, University of South Florida, Tampa, Florida

Most students in the upper elementary grades have played and enjoyed card games. *Geo-gin* is a card game that makes use of some important geometric concepts such as spatial perception, identification, and discrimination. It is designed for groups of two to four students.

Materials needed

Two sheets of poster board (preferably, one side white and the other side a bright color)

Four sheets of construction paper, one each of four different colors

Magic markers in four different colors (preferably, the same four colors as the construction paper)

Compass, scissors, ruler, and glue.

Constructing the game

Cut out 48 three-by-four-inch rectangular cards from the poster board. Separate these cards into three sets of 16 each.

On one set of 16 cards, draw a circle (two inches in diameter) on the white side of each card. There should be four circles in each of the four colors. On another set of 16 cards, draw a two-inch square on the white side of each card. There should be four squares in each of the four colors. On the last set of 16 cards, draw a parallelogram with two-inch sides on the white side of each card. There should be four parallelograms in each of the four colors.

In the upper left-hand corner of each card, write the letter "T" in the color of the figure. This marks the top of the card.

On each color of construction paper, draw the three different figures. Cut each of the construction-paper figures into four parts as in figure 1.

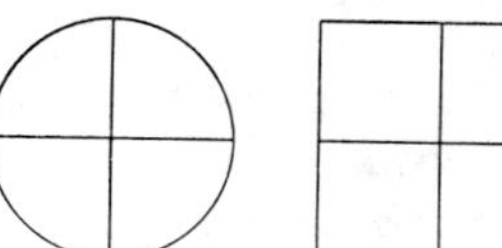
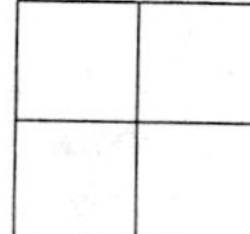
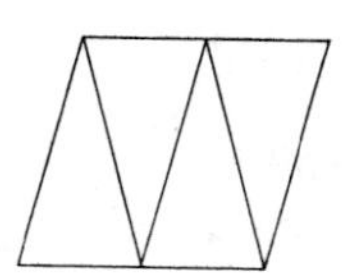

Fig. 1

Now glue one fourth of each color of the construction-paper figures in its respective position and in the corresponding color on each card. Only one piece is glued on any one card. Figure 2 shows the placement of the four parts of a parallelogram on each of four cards. On each card, draw a dotted line to show the other three fourths of each figure.

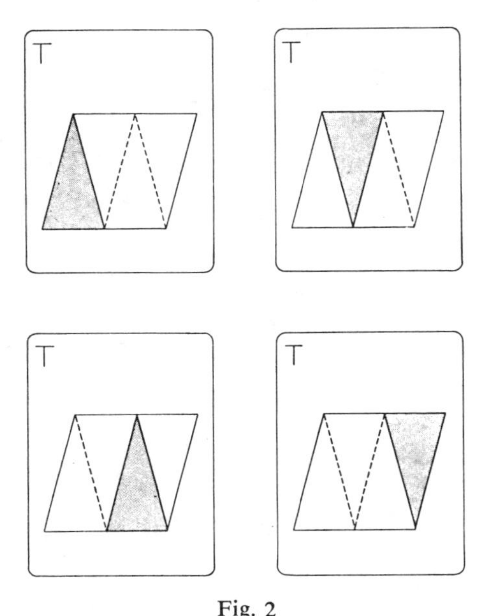

Fig. 2

(You can also decorate the colored side of each card with the name of the game and an appropriate design to make it look "professional" as in figure 3.)

Fig. 3

Playing the game

The deck is shuffled and one person starts the game by dealing eight cards to each player. The dealer then turns over the top card in the deck and places the remaining cards face down on the table. One person should be selected to keep score with a pencil and paper.

The object of the game is to obtain two complete figures. A complete figure consists of four cards of the same figure, with a different portion of each figure colored—all four parts would fit together to make a complete figure. The cards must all be turned so that the "T" is in the upper left hand corner. In making a complete figure, it is not necessary to use only one color. Since this is very hard to do, extra points are given if a figure is completed in only one color.

Play begins as the player left of the dealer decides whether to use the card turned face up or to draw the top card on the face-down pile. A player must decide which two figures he will attempt to complete, but he may change his mind at any point of the game, since he is the only one who sees his hand. For every card he adds to his hand, a player must discard one; therefore, he should have eight cards in his hand at the completion of his turn.

After a player has put down his discard, the player on his left then decides whether to use the last card placed face up or to pick up the top card in the face-down deck. Play continues in a clockwise manner until one player wins by getting two complete figures in his hand. The winner calls out "Geo-gin" and lays down his hand. He scores ten points for winning and an additional five points for each figure he completed using only one color. Any other player that has a completed figure in his hand in *one color only,* scores five points.

The player on the left of the dealer then shuffles the cards again and the same procedures are resumed until one player obtains 50 points.

Paper folding and cutting a set of tangram pieces

STEVEN S. DICKOFF
Montgomery County Public Schools, Rockville, Maryland

Steven Dickoff is an elementary mathematics teacher specialist in the Department of Supervision and Curriculum Development in Montgomery County, Maryland. He has led many in-service workshops for elementary teachers in the states of Maryland and New York.

In a recent teacher-training workshop conducted by the author, a question arose concerning a method for duplicating the seven pieces of the ancient Chinese tangram puzzle without having to trace the pieces of another puzzle. After some thought about the relationships of the pieces in the puzzle to one another, the following paper-folding and cutting method, illustrated diagrammatically, was conceived.

Start with a rectangular sheet of paper as in figure 1 (usually $8\frac{1}{2} \times 11$). Fold

A B
D C

Fig. 1

edge AD to coincide with edge DC as in figure 2. Cut off the excess, figure $EBCF$, and discard it. Unfold the shape. Square

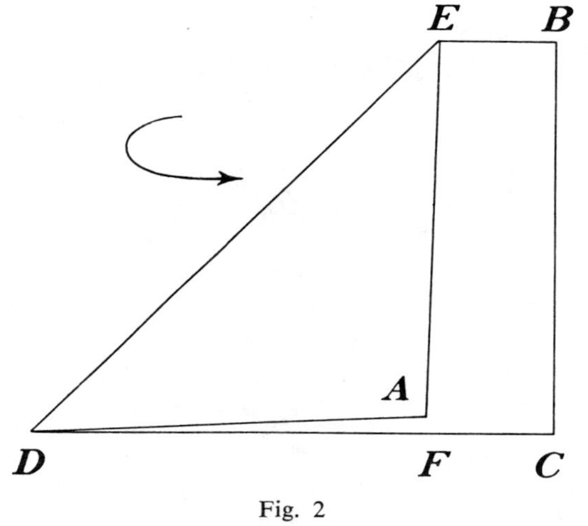

Fig. 2

$AEFD$ remains (fig. 3). Cut along fold ED. Congruent triangles AED and DEF

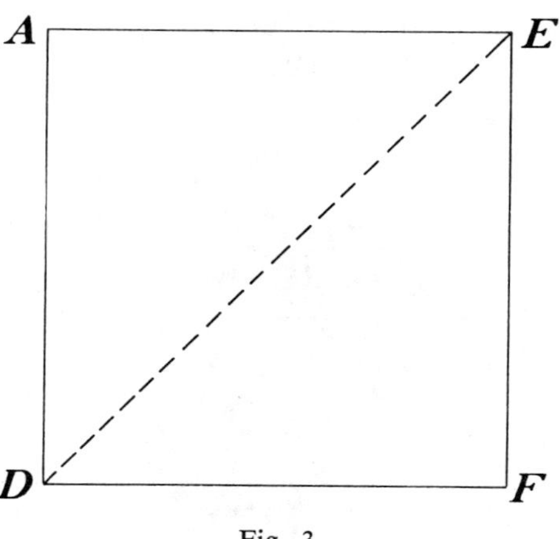

Fig. 3

result (see fig. 4). Fold each of the tri-

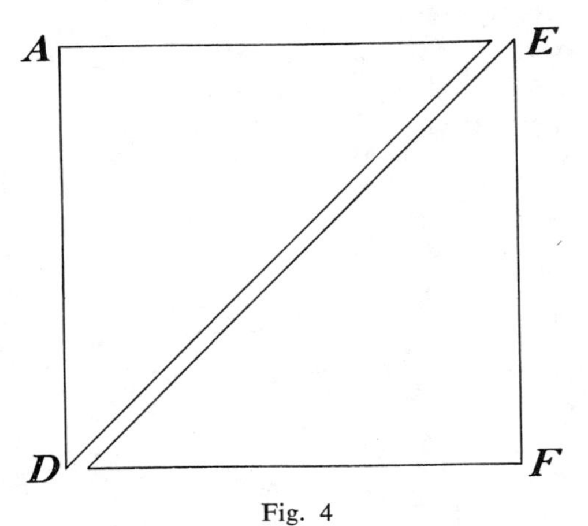

Fig. 4

angles in half as in figure 5. Unfold the

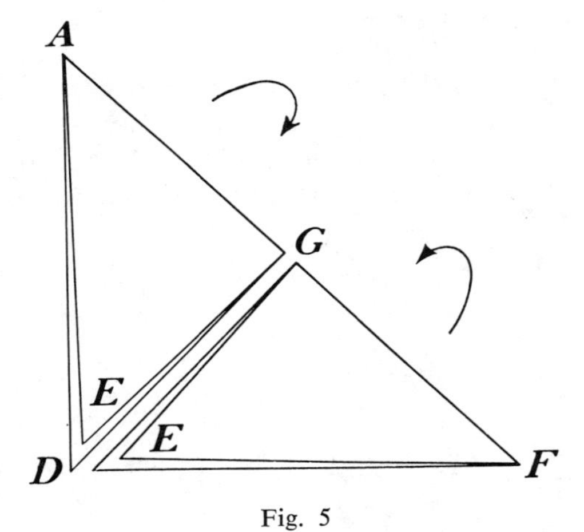

Fig. 5

two triangles and cut along fold *AG* only.

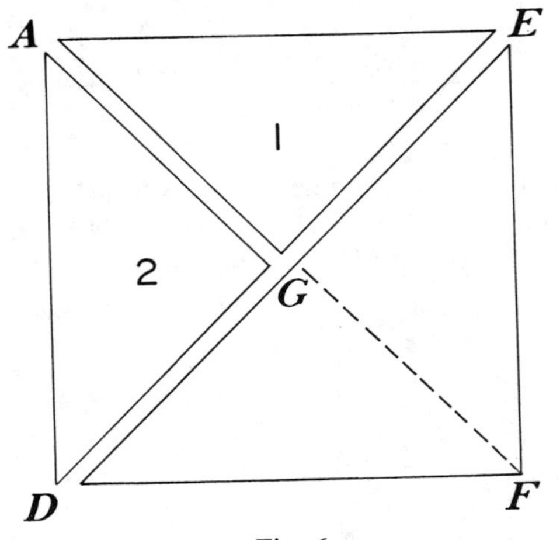

Fig. 6

Place pieces numbered "1" and "2" in figure 6 aside. They are the first two tangram pieces. Fold triangle *EFD* so that point *F* coincides with point *G* (see fig. 7).

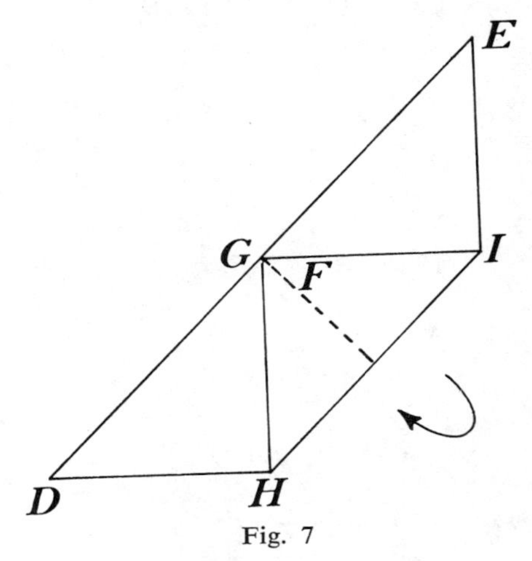

Fig. 7

Unfold triangle *EFD* and cut along fold

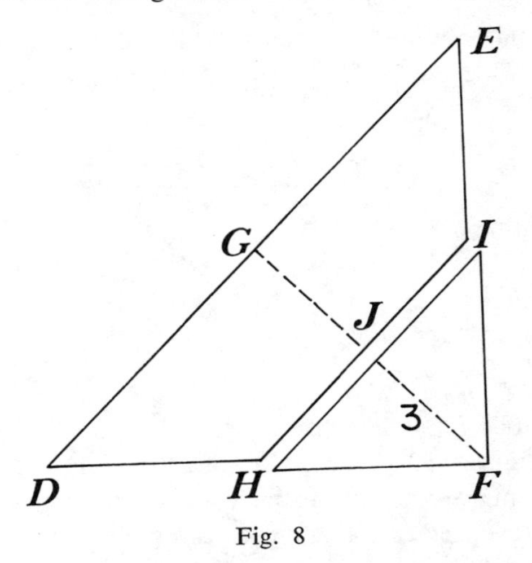

Fig. 8

HI only. Set piece numbered "3" in figure

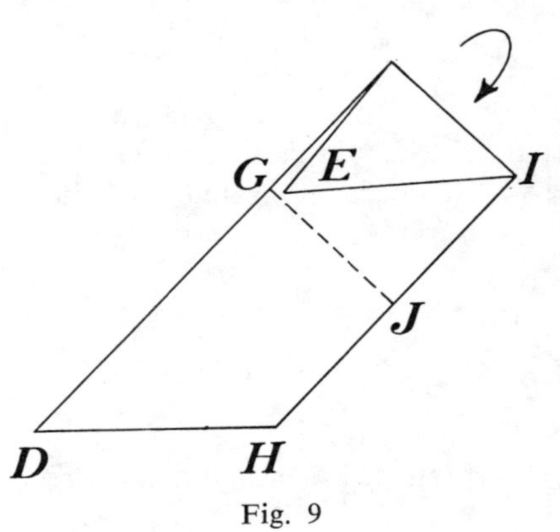

Fig. 9

8 aside. Fold figure *DEIH* so that point *E* coincides with point *G* as in figure 9. Unfold figure *DEIH*. Cut along folds *IK* and *GJ*. Set pieces numbered "4" and "5" in figure 10 aside. Fold figure *DGJH* so

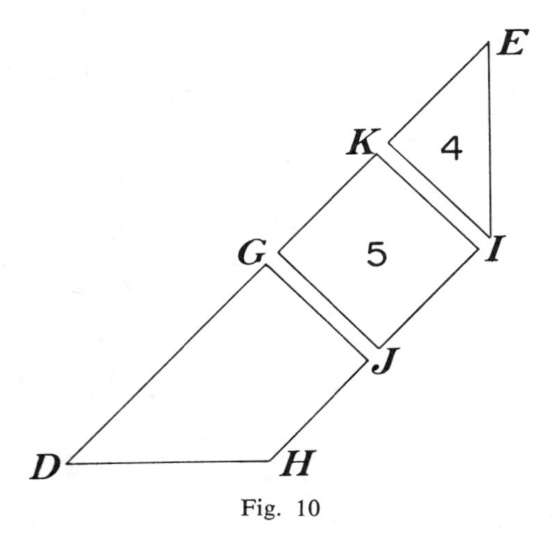

Fig. 10

that point *G* coincides with point *H* (see fig. 11). Unfold figure *DGJH*. Cut along

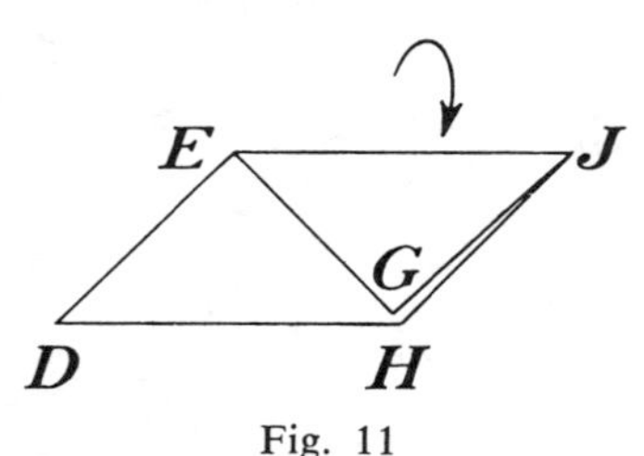

Fig. 11

fold *JL*. Pieces numbered "6" and "7" in figure 12 are now formed. Thus, the tangram puzzle is now complete with all seven pieces. The pieces can now be placed back together, as in figure 13, to form the square in figure 3.

The Chinese tangram puzzle has always been popular with children, who, once they have solved it, are delighted to "fool their friends." This same puzzle may also be sawed from a piece of ¼-inch plywood or

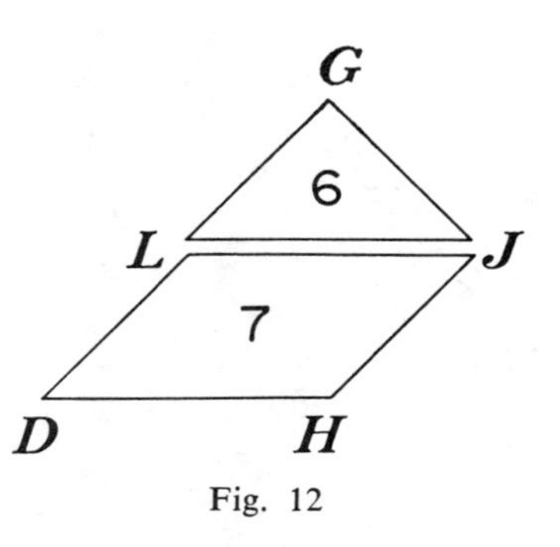

Fig. 12

Masonite and the pieces kept in a shoebox, along with different puzzle shapes on activity cards, for the children's use. Any way you use it, it's sure to capture the interest of the children as it has for ages.

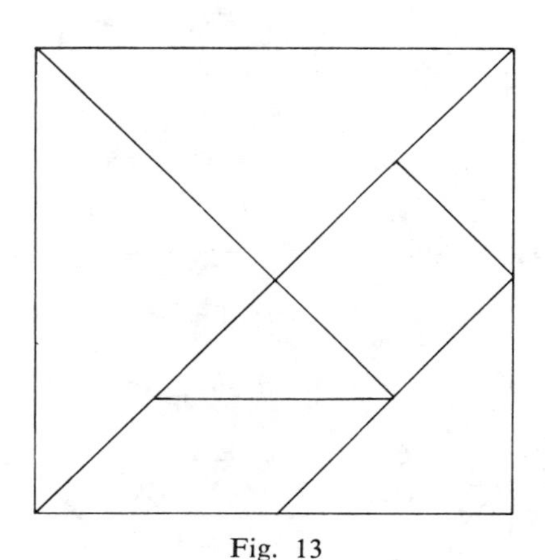

Fig. 13

Editorial Comment.—Tangramlike puzzles for basic area formulas are easy to make. For example, to show that the area of a parallelogram and the area of a rectangle are related, start with a parallelogram of two pieces.

Rearrange the pieces to make a rectangle.

A game with shapes

DAISY GOGAN

Daisy Gogan is working on her doctoral program at Teachers College, Columbia University. She has had experience in teaching high school, serving as chairman of the mathematics department at Northern Highlands Regional High School in Allendale, New Jersey, and as an assistant in the mathematics department at Teachers College, Columbia University.

During their lunch period, Tom and his friend Greg invented a new game they called "Shapes." On a sheet of graph paper they took turns filling in one square at a time with a big X to see whether they could form various shapes. They made these rules:

1. They would toss a coin to see who would start the first shape. After that they would take turns starting.
2. The "starter" could fill in any square he wished.
3. Each person, as his turn came, could then fill in any square next to one already filled in.
4. They would stop after filling in five squares.
5. When the five squares were finished, they would examine the result. If the shape were a new one, the boy who had finished it would win that point. If the shape were just like one already made, the boy who finished it would lose and the point would go to the other person.

Tom won the toss to be the first starter, and they filled in the squares quickly:

It was Tom's point, since he had finished the shape, and of course since it was the first shape it was a new one.

Greg started the second shape. When four squares had been completed, it was his turn again for the fifth square. He made sure that he filled it in so as to make the final shape a new one. He won that point.

The boys continued. The game went very quickly for several shapes, since they had little trouble making them different. Tom had just made a point by finishing this shape:

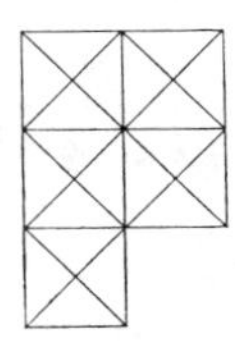

and the boys were working on the next round.

Greg placed the fifth square so that he had this shape:

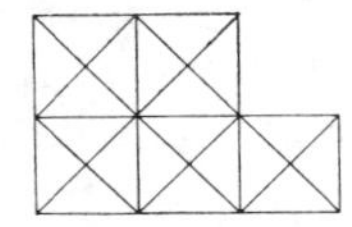

"My point," he said.

"Oh, no!" replied Tom. "That shape is just like the one I just finished."

"How can you say that, Tom? Why, this one uses three columns of squares, and the other one takes only two columns."

Tom replied, "If you turn your paper clockwise, you will see that the one you just finished takes only two columns. It is exactly the same as the other one. Let's use scissors and cut yours out so that we can see that it fits on mine exactly. If it fits exactly, they are the same."

Greg reluctantly agreed that Tom was right, and the point went to Tom.

The boys continued more slowly now. Tom was trying to keep his lead and Greg was trying to catch up.

Tom made a point by finishing this shape:

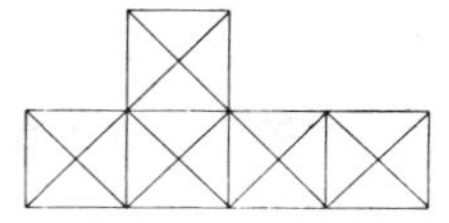

Several turns later, Greg finished this one:

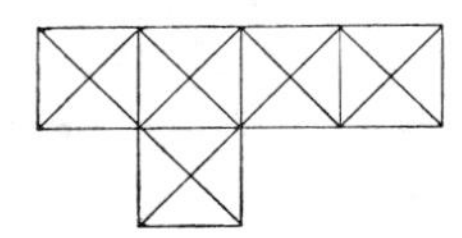

"You lose another point, Greg," cried Tom.

"How come? If you try turning this one, you'll get

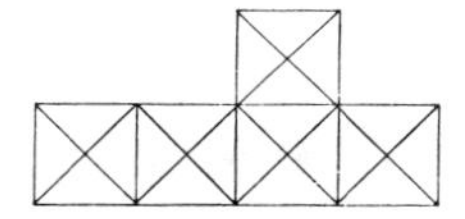

and that's different."

"If I use scissors and cut out the shape and flip it over, I can show you that yours will fit exactly on mine. It really is the same shape. Or better yet, if Jane will lend me a mirror, I can show you easily. See—yours is a reflection of mine."

"Yes, it is the same. I can see that now," said Greg. Here is how the boys used the mirror to show the reflection:

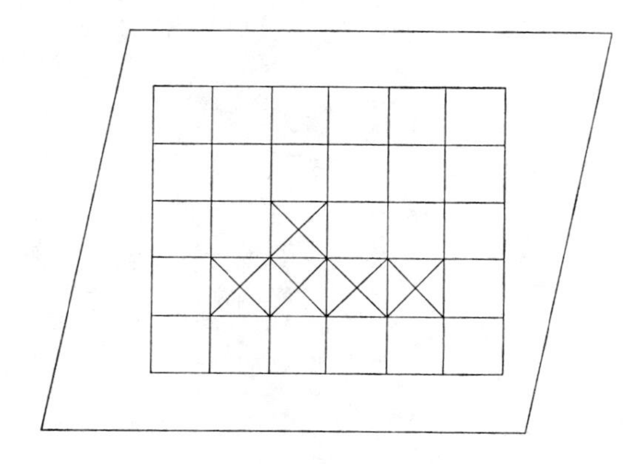

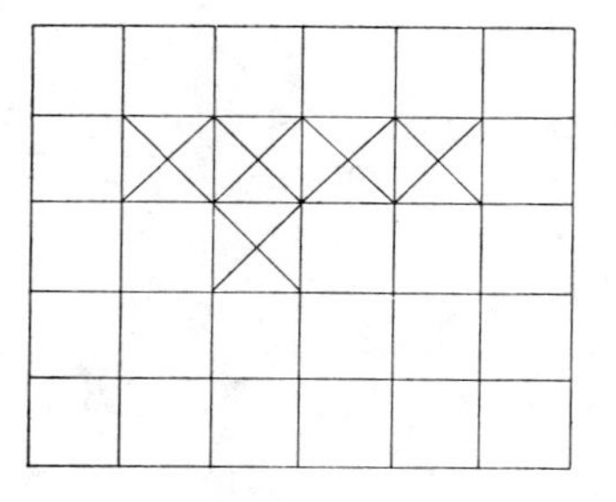

The game proceeded more slowly, as it became harder to find new shapes. But it was more fun, too, trying to trip each other up.

The game with shapes that Tom and Greg invented can be extended by having the players try to establish a relationship between successive reflections and a rotation or to find a method of detecting reflection without using a mirror.

How many different shapes can you find?

How would the game change if we used six squares?

EDITORIAL COMMENT.—Instead of using squares as the base shape, start with other geometric figures, such as triangles or hexagons. This can be done with small paper or plastic pieces. Give students a fixed number of the shapes and ask them to make as many regions as possible, using all the given regions. This encourages thinking about nonstandard area units and the recognition that different shapes have the same area.

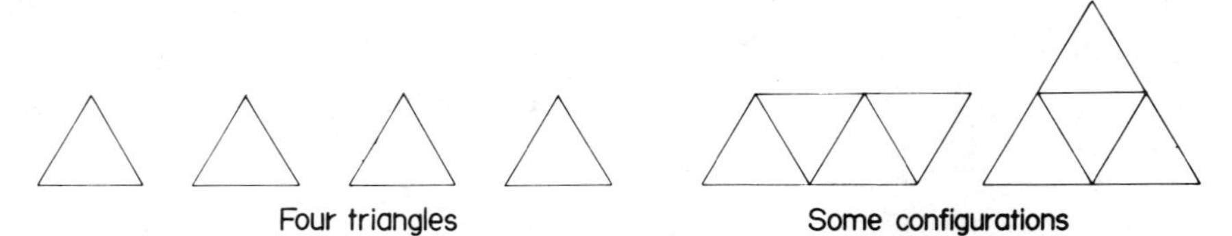

An example of informal geometry: Mirror Cards*

MARION WALTER
Educational Services Incorporated, Watertown, Massachusetts

Marion Walter is a part-time mathematics instructor at the Harvard University Graduate School of Education. She is on the staffs of Educational Services Incorporated in the Elementary Science Study and the Cambridge Conference on School Mathematics. She teaches mathematics to the students in elementary school education at the Harvard Graduate School of Education.

The need for informal geometry, especially in the earlier grades, is being recognized by educators, psychologists, and mathematicians. The Mirror Cards were created by the author to provide a means of obtaining, on an informal level, some geometric experience that combines the possibility of genuine spatial insight with a strong element of play.

The basic problem posed by the Mirror Cards is one of matching, by means of a mirror,[1] a pattern on one card with a pattern shown on another card. For example, can one, by using a mirror on the card shown in Figure 1, see the pattern shown on the card in Figure 1a?

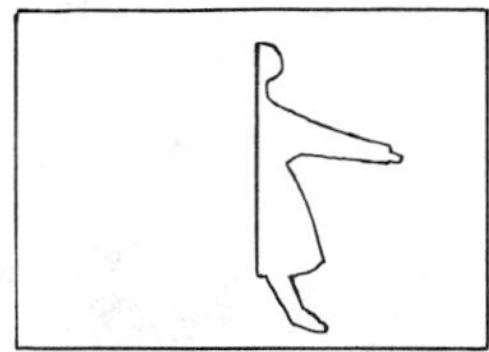

FIGURE 1

FIGURE 1a

The problems range from the simplest, such as the one shown above, to more difficult ones, such as the one shown in Figures 2 and 2a.[2] Some patterns are possible to match and others are not.[3]

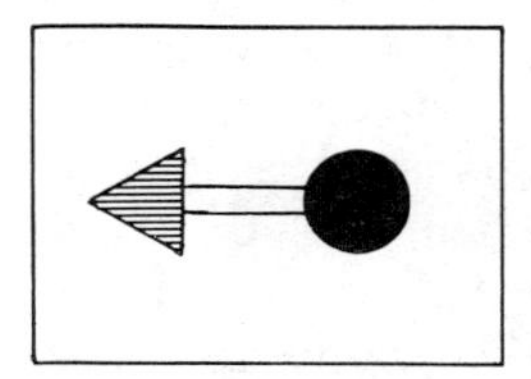

FIGURE 2

FIGURE 2a

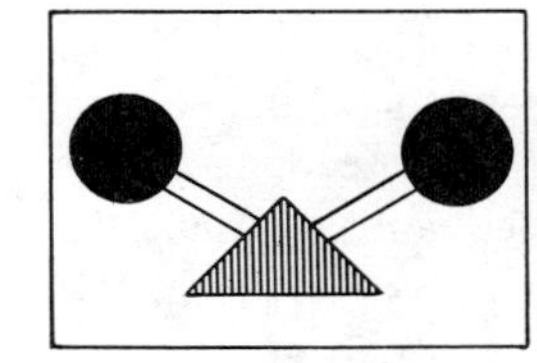

FIGURE 2b

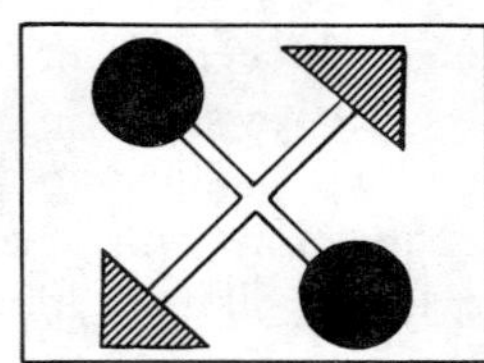

FIGURE 2c

Using the mirror on the card shown in Figure 2, which of the patterns shown in Figures 2a, 2b, and 2c can you make?

* This work was begun while the author was working during the summer of 1963 with the Elementary Science Study, a project supported by grants from the National Science Foundation and administered by Educational Services Incorporated, a nonprofit organization engaged in educational research. She would like to thank the members of the group she worked with that summer and the group in optics of the previous summer for their help and encouragement; she is especially grateful to Professor Philip Morrison, Mrs. Phylis Singer, and Mrs. Lore Rasmussen.

[1] The reader should have a small rectangular pocket mirror handy before reading on.

[2] Each box of Mirror Cards contains, in addition to mirrors, 170 cards arranged in fourteen different sets. Although the instructions for the sets vary, the basic problem is the same for all the sets and is the one described above. A trial edition of Mirror Cards was produced and copyrighted by the Elementary Science Study in June 1965. They are being used on a trial basis in over 250 classrooms around the country. The author would like to acknowledge the help received from Mrs. A. Naiman, Mrs. F. Ployer, and Mrs. J. Williams in editing the guide and producing the cards.

[3] The position of the pattern relative to the edge of the card is to be ignored.

We have noticed that the children usually find the colors and shapes pleasing and enjoy the challenge presented by the cards. They do not think of this work as "mathematics," and they often find the cards stimulating over and above the actual geometry involved. The cards may be a means of reclaiming the children who already dislike mathematics or are bored or frightened by it. The cards do not call for verbal response from the children, and no mathematical notation is needed. Closer connection with science and mathematics classes will be explored by the author in the future, since the cards can give insight into some mathematical and physical principles.

One advantage that the cards have is that the children can see for themselves whether or not they have made a pattern. They don't need to resort to authority to check whether they have solved the problem correctly. In addition, while playing with the cards they are, in effect, constantly making predictions and are immediately able to test these predictions and amend them, if necessary; and it is fun to do so! Thus, while working with the cards they should gain confidence in their own powers and learn through experience the nature of the scientific method.

While moving the mirror around on the cards, the children notice and experiment with the position of object and image in relation to the edge of the mirror. The player can decide where to place the mirror; and he soon learns that he can control its position, but that for any given position of the mirror he cannot control the position of the image!

The students also learn that a mirror does not carry out a translation. (See Figs. 3, 3a, 3b.)

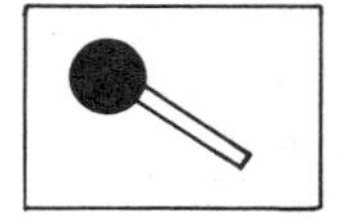

FIGURE 3

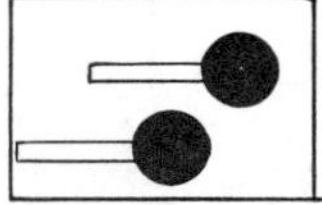

FIGURE 3a

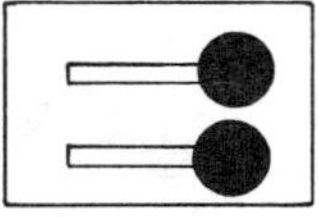

FIGURE 3b

Can one by using a mirror on Figure 3 make the patterns shown in Figures 3a and 3b? — alas, the mirror does not carry out a translation!

They learn by experience that congruence of two parts is a necessary but not a sufficient condition for a pattern to be made by use of a mirror. Most children do not know the expression "symmetric with respect to a line" or "reflection in a line." They may, nevertheless, by using the cards, gain experience that will enable them to understand the concepts that these expressions describe. This does not imply that they could give, or should be expected to give, a formal or verbal definition of these expressions. Eventually they do notice that for a pattern to be reproducible by use of a mirror, it must have two parts that lie on either side of some line and that these must "match exactly." They soon learn, for example, that the pattern shown in Figure 4a cannot be made from the pattern in Figure 4, and they probably have a good feeling for why this is so.

FIGURE 4

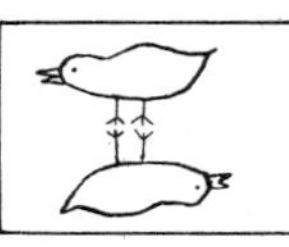

FIGURE 4a

FIGURE 4b

Pattern 4a cannot be obtained from 4. What about the pattern in 4b?

The cards provide opportunity to practice recognizing congruent figures and selecting parts of figures congruent to another.

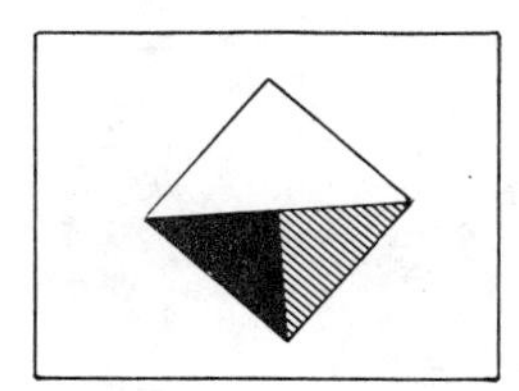

FIGURE 5

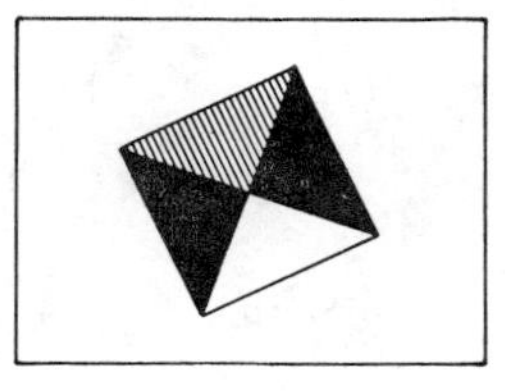

FIGURE 5a

Where must you place the mirror in Figure 5 to see the pattern shown in Figure 5a?

The children must be observant, not only about a shape and the position of that shape, but also about its colors. Some of the patterns match in shape but not in color.

They may also notice a variety of geometric properties of figures. Consider, for example, the circle. By putting the mirror on a diameter they can see the whole circle. More than that, any diameter will do and any chord not a diameter will not do. This may give young children their first feeling for a diameter of a circle, long before they know the word "diameter."

With the diamonds (see Fig. 6) they notice that there are two places where the mirror may be placed to enable them to see the whole diamond. On the other hand,

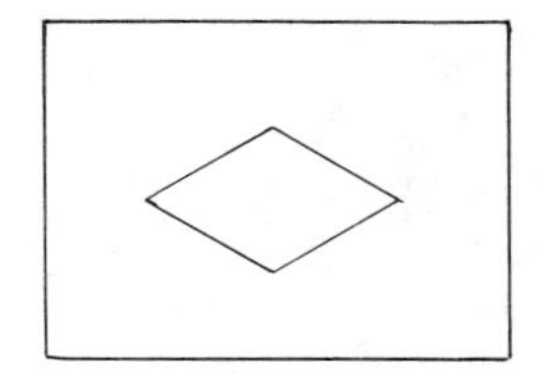

FIGURE 6

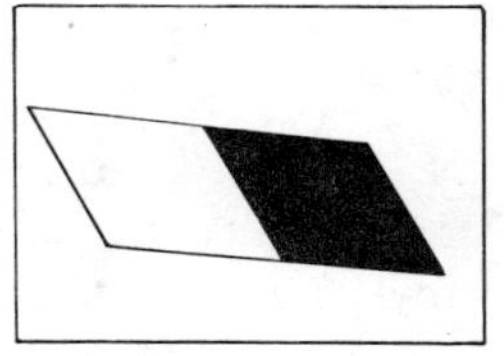

FIGURE 7

the pattern shown in Figure 7 does not have this property—to the surprise of many!

Or, again, take the triangle (see Fig. 8): the children may notice that the effect of putting the mirror along AB is in some way "the same" as that of putting it along

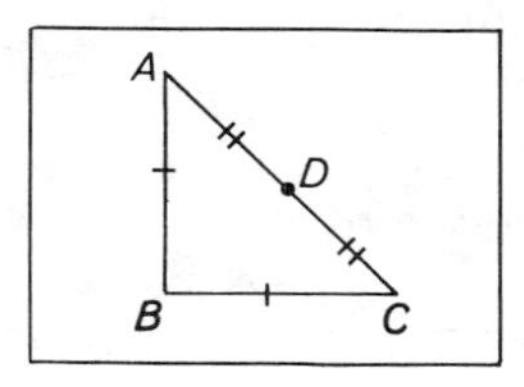

FIGURE 8

BC, but that it is quite different from that of putting it along AC. What about BD?

Other patterns on the cards, such as the ladybugs, arrows, etc., can be explored in similar ways.

For a few cards the children can obtain patterns that look somewhat like the one required but are not congruent nor actually similar in the mathematical sense. I intend to devise cards where congruent and similar patterns are obtainable, and similar but not congruent ones.

Unfortunately, none of the present cards have circles with arrows on them to show, perhaps more clearly, that the orientation gets reversed under a mirror mapping or reflection. Thus Figure 9 becomes Figure 9a.

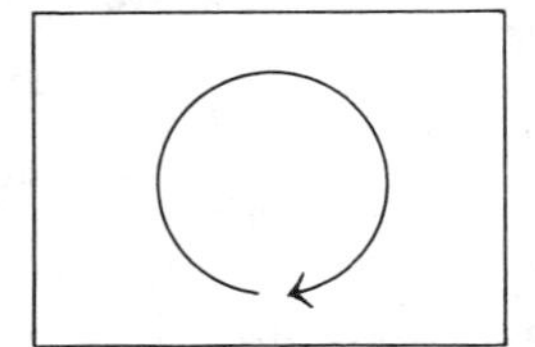

FIGURE 9

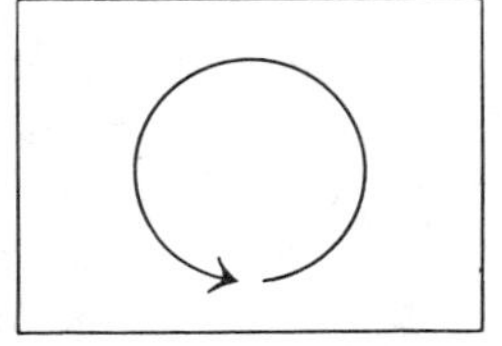

FIGURE 9a

The concept of orientation is, of course, brought out by the cards, although the arrows are not used for this purpose. Often patterns with orientation reversed and not reversed are included to make the idea more obvious. Examples taken from the ladybug and the circle set are shown below.

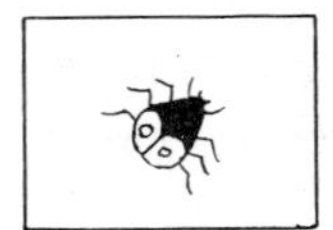

FIGURE 10

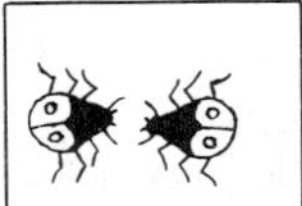

FIGURE 10a

FIGURE 10b

Can one by using the mirror on Figure 10 obtain the patterns shown in Figures 10a and 10b respectively?

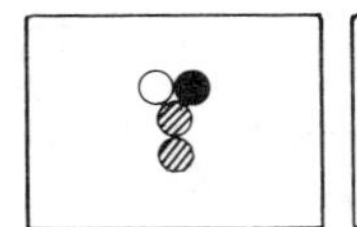

FIGURE 11

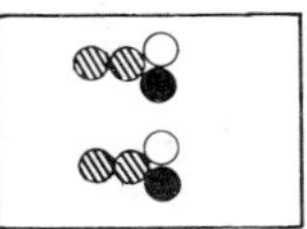

FIGURE 11a

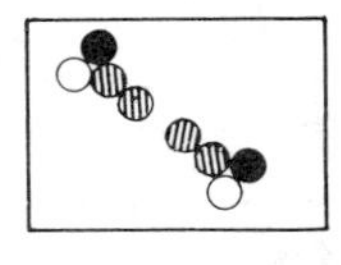

FIGURE 11b

Can one by using the mirror on Figure 11 obtain the patterns shown in Figures 11a and 11b respectively?

The fact that a mirror does not carry out a rotation in the plane is often masked by the symmetry of the figure. For example, one can make Figure 12a from Figure 12, but not Figure 13a from Figure 13.

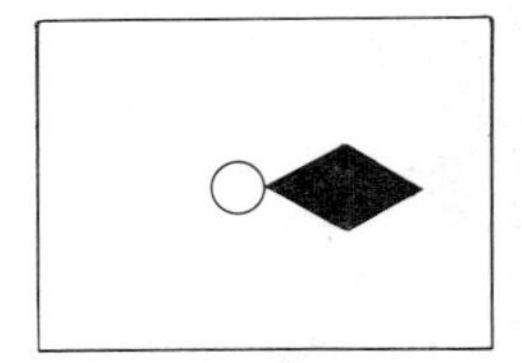

FIGURE 12

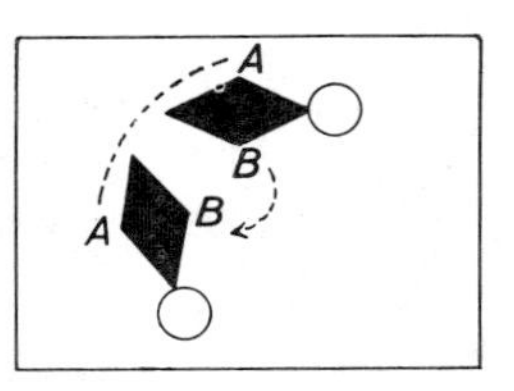

FIGURE 12a

The imagined placement of points "A" and "B" illustrates the fact that the mirror does not "rotate" the figure. Actually the mirror "flips" the image. (Points "A" and "B" are not marked on the actual cards.)

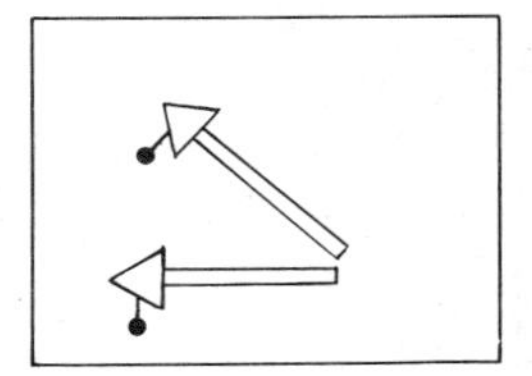

FIGURE 13

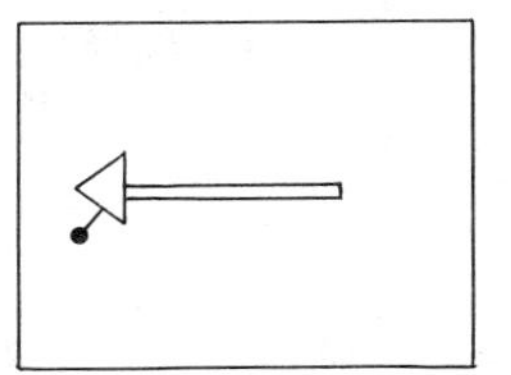

FIGURE 13a

The cards may be used at any age level. They have been used by children as young as five and by sophisticated professional scientists or mathematicians. It is interesting to note that some adults who "know

all the rules" verbally (such as "There must be a line of symmetry" or "Image distance = object distance") often have more difficulty in working through the sets than children who have not yet memorized such phrases. The one barrier to the effective use of the cards by adults appears to be an ingrained habit of respect for authority. Adults often do not want to rely on their own ability to *see* whether they have made a pattern correctly.

When the children find the problems becoming too easy, they may want to add the rule, "You may put the mirror down only once for each pattern," so that all the trial and error must go on in their heads. They may wish to make some of their own cards. When, as happens often, children are able to predict without using a mirror at all whether a pattern can or cannot be made, they will have a clear demonstration of the power of reasoning based on experience—i.e., that it is possible to predict results with confidence by thinking rather than doing! (And they are able to check their thinking if they wish.) In this way they are savoring an essential part of the nature of rational thought.

There are many questions that still need to be answered. I mention just a few. Will use of the cards make children more observant about other geometric patterns? Will it enable them to see figures within figures more easily? Does it improve their ability to visualize? Will they be able to describe patterns more clearly? Will it help or hinder children with reading difficulties?

EDITORS' NOTE. *Current information about Mirror Cards (#18418) and Mirror Cards Teachers' Guide (#18417) can be obtained from Webster Division, McGraw-Hill Book Co., New York.*

EDITORIAL COMMENT.—In addition to the measurement concepts discussed in the various articles of this section, temperature, weight, and volume are measurement systems commonly taught in the elementary mathematics program. To encourage estimation of these and other measures, or to develop the connection between a characteristic and its measure, you might construct a measurement game board.

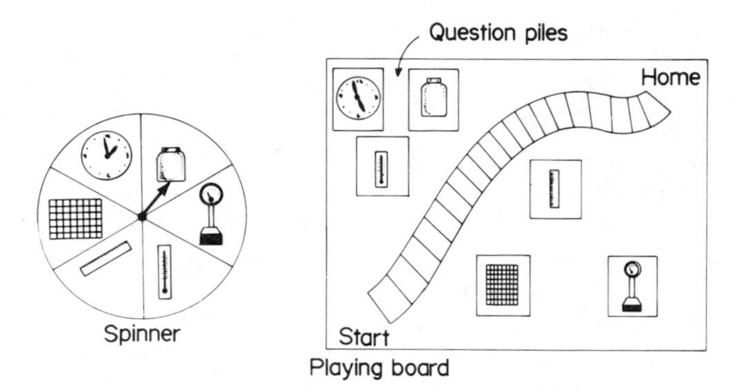

Playing board

Players spin to identify the question stack from which to draw. Question stacks exist for each of the following: time measure, volume measure, weight measure, temperature measure, linear measure, and area measure. Each question card contains a question and an indication of the number of spaces to move if the question is answered correctly. You might use questions such as these:

"What time will it be two hours from now?"

"How long is this room?"

Popsicle sticks and flying polygons

CHARLES LUND
The American School of the International Schools of the Hague, The Netherlands

In the teaching of a unit on basic ideas from geometry to children in grades 5–8, Popsicle sticks can be used to create a motivating, "hands-on," laboratory exercise. The following exercises have been used with children in both the United States and the Netherlands, and the message of identification of polygons and patterns in mathematics has been put across each time. The flying polygon puzzles and a few sample solutions are illustrated in figure 1.

THE FLYING POLYGON PUZZLES

1. Try to construct a "flying triangle" using five Popsicle sticks. (No glue!)
2. Try to construct a "flying square" using six Popsicle sticks.
3. Try to construct a "flying pentagon" using eight Popsicle sticks.
4. Try to construct a "flying hexagon" using nine Popsicle sticks.
5. What is the minimum number of Popsicle sticks necessary to construct a "flying heptagon"?
6. Is there a pattern to the flying polygon constructions? If so, what is it?

SAMPLE SOLUTIONS

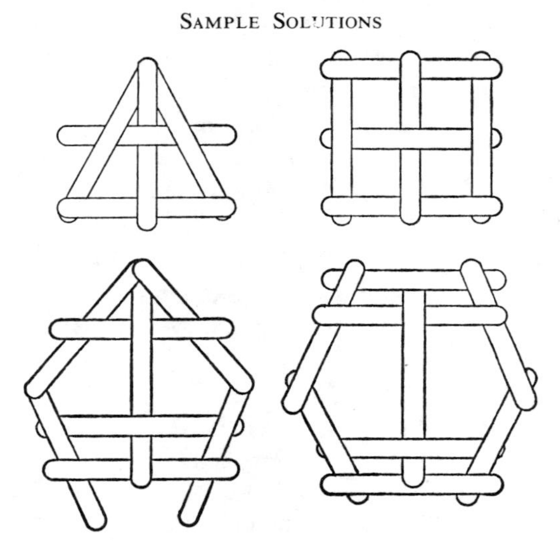

FIGURE 1

How are these puzzles used in the classroom? A dittoed copy of the directions in figure 1 and a small container filled with Popsicle sticks are placed in what I call a "thinkers' corner" near the pencil sharpener of the classroom at the beginning of the unit. Although no formal classroom discussion of the problems takes place, interest in the puzzles usually builds quite rapidly after the first day.

Students are encouraged to formulate conjectures regarding the minimum number of sticks necessary to perform each construction. Here are two interesting conjectures that have been formulated by my students in the past:

Ellen: "If a flying triangle can be constructed using five Popsicle sticks and a flying square can be constructed using six Popsicle sticks, then a flying pentagon and a flying hexagon can be constructed using seven and eight Popsicle sticks respectively." (She wasn't able to construct them but feels certain she is correct.)

Lindsay: "A flying polygon with any number of required sides can be constructed by simply adding two sticks to the previous shape. For example, a flying pentagon can be constructed by simply adding two sticks to the model of a flying square." (Six of the ten shapes he was able to construct are pictured in fig. 2.)

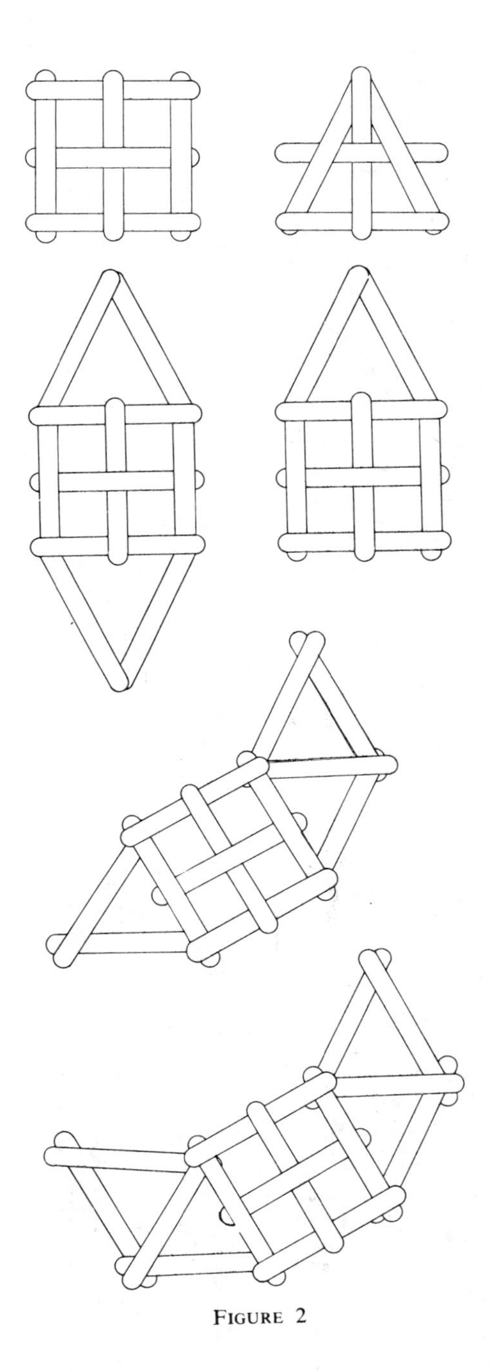

FIGURE 2

The Popsicle-stick models of polygons can also be utilized as a bridge to an exercise in direct measurement. For example, conduct a contest to see which flying polygon will sail the greatest distance. This excursion can take place on the school grounds or, if the weather is inclement, in the gymnasium. Aside from the flying polygons constructed by students, the only special equipment necessary is a 100-foot measuring tape. You may wish to have the class make that, too! By breaking the class up into teams, each member can be assigned a different task to perform. Recording, measuring distances with handmade and official measuring tapes, flying polygons, making repairs, and keeping records are a few possible jobs. All of my classes have found that the triangle construction will sail the farthest. My students say it is because "the flying triangle is the smallest and the strongest." Try this series of exercises featuring simple materials. You'll enjoy them.

The Try-Angle Puzzle

GEORGE JANICKI
Elm School, Elmwood Park, Ill.

.
. .
. . .
. . . .
. . . . .

THE above pattern of dots suggests the idea that they are to be connected. The answer is yes, but do not take your pencil or chalk off the pattern, and do not go back or retrace any line.

When you complete the pattern successfully, you win and you score 16 points.

If you fail, and fall into a trap, the game ends.

Your scoring is: 1 point for a complete triangle; 2/3 point for 2 completed sides; and 1/3 point for one side of any triangle.

In lower grades, you may want to score in terms of whole numbers; all 3 points per triangle completed, 2 points and one point for partially completed triangles.

I found the above pattern to be stimulating thinking.

It looks easy and is a challenge to critical thinking for all students. It is fun and it is arithmetic!

→

EDITORIAL COMMENT.—For something provocative related to shapes, try getting opinions on these two questions.

How many squares do you see? How many triangles do you see?

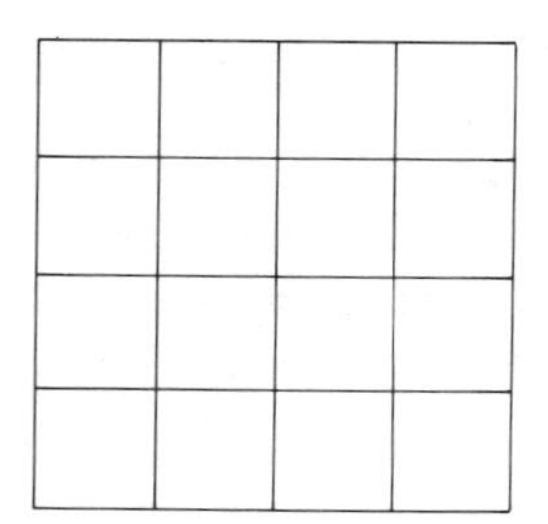

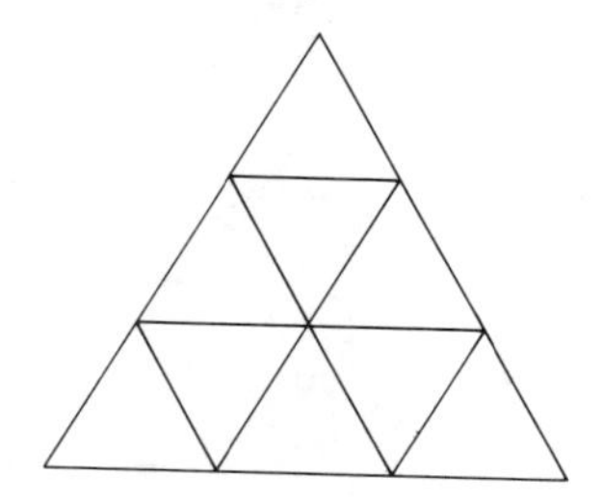

A Pythagorean puzzle

GARY D. HALL *North Judson San-Pierre Schools, North Judson, Indiana*

This puzzle was constructed as a project for a graduate course. It has been used in teaching sixth-grade mathematics.

Purpose

The purpose of this project is to teach the Pythagorean theorem to children who have no background in plane geometry. The ideas of squared numbers and the concept of a 3-4-5 right triangle are also introduced.

The child is taught these related concepts through the use of a brightly colored manipulative puzzle that guides him to form relationships involving area.

Materials

1 right triangle with sides 6 inches, 8 inches, and 10 inches.

1 square (side 6 inches) and 1 parallelogram (sides 6 inches and 10 inches; an angle equal to the smaller acute angle of the triangle described above). Each of these figures has an area of 36 square inches. They should be painted the same color.

1 square (side 8 inches) and 1 parallelogram (sides 8 inches and 10 inches; an angle equal to the larger acute angle of the triangle described above). Each of these figures has an area of 64 square inches. They should be painted the same color.

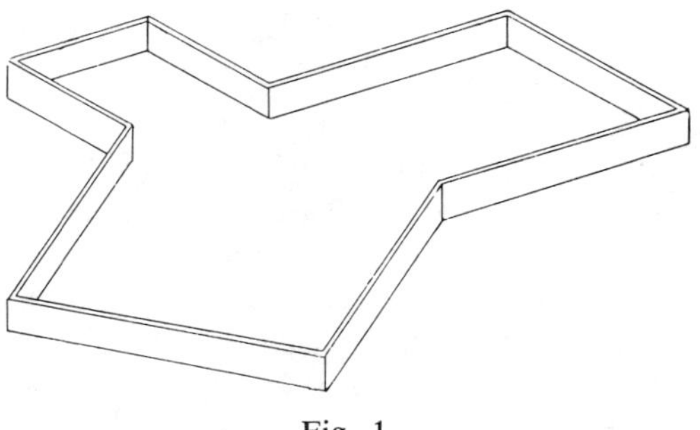

Fig. 1

1 square (side 10 inches) with an area of 100 square inches.

50 squares (side 2 inches), each with an area of 4 square inches; 9 should be painted to match the 6-inch square; 16 should be painted to match the 8-inch square; and 25 should be painted to match the 10-inch square.

1 frame like that shown in figure 1.

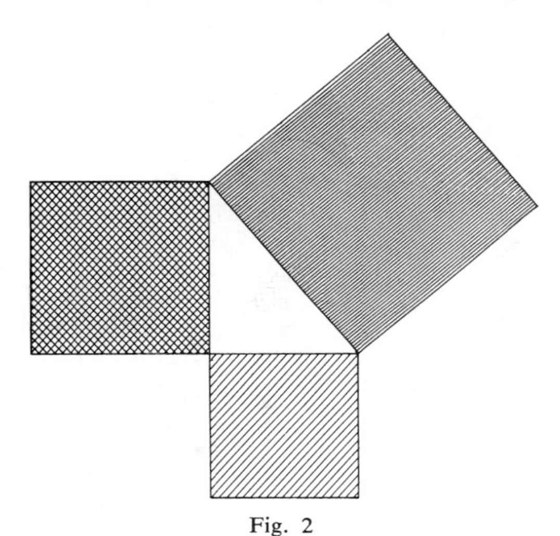

Fig. 2

Procedure

Empty the puzzle frame and give the child the basic puzzle pieces (the triangle and the three large squares). He should construct a figure such as that shown in figure 2.

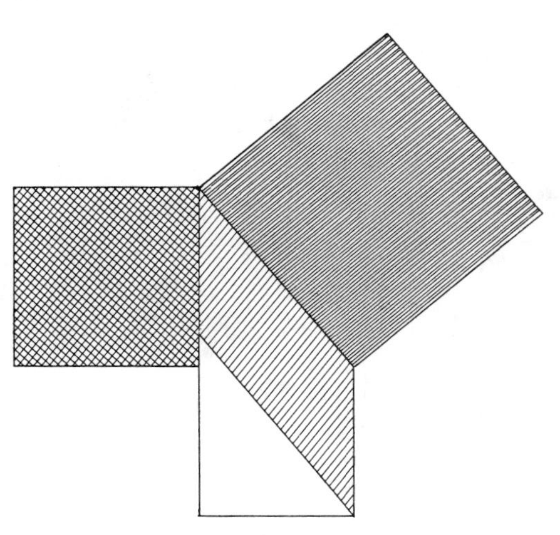

Fig. 3

Empty the puzzle again and give the child the pieces, substituting the smaller parallelogram for the smallest of the three squares. (Note: These two pieces are painted the same color to facilitate the conclusion that they are the same area.) The child should then conclude that the small parallelogram and the small square are the same size by constructing figure 3.

After emptying the puzzle the third time, give the child the largest square, the smallest square, and the triangle and sub-

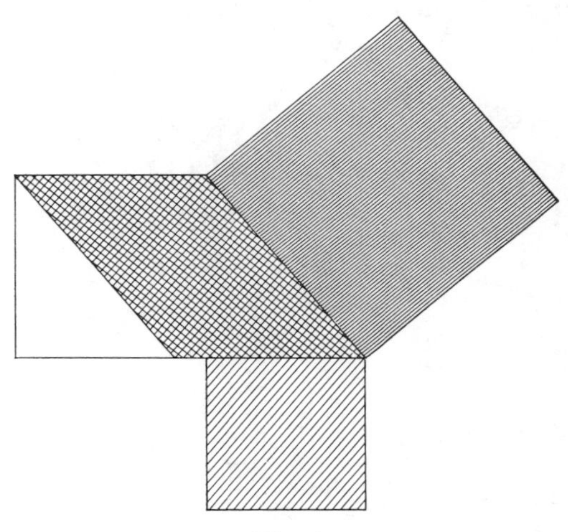

Fig. 4

stitute the larger parallelogram for the medium square. (These two are also painted the same color and are the same size.) The child should construct figure 4.

The fourth construction (fig. 5) is

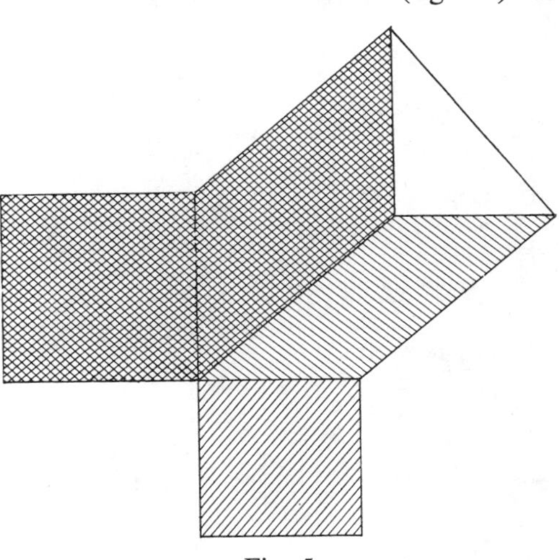

Fig. 5

formed by emptying the puzzle frame and giving the child the two parallelograms, the smallest square and the medium square, and the triangle. After making figure 4, the child should conclude that the two parallelograms are the same size as the largest square, hence the two smaller squares are also the same size as the largest square.

The next series of constructions is designed to reinforce the idea of the Pythagorean theorem and to introduce the idea of squared numbers and the concept of a 3-4-5 right triangle.

Empty the puzzle and put in the fifty small squares. The nine squares painted to match the six-inch square go in the small compartment, the sixteen squares painted to match the eight-inch square go in the middle-sized compartment, and the twenty-five squares painted to match the ten-inch square go in the largest compartment. Also, put the triangle in the middle. Have the child take the squares from the large compartment and fill up the two smaller compartments. Then have him take the squares from the two smaller compartments and construct various designs, such as those

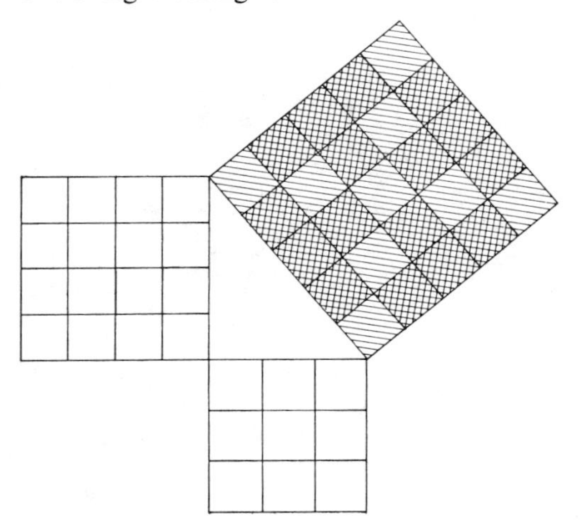

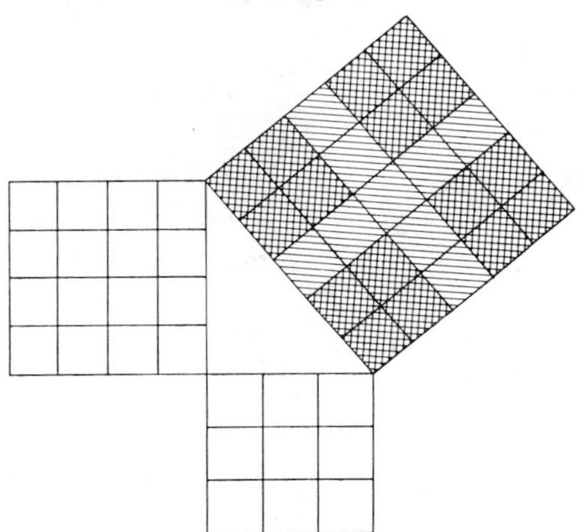

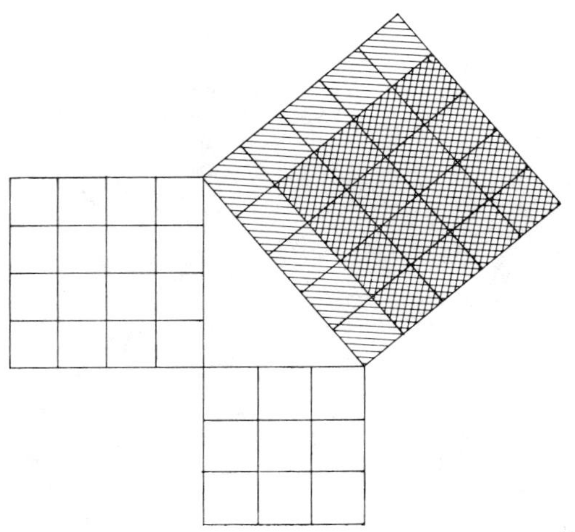

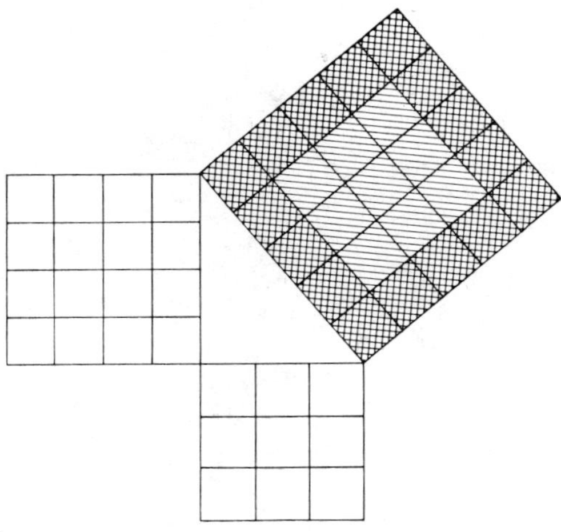

Fig. 6

shown in figure 6, in the large compartment.

Use of the puzzle would not necessarily require a rigid lecture type of presentation such as has been outlined. If students are merely allowed access to the puzzle, many of them will make interesting discoveries in their free time.

EDITOR'S NOTE. Puzzles—good ones—are helpful in developing analytical thinking. They should be readily accessible in all classrooms. They are particularly appropriate in mathematics-laboratory settings where children make original discoveries and solve quantitative problems individually.—CHARLOTTE W. JUNGE.

Tic-tac-toe—a mathematical game for Grades 4 through 9

ROBERT A. TIMMONS *Commack, New York*

Here is a game that can be played with equal enthusiasm in Grades 4 through 9. The game changes only in the amount of strategy used by the students in the upper grades.

The only prerequisite for the students is that they have had the concept of negative and positive numbers introduced to them before the game and are familiar with the ordinary game of tic-tac-toe.

The knowledge of negative and positive numbers would not have to be very great. It is sufficient for them to simply be aware of their existence.

X	O	O
	X	O
	X	X

The game is played with the entire class. An overhead projector with a prepared

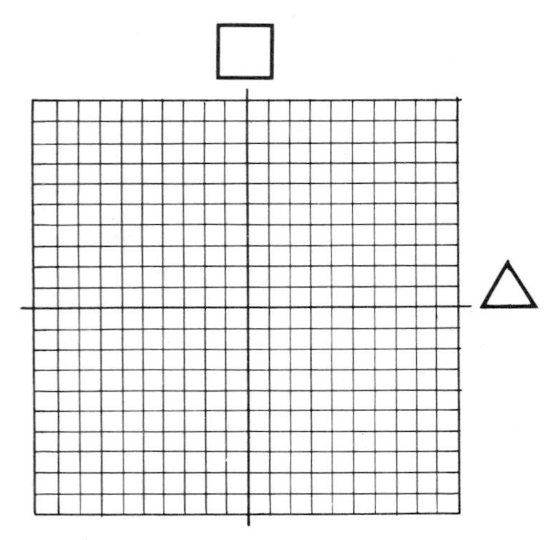

grid makes things easy. However, a grid drawn on the board will do just as well.

The instructions for the game are purposely short and simple. It will be up to the participants to fill in the gaps.

The class is divided into two teams (e.g., boys versus girls). The teams do not have to be of equal number or ability. One pupil is assigned the job as a recorder. He or she may or may not participate in the game. His job is to record, in two columns on the board, the numbers that are given to them by the students. E.g.,

	△	□
X	2	3
O	4	1
X	3	0

The teacher then gives the following instructions to the students:

1. This is a game of tic-tac-toe, but in this game in order to win you must get five "X" or "O" in a row.
2. In order to tell me where to place your "X" or "O" you must give me two numbers. *Each* number must be equal to or less than 10. The recorder will write these numbers on the board and I will place your "X" or "O" in the correct place on the grid. Watch me closely and see if you can understand how I place them.
3. Once you say a number you may not change your mind. Think before you tell us your numbers, but if you take too long you will lose your turn.
4. You are not allowed to help your team members. (This rule can be altered at the teacher's discretion.)

The game follows the rules of coordinate axes with the students supplying the two variables. The teacher should be careful that he or she does not count by pointing to the lines but rather simply placing the marks in the correct place.

The first couple of games played will most likely be played and won all in the first quadrant.

I have been surprised each time I have played this game at the speed at which the students discover how to locate the points. Frequently it is the student who is having difficulty with his regular program who is the first to discover it.

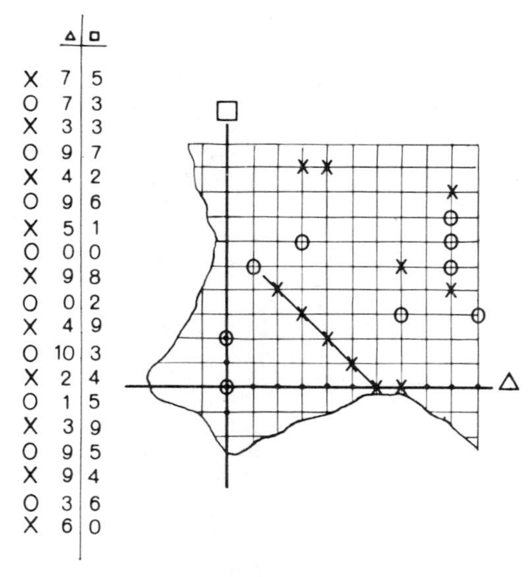

If a student should happen to give a coordinate that has been already given or is not in the limits previously set, he is informed that he cannot go there and that he loses his turn.

In order to force the game out of the first quadrant, the teacher may reduce the limits set in rule #2 to "numbers equal to or less than five." The game will then

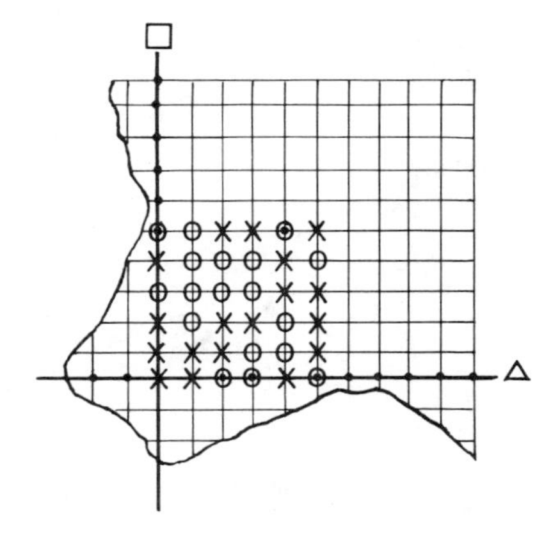

quickly come to a stalemate with the first quadrant completely filled in.

At this point the students will urge the teacher to tell them how they can get out of the first quadrant. The teacher should not give in but rather keep on promoting them on with questions like this:

"Give me the right pair of numbers and it will get you out of that corner. Try a different type of number. Think back. Didn't we learn about any other types of numbers?"

Sooner or later someone will come out with a negative number or a pair of negative numbers. The game can then continue with the previous limits or any limits set by the teacher.

One of the difficulties with this game is that the children would like to play it all the time. I have yet to find a class that tires of it.

I believe the activity originated in the Madison Project. A complete description of it can be found in *Discovery in Mathematics, A Text for Teachers,* by Robert B. Davis. (Reading, Mass.: Addison-Wesley Publishing Company, 1964).

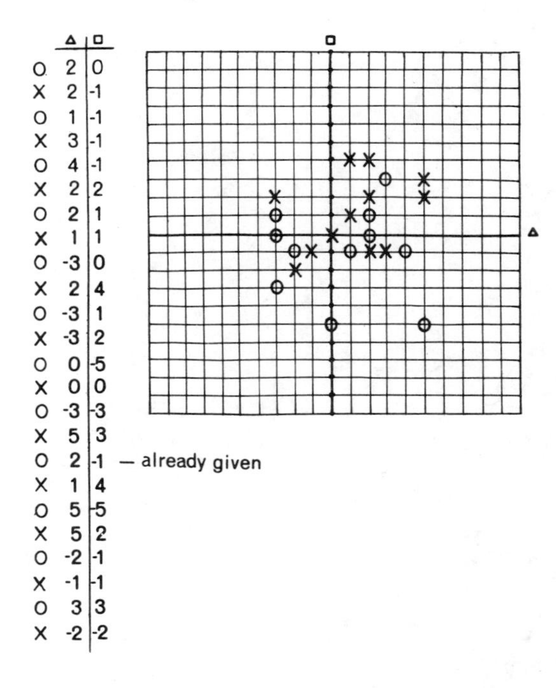

The value of this activity and others like it is the enthusiasm it generates. It allows all to participate no matter what their degree of competency, and lends itself to the discovery method of teaching with little effort on the teacher's part. If a student is unable to comprehend how to locate the points, there is no loss in the mathematics sequence. If on the other hand, he becomes proficient in locating points in all four quadrants, he has been brought to the threshold of analytical geometry.

Hide-a-region—$N \geq 2$ can play

JEAN S. OVERHOLSER
Oregon State University, Corvallis, Oregon

Hide-a-Region is a game that can be played by two or more persons, from the first to the twelfth grade. Its purpose is to give practice in locating points on a grid, and in the concept of the area of a region.

In its simplest form, all players are given graph paper or a grid. One group decides on the location of a square region on the grid, with the vertices at ordered pairs of whole numbers. In figure 1, a region of area 16 is shown on a ten-by-ten grid. The other group tries to locate the region by calling out ordered pairs, while the group that has hidden the region calls out "Inside!" or "Outside!" in response to each trial. The region is located when the opposing team has named all four vertices. A tally is kept of the number of guesses.

It is now the turn of the second team to hide a region. After $2n$ games, the winning team will have the fewest number of guesses.

For a variation, a rectangular region of

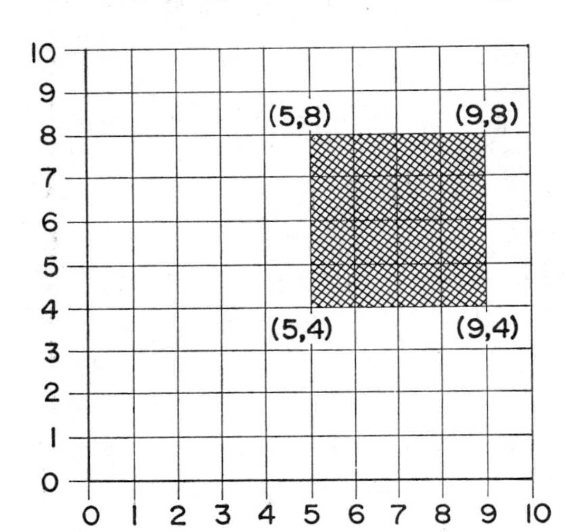

FIG. 1.—A "hidden" square region of area 16.

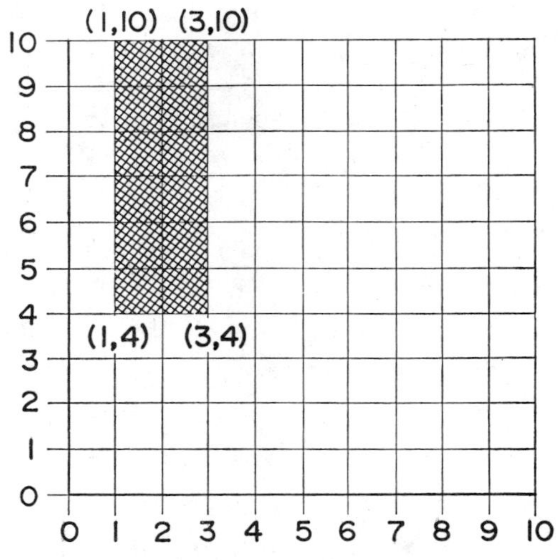

FIG. 2.—A "hidden" rectangular region of area 12.

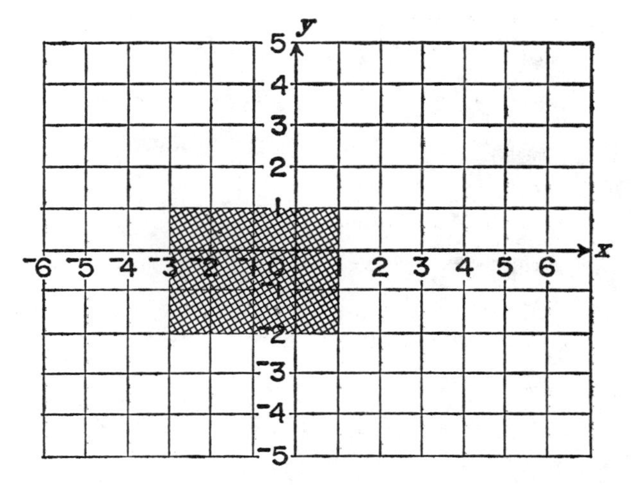

FIG. 3.—A "hidden" rectangular region of area 12.

a given area can be hidden as shown in figure 2. The group who hides the region says, "Rectangular region, area 12." The solution is (1,4), (3,4), (3,10), (1,10), in any order.

For another variation, the team that hides the region could specify that the boundaries are excluded. Then, if the opposing team calls out a point on the boundary, the team that hid the region can say, "Outside!" Here the vertices determine the region, although they are outside the region.

After negative numbers and three more quadrants are introduced, the area available for hiding a region is expanded. Figure 3 shows a hidden region of area 12 with vertices in all four quadrants. The group that has hidden the rectangular region says, "Rectangular region, area 12, hidden in the region where the x-coordinates go from negative 6 to positive 6, and the y-coordinates go from negative 5 to positive 5."

A further variation would involve putting the vertices at points determined by ordered pairs such as (1½, 2¾), (9½, 2¾), (9½, 1¾), (1½, 1¾) to determine a rectangle of area 8.

EDITOR'S NOTE.—Is it possible that some regions could have the same area but have different dimensions? It occurs to me that the region described as "rectangular, area 12," might be a region 2 by 6 units as well as 3 by 4!—CHARLOTTE W. JUNGE.

EDITORIAL COMMENT.—An appealing variation on Hide-a-Region is Hide-a-Name. And if you make the name hidden that of one of the children, it is even more exciting!

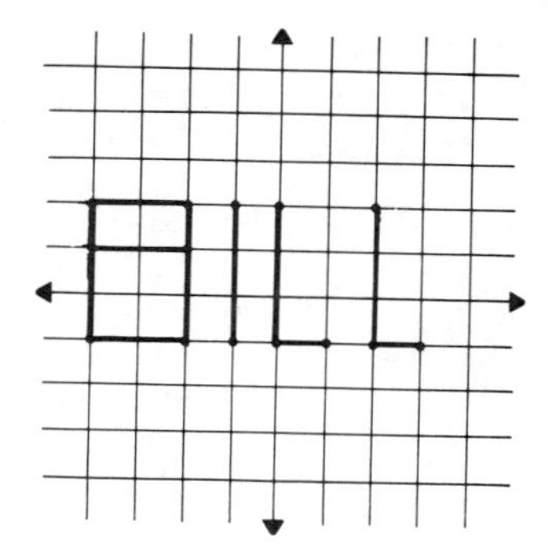

Cartesian coordinates and battleship

WILLIAM R. BELL
Boca Raton Middle School, Boca Raton, Florida

Mathematical games are a way of relieving the drudgery of practice for some students, and this adaptation of an old and familiar game (which is often played surreptitiously by students) can also be justified as being educational. In the version of Battleship described here, the basic difference is that the intersections of grid lines are named instead of the squares between the grid lines. The game is best introduced after a study of integers and as a prelude to graphing equations of lines.

The game is played on the standard two-dimensional grid. (See fig. 1.) Students must understand that all points on the grid are identified by a pair of numbers (x, y), where x is counted horizontally and is always named and located first, and y is counted vertically and is always named and located second. A few examples—(1, 1), (2, 2), (−2, 2), (1, −1)—may be appropriate. Limiting the size of the grid to six units in each direction from zero is best if the game is to be completed within the normal class period of forty or fifty minutes.

Each player should have a single sheet of paper with two grids drawn on it. The player spots his own battleships on one grid, and on the other he records his shots at his opponent's forces. Each player is allowed three battleships. A battleship consists of three adjacent (horizontally, vertically, diagonally, or on a corner) points.

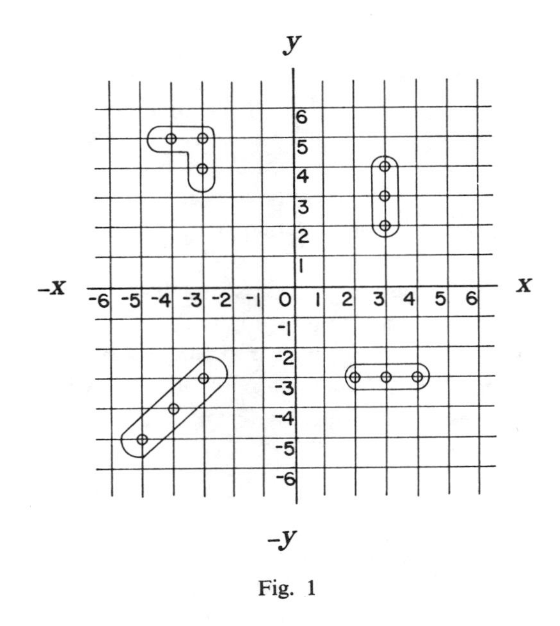

Fig. 1

Players take turns firing volleys of three shots. (The decision regarding who fires first can be made by any method.) When hits are made, they are acknowledged immediately. A game ends when a player has lost all three of his battleships.

During a game, players must be placed in such a way that they cannot see their opponent's sheet or the opponent marking his sheet. Notebooks placed on end can serve as improvised walls.

Be prepared to be challenged by your students. You will be pleasantly surprised at their ability and determination to beat the teacher at his own game.

EDITORIAL COMMENT.—Rather than using battleships on the grid, you might consider placing fish on the ocean and have a "Fishing Rodeo." Whales might be five connected locations, mackerel could be four points, and so on to minnows, which are single points.

Holiday plot-dots

GARY A. DEATSMAN

Moorhead State College, Moorhead, Minnesota

Here is an activity that will interest and challenge third graders and advanced second graders. By carefully following written directions, each child plots ordered pairs of natural numbers to get points on a coordinate system. Each point is labeled with a number and when the plotting is done the child has constructed a follow-the-dots puzzle which he can then complete. Directions for three of these "plot-dots" are given here. The first turns out to be a Halloween pumpkin and the second a Christmas tree. The third, which is shown in figure 1, is a Valentine heart.

To do one of the plot-dots, each child should be supplied with an instruction sheet and a piece of special graph paper. Half-inch squares must be used if the figure is to fit properly on 8½-by-11-inch paper. The coordinates should be on the paper. The graph paper can be teacher-made. I made mine very easily by drawing the coordinates on a piece of ordinary graph paper and then making a spirit duplicating master.

Accuracy is very important if the pictures are to look right. Giving each child an instruction sheet is essential; trying to read the instructions aloud to the class will result in chaos. I would encourage the children to work in pairs or small groups and to check each other's work. It seems to be futile to try to get them to write *lightly* so they can erase easily if they make an error, but maybe it's worth a try. Some children confuse "over" with "up," so some preliminary practice in this area may help. Sometimes demonstrating the plotting of some points on the blackboard helps. If your pupils are like the children I worked with, many of them will finish their work with complete accuracy and get very nice

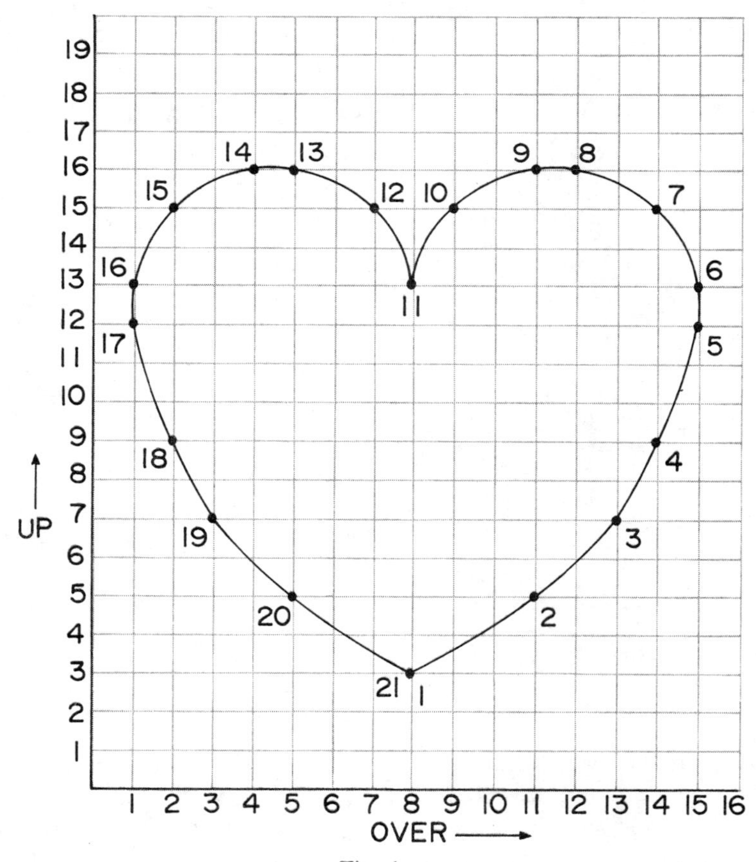

Fig. 1

results. A few may get discouraged and quit, so I would try to keep competition at a low key.

When a child completes a plot-dots he should then be allowed to add to his picture—draw a face on the pumpkin, draw decorations on the Christmas tree, or write something appropriate on the heart. He then can color his creation.

Although I can't prove it, it seems reasonable to me that experience with plot-dots may help prepare children for later work with linear measurement and graphing. In any case, it's fun.

Directions for Halloween Pumpkin Plot-Dots

1. Go over 14 and up 10 for dot number 5.
2. Go over 12 and up 13 for dot number 7.
3. Go over 4 and up 4 for dot number 19.
4. Go over 6 and up 16 for dot number 11.
5. Go over 9 and up 14 for dot number 9.
6. Go over 4 and up 13 for dot number 14.
7. Go over 14 and up 7 for dot number 4.
8. Go over 10 and up 3 for dot number 1.
9. Go over 2 and up 7 for dot number 17.
10. Go over 3 and up 12 for dot number 15.
11. Go over 8 and up 16 for dot number 10.
12. Go over 7 and up 14 for dot number 12.
13. Go over 2 and up 10 for dot number 16.
14. Go over 3 and up 5 for dot number 18.
15. Go over 6 and up 3 for dot number 20.
16. Go over 10 and up 14 for dot number 8.
17. Go over 12 and up 4 for dot number 2.
18. Go over 6 and up 14 for dot number 13.
19. Go over 13 and up 5 for dot number 3.
20. Go over 13 and up 12 for dot number 6.
21. Go over 10 and up 3 for dot number 21.

Directions for Christmas Tree Plot-Dots

1. Go over 8 and up 1 for dot number 1.
2. Go over 8 and up 3 for dot number 2.
3. Go over 10 and up 12 for dot number 9.
4. Go over 5 and up 12 for dot number 16.
5. Go over 7 and up 18 for dot number 12.
6. Go over 9 and up 15 for dot number 11.
7. Go over 11 and up 6 for dot number 4.
8. Go over 3 and up 6 for dot number 20.
9. Go over 4 and up 9 for dot number 18.
10. Go over 13 and up 3 for dot number 3.
11. Go over 6 and up 3 for dot number 22.
12. Go over 8 and up 15 for dot number 10.
13. Go over 5 and up 15 for dot number 13.
14. Go over 11 and up 9 for dot number 7.
15. Go over 12 and up 6 for dot number 5.
16. Go over 1 and up 3 for dot number 21.
17. Go over 3 and up 9 for dot number 17.
18. Go over 6 and up 1 for dot number 23.
19. Go over 10 and up 9 for dot number 6.
20. Go over 9 and up 12 for dot number 8.
21. Go over 6 and up 15 for dot number 14.
22. Go over 4 and up 12 for dot number 15.
23. Go over 2 and up 6 for dot number 19.

Directions for Valentine Heart Plot-Dots

1. Go over 13 and up 7 for dot number 3.
2. Go over 8 and up 13 for dot number 11.
3. Go over 8 and up 3 for dot number 1.
4. Go over 1 and up 12 for dot number 17.
5. Go over 15 and up 12 for dot number 5.
6. Go over 15 and up 13 for dot number 6.
7. Go over 12 and up 16 for dot number 8.
8. Go over 2 and up 9 for dot number 18.
9. Go over 5 and up 16 for dot number 13.
10. Go over 2 and up 15 for dot number 15.
11. Go over 14 and up 15 for dot number 7.
12. Go over 11 and up 16 for dot number 9.
13. Go over 11 and up 5 for dot number 2.
14. Go over 5 and up 5 for dot number 20.
15. Go over 3 and up 7 for dot number 19.
16. Go over 9 and up 15 for dot number 10.
17. Go over 1 and up 13 for dot number 16.
18. Go over 4 and up 16 for dot number 14.
19. Go over 14 and up 9 for dot number 4.
20. Go over 7 and up 15 for dot number 12.
21. Go over 8 and up 3 for dot number 21.

Stick Puzzle

If you have 12 sticks of equal length arranged as in the diagram below, show how you can make the following rearrangements.

1. Remove 4 sticks and leave 2 squares.
2. Remove 4 sticks and leave 1 square.
3. Change 3 sticks and have 3 squares.
4. Remove 2 sticks and leave 2 squares.

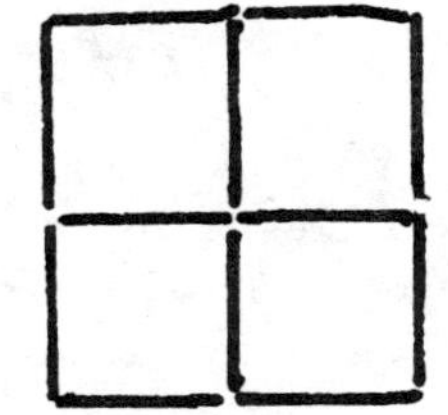

A Game of Squares

George Janicki
Elm School, Elmwood Park, Ill.

. . . . .
. . . . .
. . . . .
. . . . .

The pattern above suggests a game of dots to any bright student. In my classes, I used this puzzle as follows:

First set a time limit of one minute. (This makes it very exciting!)

The idea is to start at any place and to connect each dot, and to try to complete as many whole squares as possible.

You do not permit any diagonal connections. NEVER TAKE PENCIL OFF THE PAPER!

You cannot retrace or cross any previous line.

When you end up in a blind alley, the game is finished.

The scoring is: 1 point for a completed square; 3/4 point for 3 sides of a square completed; 1/2 point for two sides completed; and 1 point for one side since you might possibly have such an arrangement.

The average score is: 7 points; the real bright students reach 8 3/4. This game is really fascinating.

I recommend it highly to any arithmetic teacher in grades 4, 5, 6, 7, 8.

(You can change the scoring points for lower grade students to all whole numbers: 4 points, 3 points, 2 points, and one point since they may be unfamiliar with adding unlike fractions. In such cases, 48 is top score; see how high they can make using these scoring rules.)

Cross Figure Puzzle—Measures

George Janicki
Elmwood Park, Illinois

ACROSS DEFINITIONS

1—$\frac{1}{8}$ of a mile in feet
4—1/10 of a mile in feet
7—A certain type of fire alarm
8—$14\frac{1}{2}$ minutes in seconds
9—A dozen
10—Number of ounces in 1 pound
11—Baker's dozen
14—Ounces in pound (Troy system)
16—A full day in hours
17—6 feet 5 inches expressed in inches
19—$466 \times 1 + 0 =$
21—Largest 3 place whole number
22—A 1/12 of a mile in feet
23—$19 \times 3 \times 4 \times 4 =$

DOWN DEFINITIONS

1—Product of 23 and 27
2—$38\frac{1}{4}$ pounds changed to ounces
3—A penny in decimal form
4—5 feet less 2 inches in inches
5—$100 \times 2.71 =$
6—$3212 \times \frac{1}{4} =$
12—One hour in minutes
13—3 feet in inches
14—A gross
15—22 dozen is how many units?
17—An odd number
18—An even number
20—5 feet tall is how many inches?
21—Largest 2 place whole number

8

Reasoning and Logic

Reasoning and logic are pervasive throughout all mathematics. The canons of logic form the bases of the deductive nature of abstract mathematics. Students naturally acquire logical prowess through reasoning in everyday affairs. This occurs throughout the elementary school years. Formal instruction in logical ideas frequently occurs in contemporary mathematics programs in the late elementary and junior high school years.

Most mathematical games require some degree of reasoning or logic (although some rely more on chance factors). Some games are designed specifically to encourage the development of reasoning, and others depend heavily on the development of strategy. This section contains illustrations of games of both types.

The lead article, "Rainy-Day Games," illustrates four games designed to help students "learn an essential process of mathematics, namely, asking good questions and piecing information together to draw a conclusion." Be sure to follow the authors' suggestions about supplying props to the students—it will help the students put structure on the situations.

Ruderman's "Nu-Tic-Tac-Toe" is a variant on checkers, requiring strategies for timing moves, blocking, and positioning. The tic-tac-toe array becomes the playing board for the hexapawn game described in Ackerman's article, "Computers Teach Math." (You don't really need a computer for this game.) The article provides a good example of a procedure for analyzing games having a finite number of possible moves. Such games and their analytical procedures demonstrate clearly that there is really much more than motivation available in the use of classroom games.

The game Kalah has been around for thousands of years. It has been used in different versions in many cultures, particularly for the development of quantitative judgment. It is a game in which victory depends solely on reasoning rather than on chance. Haggerty's description of the game provides both a clear exposition of the rules and some interesting historical notes. For those with limited resources, be sure to note the inexpensive way to construct the Kalah board in the follow-up article by Brill.

Other inexpensive strategy games are described in Massé's "Drill Some Fun into Your Mathematics Class." Massé gives directions for playing as well as for constructing the game boards. In addition to the strategy features, these games are useful for practice with addition-subtraction relationships and number-theory concepts.

The last two articles are versions of the "Whodunnit" puzzles or riddles. "Jupiter Horse Race" is a student-created puzzle—a reminder that our students can be a rich source of legitimate mathematical activity. "Paper, Pencil, and Book," like Kalah, is a game of ancient origins. This one dates from medieval times. Games survive not only because they provide diversions for man but because the best ones whet and test his intellectual appetite.

Rainy-day games

ROBERT C. GESSEL, CAROLYN JOHNSON, MARTY BOREN, and CHARLES SMITH

At the time this article was written, all four of the authors were Miller Mathematics specialists with the Fullerton, California, Elementary Schools. Since then, both Robert Gessel and Marty Boren have left Fullerton. Robert Gessel is currently teaching mathematics in the Audubon Junior High School in Cleveland, Ohio.

There are many occasions when "rainy-day games" have their special value and usefulness, and, of course, rainy-day games may even be played on sunny days too. The games described here give children a chance to ask questions that enable them to obtain information; the information is then used to reach a conclusion.

In playing the games, children frequently ask redundant or useless questions. They realize their questions are useless when no information is gained. Consequently, the children listen to classmates' questions to learn how to ask questions that will elicit useful information. This type of learning experience provides an opportunity for the children to develop an intuitive feeling for deductive reasoning.

One cannot say that because of playing these games students will improve their arithmetic skills. However, they will learn an essential process of mathematics, namely, asking good questions and piecing information together to draw a conclusion. The children will be actively involved in an experience that encourages deductive reasoning. Finally, the games are fun!

Pico-Fumi

When the game of "Pico-Fumi" is presented to a class for the first time, it is best to begin with the definitions of the terms written on the chalkboard for easy reference as follows:

Pico means that one of the digits is right, but it is in the wrong place.
Fumi means that one of the digits is right, and it is in the right place.

To illustrate how the terms are used, you might pick the number 43 and say, "I have selected a number between 9 and 100; guess the number." If someone guesses that the number is 63, your answer would be, "That is one fumi," since the 3 is a correct digit in the right place." You would then record 63 on the board with "one fumi" next to it. Suppose the next guess is 74. You would answer, "That is one pico," since the 4 is a correct digit but in the wrong position." Under the 63 you would record the 74 with "one pico" written beside it. If someone guessed 98, you would call it nothing, since it is neither pico nor fumi. The 98 would be recorded with "nothing" beside it under the 74. More examples may be given, if it seems necessary, to further establish the answers that may be given.

To actually play the game, the class is divided into two teams, and a member from each team, say Thad and Wendy, is picked to write a "secret" number between 9 and 100 on a slip of paper. They come to the front of the room. Since the teacher should record the answers for the first few games, you look at their numbers; suppose Thad's is 25 and Wendy's is 61. Thad picks a member of his team to guess Wendy's number. If the person guesses 56, then you record "one pico" next to the 56 under Wendy's name. Wendy then chooses a member of her team to guess Thad's number. If the person guesses 46, a 46 with "nothing" beside it would be recorded under Thad's name. The game continues until one team gets two fumis, which will be the correct number. The progress of a game might be recorded on the board in a table like that shown in figure 1.

Thad's Number			Wendy's Number		
nothing	←	46	56	→	1 pico
1 fumi	←	35	63	→	1 fumi
2 picos	←	52	67	→	1 fumi
2 fumis	←	25	69	→	1 fumi

Fig. 1

At the end of a game, another person from each team is picked to write down a secret number. Thus, a new game is started. After one or two games, the students can record the guesses themselves.

A few suggestions may be helpful. Play the game with a friend first in order that you clearly see some of the strategies of the game. Do not expect the students to develop good strategies at the start.

Hats

The game of "Hats" requires six hats and a paper bag big enough to hold them. Three of the hats are of one color and three are of another. For the purposes of this explanation, let us assume that three of the hats are red and three are black.

To play the game, select three students and arrange them in a circle so that each may see the other. With the six hats all hidden in the bag, ask the students to close their eyes. Then remove three of the hats and place one on each head. Tell the students to look carefully at the others but caution them not to give away what they see.

Ask them to raise their hands if they see a red hat. The game is then to see if each student can figure out what the color of the hat he is wearing is.

Name Game

The "Name Game" is a fun game that allows children more practice in logic and good question asking. It discourages wild guessing and ragged thinking.

The Name Game begins when the teacher selects a child. No one in the class, including the child himself, knows who has been chosen. The teacher gives a clue. The children use the clue to ask a question that will yield more information. If the question uses the clues the teacher has given, the yielding of information continues until the chosen child has been determined. The following is an example of how the game proceeds.

TEACHER: The person I have chosen is not a girl.

BOBBY: Is it Charles?

TEACHER: No, but you used my clue in your question; so you get another clue. The person sits on the right side of the room.

LAURA: Is it Sherry?

TEACHER: No. Sherry sits on the right side of the classroom, but is Sherry a boy? Did you use both the clues? Sorry, no new clues.

MARGARET: Is it Danny?

TEACHER: No, although Danny is a boy and sits on the right side of the classroom. Good guess. The person sits in row one.

JOSE: It's Bill!

TEACHER: Yes, how did you know?

JOSE: In row one on the right side of the room, Bill is the only boy.

TEACHER: Very good thinking, Jose.

Number on the Shoe

The entire class can play this game. The teacher chooses a student with a sturdy shoe sole and asks the student to think of a number between one and one hundred. The teacher records the number on the student's sole with a piece of chalk. The class then tries to guess the number by asking questions that can be answered yes or no.

The students usually begin by asking questions like "Is it 73?" But eventually someone asks, "Is it in the thirties?" The latter type of question should be encouraged by the teacher because it takes in a greater range of numbers.

The object of the game is for the students to ask the minimum number of questions to determine a classmate's number. During the game students have an opportunity to listen, to verbalize, and to use and build on information gotten from students' previous questions. The teacher may want to record data from the children's questions on a number line on the chalkboard.

These games give children an excellent opportunity to develop game strategy and game planning. Overall, they offer a nonthreatening kind of total-class involvement by which deductive reasoning is exemplified through a question–answer–information–further-deduction type of sequence.

ic
Nu-Tac
oe

HARRY D. RUDERMAN *Hunter College High School, New York City*

Games like checkers and chess offer the child an opportunity to think ahead by considering alternate moves for himself as well as for his opponent. Unfortunately, in this fast-moving period many are reluctant to sit down and ponder over such games, mainly because it takes too much time.

The game described here is a type of tic-tac-toe game that has ingredients of strategy common to checkers and chess: timing, blocking, position, anticipation. Moreover, children, as well as adults, learn this game in less than one minute, get to grips with the challenge very quickly, and seem to derive considerable enjoyment in playing it. It is a game that I invented recently and would like to share with others.

Directions for playing the game

1 Start with four pieces marked with crosses and four pieces marked with circles, and arrange them as shown on the twenty-square board in Figure 1.
2 Two players play. One moves the pieces

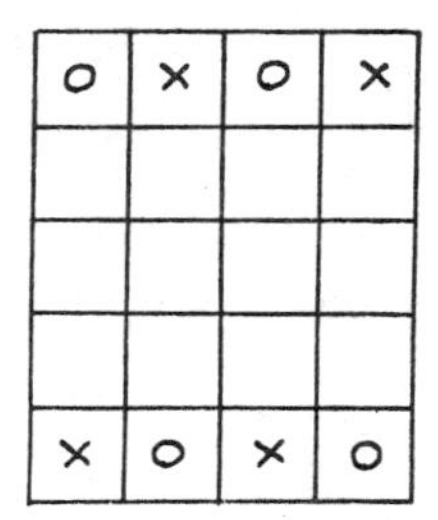

Figure 1

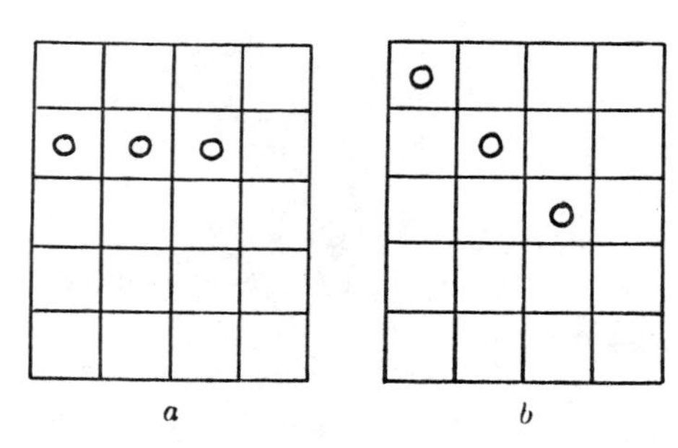

Figure 2

with crosses, the other moves the pieces with circles, taking turns, first one player then the other.

3 A move consists of pushing your own piece into an adjacent vacant square up or down, to the right or to the left, but NOT diagonally. There is no "jumping" or "taking" in this game, and if a square is occupied, no other piece may be moved into this occupied square.

4 The object of the game is to place three of your pieces in the same line: vertically, horizontally, or diagonally without any intervening vacant squares. Examples of wins for the circles are shown in Figures 2*a* and *b*. Examples of no wins for the circles are shown in Figures 3*a* and *b*.

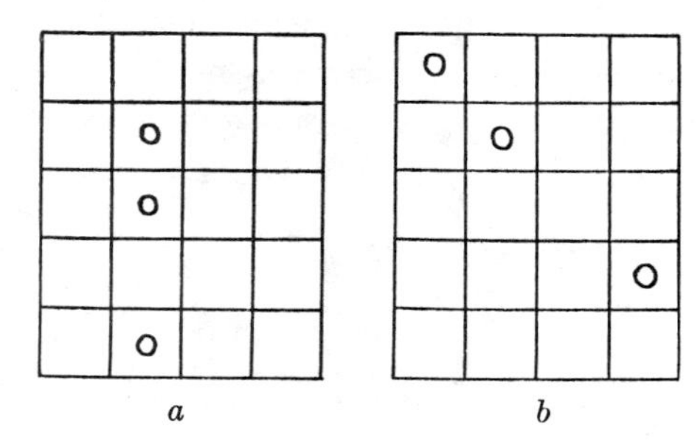

Figure 3

A variation

One player places all eight pieces on the board into squares in any arrangement he chooses, one to a square. The other player makes the choice of moving first or second.

EDITORIAL COMMENT.—A quick but intriguing game is the game of NIM. There are many variations on this game. One of the simplest involves three containers of beads. Container A has five beads, container B has four beads, and container C has three beads.

Two persons play, alternately taking beads from the containers. When it is a player's turn, he may take as many beads as he wants from a single container as long as he takes at least one bead. The person to take the last bead is the winner. (The strategy for winning uses concepts from binary numeration.)

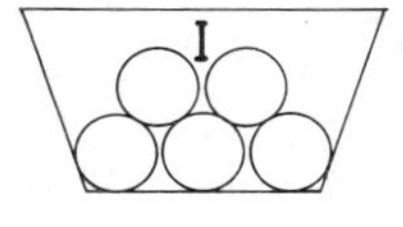

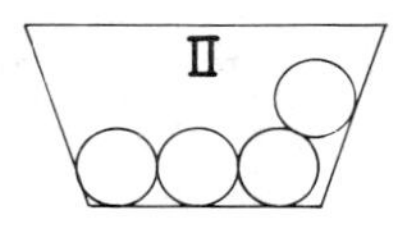

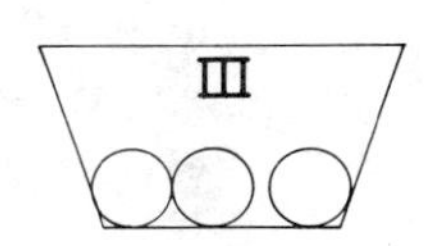

Computers teach math

JUDY ACKERMAN

West Boylston, Massachusetts, School System, Worcester, Massachusetts

In order to maintain a high degree of interest in my sixth-grade mathematics classes, I decided to introduce a year-long class project. Each child belongs to a committee and is able to help with the work.

Our project for the year is to learn about, and, build game-learning computers. The classes are working with the simplest kind of game-learning computers—ones that play games that can be completely analyzed.[1] The ultimate goal is to design and build a game-learning tick tack toe computer. The first phase of this project involved studying and building a small game-learning computer for hexapawn,[2] a game designed by Martin Gardner.

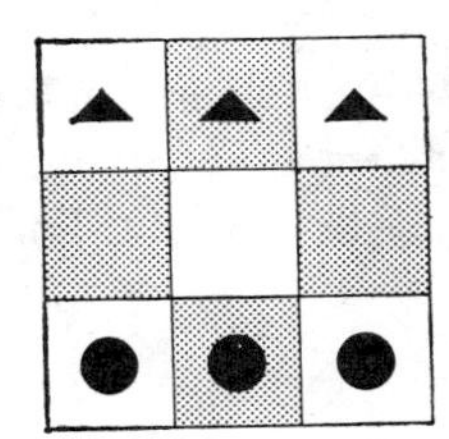

FIGURE 1

The game of hexapawn is played on a 3-by-3 board with three markers for each player. (See Fig. 1.)

The markers are moved in the same way as the pawns in chess:

1. A marker may be moved forward one square to an empty square.

2. A marker may be moved forward diagonally to the left or right to capture an enemy marker. The captured marker is taken out of play. The game is won in three ways:

1. By moving a marker to the opponent's side.

2. By capturing all of the opponent's markers.

3. By planning moves so that a point is reached in the game where the opponent is unable to move.

As the class learned to play hexapawn, they observed that a draw was impossible and that there was a decided advantage for one of the players. (See if you can discover which player has the advantage.)

When the class became familiar with the game, they charted all the possible combinations of moves that could face the computer during any game if the computer went second. (There are only twenty-four possible combinations.) For each combination, arrows were drawn in different colors to illustrate the choices that the computer faces on each move (Fig. 2). The "brain" of the computer consists of twenty-four cups which were suspended

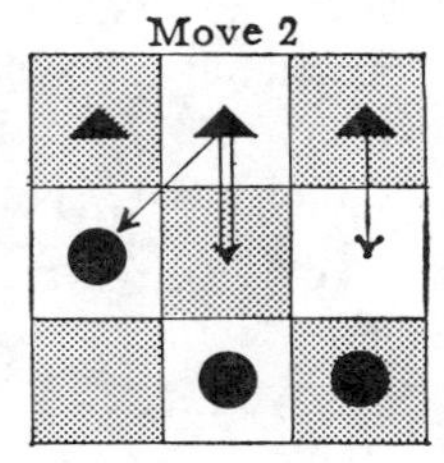

A combination diagram
FIGURE 2

[1] C. E. Shannon, "Game Playing Machines," *Journal of the Franklin Institute*, CCLX (December 1955), 447–53.

[2] Martin Gardner, "Mathematical Games," *Scientific American*, CCVI (March 1962), 140–43.

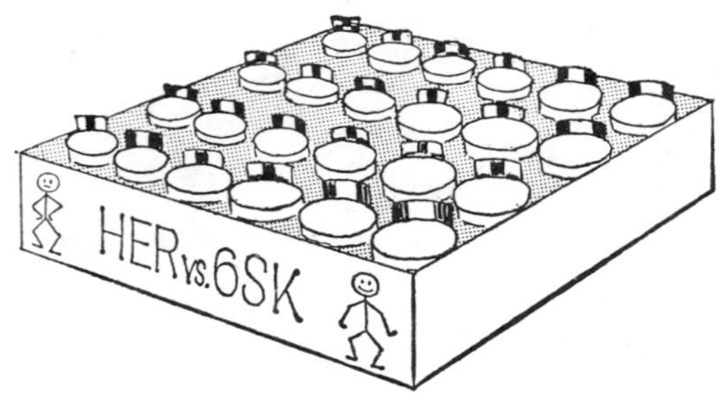

The hexapawn computer
FIGURE 3

in the side of a cardboard carton (Fig. 3). The combination diagrams were stapled to the cups. All the possible second moves, fourth moves, and sixth moves were grouped together.

The "memory" was made up of marbles placed in the cups. The number and colors of the marbles in each cup were determined by the number and color of the arrows on each combination diagram.

The learning took place as the class played the computer. The human player always started first. The operator then located the cup that contained the present board position. He randomly removed one marble from that cup and made a move for the computer that corresponded to the color of the marble. The game continued until there was a winner. When the computer won, it was rewarded by replacing all the marbles removed during the game and by adding an extra marble to correspond to the marble removed on the last move of the game. If the computer lost, it was punished by permanently removing from its memory the marble that corresponded to its last move of the game. The other marbles removed during that game were replaced.

As the members of the class took turns playing the computer, the rest of the class was busy keeping records of the results of each game. In this way there was an excellent opportunity to introduce a unit on graphing. The class found that the line graph was one of the more effective ways of illustrating the results. (See Fig. 4.)

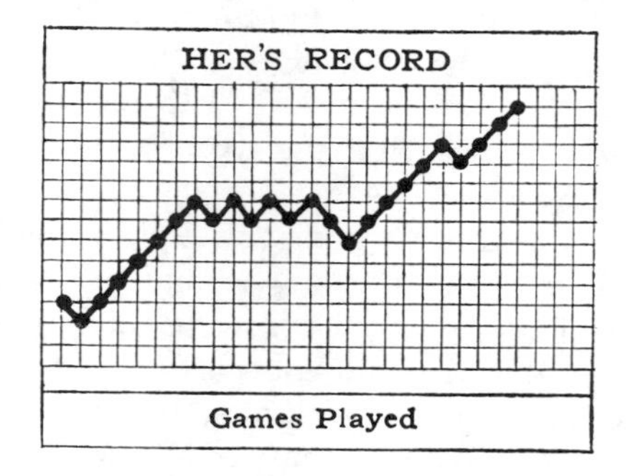

FIGURE 4

The experience that my classes have gained from building the hexapawn game-learning machine has given them many ideas for the design and construction of a ticktacktoe-learning computer. They are most enthusiastic and willing math students who look forward to coming to class.

EDITORIAL COMMENT.—Be sure to make a file of all the "brainteasers" you run across. Put each one on a separate file card and keep in your puzzle corner. Students can pull out cards to work on when they have a spare moment. For example:

Cut a circular pie into *eight* pieces by making only *three* cuts.

KALAH—an ancient game of mathematical skill

JOHN B. HAGGERTY *Public Schools, Melrose, Massachusetts*

Mr. Haggerty is an instructor of mathematics in the Calvin Coolidge School in Melrose.

A large number of elementary and secondary teachers of mathematics have found it advantageous to use games as part of their instruction. The use of these games is much more than a mere whim upon the teacher's part. It has been shown that games, when properly used within the classroom environment, do actually produce an improvement in the ability known as quantitative judgment.

We realize that this term, quantitative judgment, is inclusive; there are many facets involved. The important point is, however, that the pupil is led by involvement in the game to make a decision based upon the relative position of his opponent.

The past few years have seen a sudden upsurge in the use of mathematical games in the classroom. More teachers than ever before are becoming interested in the use of games in teaching. More people than ever before are playing these games. Games such as chess and checkers are interesting, and do involve to some extent the quality which this article describes, namely, quantitative judgment. There are games which do even better in this regard. For several years we experimented with various games in the classroom. What we were looking for were games in which the element of chance was minimal or absent altogether. These are extremely hard to find and, in many cases, are not altogether practical from an educational point of view.

Quoting from notes which the author made in a class in advanced statistics at Harvard University in 1947, "Most games now known contain at least some elements of probability. To the best of my knowledge there are but a handful of games which do not, and most of these are listed in *Recreations Mathematique*, Volume III, page 105 . . . written by Lucas."[1]

In the October, 1963, issue of *Scientific American*, page 124, two of these games are described, together with two games of more modern origin. The two older games are The French Military War Game and Tafl, a game of Scandinavian origin. These games combine the characteristics of extreme simplicity and unusual strategy based upon immediate decision depending solely upon the last move by the opponent.

We have used all these games throughout the system in Grades 1 through 8 with some limited success. The games lacked in most cases an essential element which we felt was highly desirable: cultural significance and a relationship to the historical development of the number systems, bases, and systems of numeration.

Dr. Kelley did not know back in 1947 that there was, indeed, a game which possessed all of the essential elements to promote the development of pure reason in the form of quantitative judgment without a trace of chance or probability as it is known in most modern mathematics books and courses. Further, and equally important, was the fact that the game was as old as civilization itself and had been continuously played throughout the Near and Far East for seven thousand years.

The name of this fascinating game was Kalah. While we were busy taking notes

[1] Truman L. Kelley, "Advanced Statistics," Harvard University, 1947.

in Dr. Kelley's classes, less than fifty miles away from us, in Holbrook, Massachusetts, 67-year-old William Champion was developing a modern version of this ancient game after a lifetime of personal historical, as well as archaeological, research throughout the countries of the eastern half of the world.

It is a fact of recorded history that many games which were popular at one time or another became lost in antiquity. Many of these games never appeared again. We only know about them through archaeological research. So far as we know, the game of Kalah is the only game continuously played in widely separated parts of the world on at least three continents from the time of the first civilized country, Sumeria, down to the present.

The following quotation from the June 14, 1963, issue of *Time* gives us a vivid account of the background and ancient origin of this game.

Carved on a vast block of rock in the ancient Syrian city of Aleppo are two facing ranks of six shallow pits with larger hollows scooped out at each end. The same design is carved on columns of the temple at Karnak in Egypt, and it appears in early tomb paintings in the valley of the Nile. It is carved in the Theseum in Athens, and in rock ledges along caravan routes of the ancient world. Today the same pits and hollows are to be found all over Asia and Africa, scratched in the bare earth, carved in rare woods or ivory inlaid with gold.

In 1905, the year he graduated from Yale, William Champion read an article about an exhibit of African game boards at the Chicago Exposition of 1893 in which the author noted that Kalah "has served for ages to divert the inhabitants of nearly half the inhabited area of the globe." Fascinated by the failure of such a pandemic pastime to catch on in the U.S. and Europe, Champion began tracing its migrations and permutations.

He found an urn painting of Ajax and Achilles playing it during the siege of Troy; he found African chieftains playing for stakes of female slaves, and maharajahs using rubies and star sapphires as counters. He finally traced it back some 7,000 years to the ancient Sumerians, who evolved the six-twelve-sixty system of keeping numerical records.*

These people inhabited the fertile valleys between the Tigris and Euphrates rivers in that part of the world presently known as Iraq. This has long since been accepted in archaeological circles as the birthplace of civilization.

After having searched for several years for the game which was most adaptable to the modern math approach, I was naturally intrigued by the *Time* article. A letter to both the periodical staff and the author drew unexpected responses. *Time* granted unqualified authority to quote at length from the article entitled "Pits and Pebbles," but most pleasant of all was the surprise of receiving a call from the subject in person from his home in Holbrook, Massachusetts. He thanked me for my letter and invited me to meet him at the Boston Public Library where the game was being granted prime exhibit space during the month of September in the Sargent Gallery which is devoted exclusively to subjects of documented archaeological and cultural significance.

Naturally, I went to meet this remarkable man, who at the age of 83 can still run at full speed up three flights of stairs. He spelled out for me the fascinating story of his travels over much of the earth's surface in his quest of the ancient origin of this game which is still widely played in Semitic countries of the Near East and Africa.

He told me he first became interested in the game that his porters on an African trip were playing in the sand. The place was the Makahlai Desert of Africa. These Arabs called it Mancalah, or "The Game of Intelligence."

"Two players sit behind the two ranks of six pits on the board between them.

Each pit contains three (for beginners) or six "pebbles" (which may be anything from matches to diamonds). Purpose of the game is to accumulate as many pebbles as possible in the larger bin (kalah) to his right. Each player in turn picks up all the pebbles in any one of his own six pits and sows them, one by one, in each pit around the board to the right, including, if there are enough, his own kalah, and on into his opponent's PITS (but not his kalah). If the player's last counter lands in his own kalah, he gets another turn, and if it lands in an empty pit on *his own side*, [only] he captures all his opponent's counters in the opposite pit and puts them in his kalah together with the capturing pebble. The game is over when all six pits on one side or another are empty. It is not always an advantage for a player to go "out," since all the pebbles in the pits on the opposite side go into the opponent's kalah. The score is determined by who has the most pebbles. Each player cleans out his own kalah at the end of the game and replaces three pebbles in each pit on his own side. All pebbles left over represent the margin of victory."*

This method of tallying enables even very young children to play the game and, although they are unable to count formally, they know whether they won or lost by this tally. If they end with excess counters after loading each pit, they know intuitively that they have won. It is precisely this intuitive outcome of the game which we feel is one of the important outcomes of playing the game.

A sophisticated player learns not to accept all short-term advantages, however tempting. Sometimes the early gain is wiped out in the later stages. It is this fact which adds to the fascination of the game. We would invite the reader to play one full game to check this fact for himself.

Figure 1 on page 329 is a sketch of the layout of the board. These pits can be scooped out of the ground, and ordinary pebbles can be used in playing, much as in the game of hopscotch. The mathematical implications are, however, much more important than those of hopscotch.

The modern version of the game is played upon a wooden game board which is made in the loft of a factory in Holbrooks, Massachusetts. Some 24 styles of boards in all are turned out. The simplest board is routed out of ponderosa pine by automated machining. The counters are a species of Italian bead which are extremely durable and adaptable to vegetable dyeing processes. The most elaborate of these boards is carved in the form of an original Nile skiff, a craft which is so common to that area of the world.

The city of Boston's 160 playgrounds have 1600 of these game boards. Los Angeles, Cleveland, Chicago, and New York playgrounds are also equipped with this highly interesting educational-diversionary aid.

The most frequent question asked by parents is, "Will this game increase my child's mathematical ability?" Our most frequent answer is not lightly stated: "Kalah is the best all round teaching aid in the country." We are not alone in this belief. The Harvard Graduate School of Education also thinks highly of the game and is publishing a teaching manual to cover not only the rules for play but also the ancient cultural origins of the game. Also discussed in the Harvard manual are the implications of the game for the teaching of "modern mathematics."

One must again recognize that the term "Quantitative judgment" is a general term which has several facets. Among these is one which, for want of a better term, might be defined as a "quality of the mind to make a specific decision based upon the array of the opposing position. This decision is entirely intuitive and is based upon a grasp of special relationships within the scope of that which is loosely defined as 'human intelligence.' "[2]

* Courtesy *Time*. Copyright, Time, Inc., 1963.

[2] Jerome S. Bruner, *The Process of Education* (Cambridge: Harvard University Press, 1961) p. 64.

This ability seems to be best fostered in children by providing them with problem-solving situations which require them to use whatever natural intuitive skill they possess. In this sense the game of Kalah is purely mathematical, no element of chance enters in, and the basic rules are so simple that even a young pre-school child can play the game after a short demonstration.

We have found that this ancient game of Kalah not only induces very young children to think quantitatively but also develops intuitive decision-making so necessary in problem-solving. It also helps pupils think computatively. The method of play—distributing counters one by one to the right—confirms and structures the habit of moving from left to right as in reading and writing.

Small children learn to count without using their fingers; to distinguish units from multiples; to assess special array in a physical order to objects as well as to acquire other mathematical concepts in a natural way while playing an interesting and absorbing game.

The four fundamental processes in math are all used in determining the number of counters in the twelve pits at any given time. This attribute is also used in computing the score. Pure reason is vital to victory. The strategic move must be immediately evident to the player involved.

There is absolutely no element of chance in this game. It is purely a game of skill, as it has always been. The opponent moves, and the player responds intuitively to the situation at his own level of sophistication.

The number of counters in each pit can be increased to six or reduced to one for pupils of widely varying mental ages. The use of increased numbers of counters creates so many variables that no mathematician could compute the combinations. No doubt some of today's larger computers could be programmed to do this, but the probabilities are so great against there being two identical games in a lifetime.

The following demonstration game has been found effective for teaching even very young pupils how the game is actually played. An interesting projection of this game, which the classroom teacher will find useful in introducing the game to large numbers of pupils in Grades 1 through 12, will be described later in the article.

In the game illustrated Player A began by moving the three pebbles in his pit A4, ending in his Kalah and thus earning another move, which he used to play from pit A1, ending on empty pit A4 and thereby capturing B's men. By similar moves and captures, A, by the fourth turn, has become pebble-proud with eleven in his kalah to a pathetic one in B's (see Fig. 1). The sequence of moves in the sample

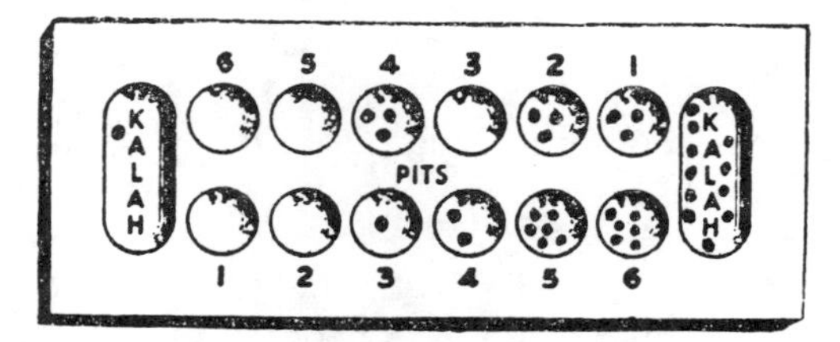

Figure 1. *The board after* A1

game is A4, A1–B6–A3, A2, A1–B4, B6, B1–B3–A6–B2, B1. By the fourth turn A is dangerously concentrated in the two pits, A5 and A6. B, seeding six pebbles on his own side, forces A to start distributing his hoard around the board. By the eighth turn (see Fig. 2) A still has twelve in his Kalah to five in B's, but B moves the five

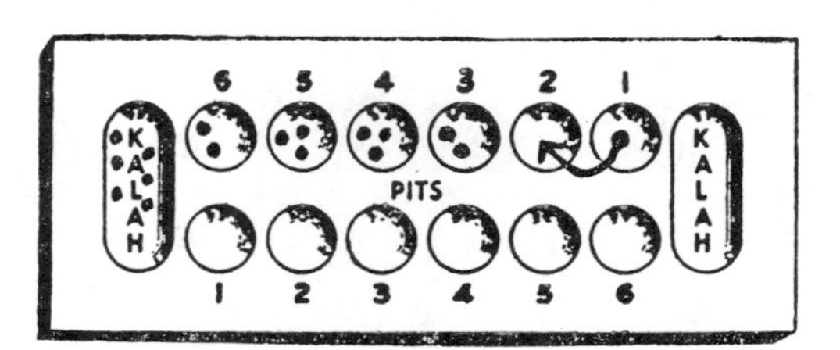

Figure 2. *B's winning move*

pebbles in B2 and then has only to move the single pebble in his pit B1 to capture A's seven remaining pebbles, ending the game and winning it by a score of 24–12.

We have found that the use of a demonstration game such as that just described is the most practical way to introduce the game to a group of pupils. In our own case two applications of this method were adopted at different levels of instruction.

For groups of pupils in Grades 4 and up we used an overhead projector with the Kalahs and the pits drawn on the platen overlay. We used corn seeds as counters which appeared on the screen as irregular opaque pebbles. As these objects were manipulated upon the platen, each move was clearly visible on the overhead screen to the watching group.

With younger pupils who have a much shorter attention span we found that an actual demonstration game on a teaching board which used separate colors for each side of the board was more interesting. The game was taught developmentally while the children observed, asking and answering many questions. This demonstration was followed by several practice games between pupils selected from the class. The two basic rules of the game were structured by discussion and additional demonstration. We then paired off the whole class at boards previously set up. Considerable interest resulted, and it seemed to be entirely spontaneous.

It has long been felt in professional circles that experiences in problem-solving and the development of intuitive powers in solving problems have been sparingly used in pre-school and Grades 1 and 2. One of the difficulties has been that many young pupils do not read well at all. Some teachers have attempted to surmount these difficulties through extensive use of diagrammatic and pictorial representation. This procedure, while useful in some situations, is generally not the way to start. We think now that a mathematical game such as Kalah has more value.

We feel that the pupils' inability to read is no justification for omitting problems from these early programs. The teacher can present these situations orally to the pupils. The game of Kalah is one of the best nonverbal means of accomplishing this end.

In addition to its value as a diversion and as a means of developing the intuitive abilities so important to problem-solving, there is another outcome equally valuable. This outcome is the recognition of the close identification of the game throughout the history of civilization with the development of systems of numeration and the concept and ideas of number. This outcome alone would make it a valuable activity for pupils in the mathematics laboratory.

At the very least it makes the classroom a more interesting place. Pupils are given an opportunity to identify themselves closely with a game which certainly has a rich cultural heritage.

We do not wish to imply, however, that mathematics can be completely taught by means of a game or any series of games. We merely think that this game of Kalah is one way of producing some desirable outcomes in the teaching of mathematics since pupils usually play it with great interest and enthusiasm.

The understanding, enthusiasm, and interest that Kalah engenders is not at all remarkable when one considers that the game has persisted for seven thousand years in widely separated areas all over the globe. We tend to agree with the ancient historian Herodotus, who was quoted as having said, "A cultural heritage may be lost once and yet be retrieved by someone of persistence and dedication, but twice lost it may prove to be irretrievable."[3] As a result of our own brief experience with this fascinating game we certainly hope that other school pupils throughout the country will soon be given a chance to play it.

[3] *World's Great Events Series*, IV, 166.

A project for the low-budget mathematics laboratory: the game of kalah

RANDALL L. BRILL

A teacher of sixth-grade mathematics at St. John Vianny School in Northlake, Illinois, Randy Brill has studied with Dr. James Lockwood at Northeastern Illinois University.

Teachers often see a kit or game that they feel would be ideal for use in the mathematics-laboratory activities of their classrooms but find it financially out of reach. By using a little imagination, they very often can make similar kits or games and at a much lower cost.

Kalah, a fascinating count-and-capture game, is easily made and offers endless opportunities for the student to develop his basic arithmetic skills while having fun. It involves the players in a strategic and logical pattern of thinking.

We are told that this game, which was first played in ancient Africa, was found to be so simple and fascinating that it became a popular form of gambling. It is said that men even determined their empires and harems over a game of Kalah. Although no longer used for such purposes, it still holds a great fascination for young and old alike and has spread to many parts of the world.

Following is a list of materials needed to make this game:

1. One piece of stiff cardboard approximately 8″ × 24″
2. Eighteen paper cups (the 4 oz. size is best)
3. Glue and cellophane tape
4. Dried beans to be used as counters

(You may want to vary the sizes or the materials to fit your situation. For instance, egg cartons are a possible substitute for the cardboard and paper cups, and marbles or beads might be used as counters.)

Cut all the paper cups so that they are ½ inch to ¾ inch deep as shown in figure 1.

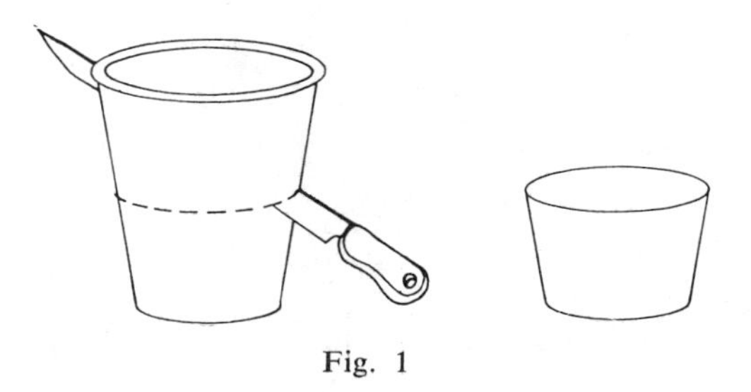

Fig. 1

Two large containers, called *kalahs,* are needed. To make a kalah, prepare three of the cups as shown in figure 2. One cup should be trimmed on both sides and the

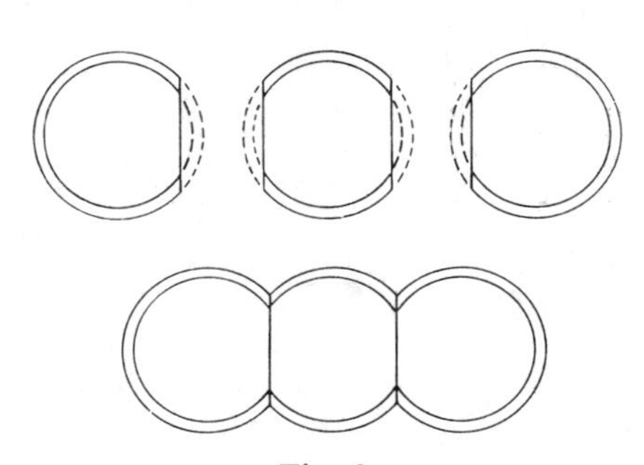

Fig. 2

other two cups should be trimmed on one side only. Taping these three cups together provides the large container that is used to gather the playing pieces during the game. (Other available materials—for example, milk shake or cottage cheese containers—can be used to form the kalahs.)

Arrange the remaining twelve cups in two rows of six on the cardboard so that the members of each pair are directly across from each other as shown in figure 3.

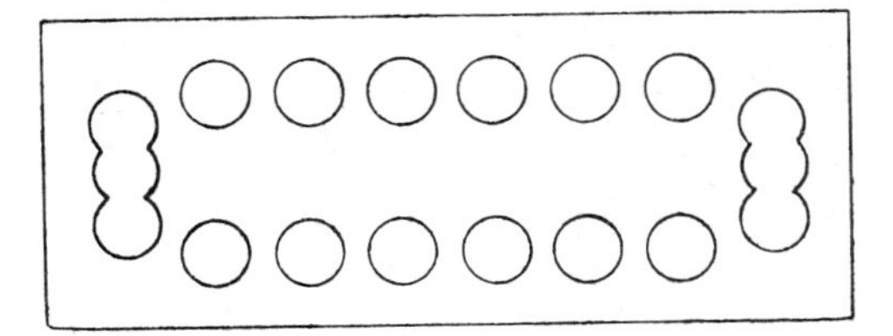

Fig. 3

The two rows should be about two inches apart. Glue the cups into place. At each end of the cardboard glue one of the kalahs. Armed with your beans, which will serve as the counters, you are now ready to play a game of Kalah.

The game is played by two players who sit on opposite sides of the playing board. Each player deposits three counters in each of the six cups on his side of the board. The object of the game is to collect as many counters as possible in the kalah, the large container, at each player's right.

The method of determining who moves first can be decided by the players. Each player, in turn, takes all the counters that are in any one of the six cups on his side of the board and distributes them one by one in each cup going to his right. If a player has enough counters to go beyond his kalah, he distributes them in his opponent's cup, skipping the opponent's kalah. Those counters now belong to the opponent.

There are two important elements that give the game its challenging strategy. If the player's last counter lands in his own kalah, he gets another turn; if his last counter lands in an empty cup on *his side* of the board, he captures all his opponent's counters in the cup opposite and puts them in his own kalah, along with his capturing piece. In figure 4, player A has emptied cup A-3 for his opening move. His last counter landed in his own kalah, and so he is entitled to another move. He now empties cup A-6 to capture the opponent's counters in cup B-4 because his own cup A-3 was empty when his last counter landed in it. All the counters from B-4 and the single counter that landed in A-3 go into his kalah. Once a counter is placed in either kalah, it remains there until the round is ended. A capture ends the move, and play goes to the opponent.

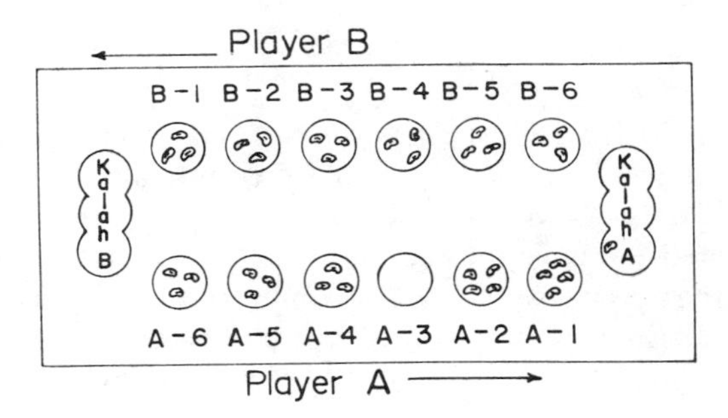

Fig. 4

A round of play is over when all six cups on one side are empty. The other player adds the remaining counters in his cups to the ones that are in his kalah. The score is determined by who finishes with the most counters. If the winning player has collected 23 counters, his score is 5 because each player begins with 18 counters. The winning score for the game can be determined according to the situation or playing time allowed, but a score of 40 is usually the goal.

As you can see, the rules of the game are quite simple, but play can become very complex depending on the abilities of the opponents. Lower grades or slow learners can begin learning to play by simply putting one counter in each cup. More experienced players or faster students can play with as many as six counters in each cup.

Three general objectives in having pupils play the game of Kalah are (1) to offer a recreational learning experience, (2) to sharpen basic computational skills and concepts, and (3) to involve students in a situation in which they use a logical pattern of strategy.

In addition to being played as an independent activity, Kalah is very effective when used as a part of a specific lesson or guided activity. It may be used to present concepts as well as to reinforce skills. Counting can be demonstrated in playing the game and tallying the score, addition and subtraction are used during play as the number of counters in the cups increases or decreases according to the moves that are made, and increasing the number of counters in the cups at the beginning of the game encourages new strategic patterns.

The board and counters can be used as a model to demonstrate other concepts, even without playing the game. The counters in a cup can become a set and the board a universe, or using simple modifications on the board, such as markings or coloring, can be helpful in discussing place value and grouping. Creative individuals may find many other variations of Kalah that will prove effective and interesting in their classrooms.

EDITORIAL COMMENT.—Many classroom games can be created using sets of objects variously called "logical blocks" or "attribute blocks." You can make an inexpensive version of these materials with file cards and the three attributes—color, number, and shape. Use two colors, red and blue; four numbers, 1, 2, 3, 4; and three shapes, ■, ●, ★. Using these characteristics, construct twenty-four attribute cards representing all the possible combinations of the three attributes. For example, you will have cards looking like these:

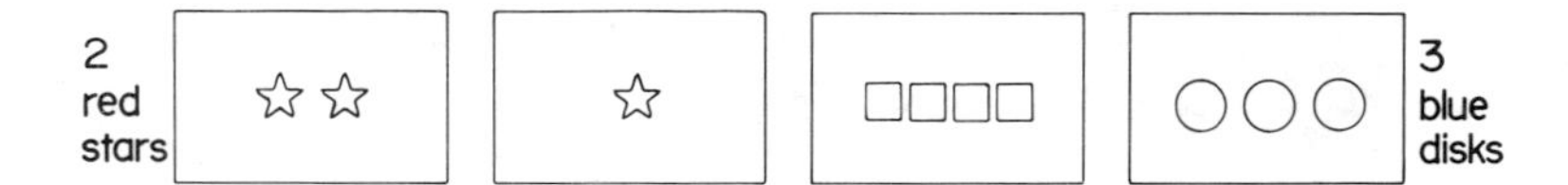

Distribute the 24 cards to a small group of children. Place one card face up on a table top. Students then take turns discarding (one card per turn) from their hands following the game rules. For example, one game involves discarding cards if they are different in one and only one way from the card on top of the discard pile. The first person to discard all the cards is the winner. Another game is based on this rule: Discard if the card is like the top card in one and only one attribute.

Students can make up their own rules. The student who made up the rule "different in two ways" soon realized that this was equivalent to the rule "alike in one way."

Drill some fun into your mathematics class

MARIE MASSÉ
Central Elementary School
West Newbury, Massachusetts

Would you like to drill some fun into your mathematics class? Put aside the usual mathematics drill and get out an electric drill, some scrap wood, some golf tees, and some dice. From these inexpensive materials you can develop a variety of mathematics games, even some that accomplish the same results as the usual mathematics drill, but in an unusually enjoyable way.

Project 1

Materials needed

A piece of wood measuring approximately 1″ by 3″ by 8″ in which two parallel rows of 10 holes each are drilled (see fig. 1), and two golf tees.

Object of the game

To land exactly on 20.

Player *A* moves his tee one, two, or three spaces; player *B* moves his tee one, two, or three spaces from where *A* landed. Play continues alternately until someone lands on 20.

The strategy, which some children see almost immediately, is to land on 16 thereby forcing your opponent to land on 17, 18, or 19, any of which leaves the next move the winning play. After discovering that 16 is a crucial point, a player must then figure out how he can control the play so that he can land there every time. Gradually it becomes clear that the multiples of four are the key to winning every time. The sly player learns that it's not only polite to let your opponent play first, it's also good strategy. The game can be varied by changing the number of holes on the board

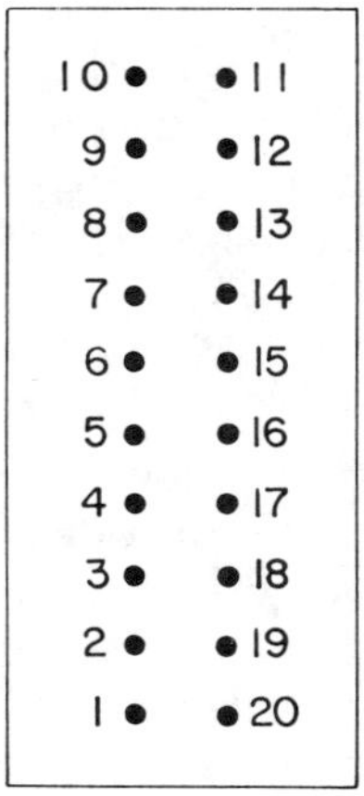

10 •	• 11
9 •	• 12
8 •	• 13
7 •	• 14
6 •	• 15
5 •	• 16
4 •	• 17
3 •	• 18
2 •	• 19
1 •	• 20

Fig. 1

or by taking moves of one through four, or more, spaces.

Project 2

Materials needed

A piece of wood measuring approximately 1″ by 4″ by 4″ in which a square array of nine holes is drilled (see fig. 2), three white tees, three red tees.

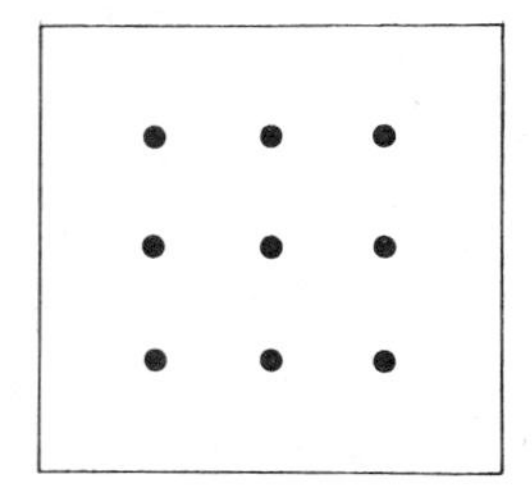

Fig. 2

Object of the game

To get your three tees in row.

Each player gets three tees. Players take turns placing tees, one at a time, on the board. When all six tees have been placed, play continues by players alternately moving tees one at a time and one space at a time in any direction until someone is able to get his three tees in a row.

Project 3

Materials needed

A piece of wood measuring approximately 1″ by 3″ by 10″ in which two parallel rows of twenty holes each have been drilled (see fig. 3) and numbered from 1 to 20; two white dice, one red die, two tees.

Object of the game

To reach 20, or beyond.

The players take turns throwing the dice. On each throw, the player moves his tee forward the total of the white dice and backward the number on the red die. The player who first finishes his play on 20 or beyond wins the game. A player does not need to end up on 20 to win.

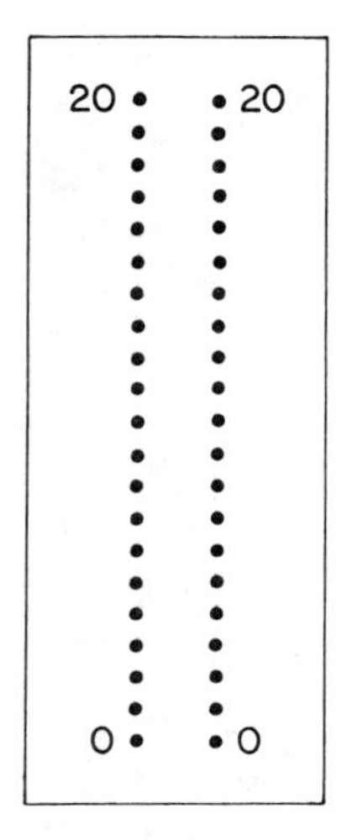

Fig. 3

Project 4

Materials needed

A piece of wood measuring 1″ by 6″ by 6″ with lines marked and intersections drilled as in figure 4, thirteen white tees, and one other colored tee.

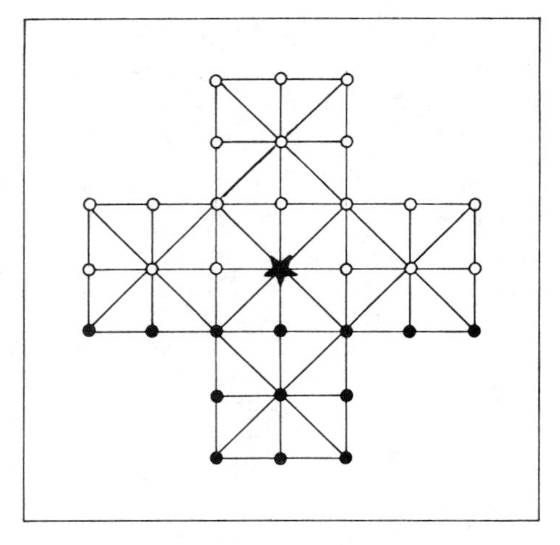

Fig. 4

Object of the game

To pen the fox (colored tee).

The name of the game is "Fox and Geese"—the white tees are the geese and the odd color, the fox. The fox is placed in in the center (see star in fig. 4). The geese are lined up on one side as shown in figure 4. This is a strategy game in which the geese try to hem in the fox. The problem is that the fox can jump any lone goose and remove him from the board as in checkers. Double and triple jumps are allowed. Clever geese should be able to trap the fox.

With these as a starter, children can design their own games. Children can mark out the boards to be drilled even though, generally, they would not be allowed to use the power drill.

Jupiter horse race

ELINOR J. WRITT

Brunswick Junior High School, Brunswick, Maine

Eighth graders at our school recently had a great deal of fun trying to solve a problem in logical reasoning that was invented by one of their classmates. Perhaps your students will enjoy solving such a puzzle, and they might possibly be motivated to create similar ones. The problem is presented here as it was submitted by the student.

Jupiter horse race *by Oliver Batson*

(Do not be put off by the weird things you see in this puzzle.)

Fill out the chart below.

Place	Owner	Stall number	Name of horse	Color of ribbon	Age	Sex	Color of hair
First							
Second							
Third							
Fourth							
Fifth							

Clues:

1. Jupiter horses all foal at the age of three, with one exception.
2. Prepon has the middle stall.
3. First place is green.
4. Curly came second in his/her heat, but did better in the finals.
5. Speedball lost to Running Thing and Prepon by an inch.
6. Mr. Smith washed his horse's ribbon, and his white socks turned pink.
7. Mr. Peters collected 5th-place money.
8. Curly is a celebrated stallion.
9. If Running Thing had come in fifth, Runner wouldn't have placed.
10. Prepon hated Running Thing, and kicked his/her wall.
11. Prepon loves Curly because of his/her polka dots.
12. Speedball was born two years before the latest leap year.
13. Prepon's ribbon is a mixture of black and blackish-black.
14. Mr. Jones' favorite color is violet.
15. Mr. Peters and Mr. Evans were seen betting together.
16. This was Prepon's first race.
17. Running Thing is the same age as his/her stall number, less one.
18. Runner foaled last month.
19. Runner is his/her mother's first foal.
20. Winners of horse shows are put in stall two.
21. Speedball lives next to Prepon.
22. Runner wishes that his/her ribbon wasn't the same color as his/her coat.
23. Runner lives next to the 3rd-place winner.
24. Purple horses always win horse shows.
25. When Prepon flies, you lose him/her against the clouds.
26. Mr. Peters bet $50 that Curly would come in second.
27. 4th place brought a blue ribbon.
28. All Jupiter horses race at the age of one.
29. Curly won a prize for one-year-olds last year.
30. The electronic picture put Prepon ahead of Running Thing a tiny bit.
31. Runner has the color of a tree trunk.
32. Mr. Evans jeeringly calls Speedball a flower.
33. Mr. Evans got $50 on a bet he made on his horse.
34. Running Thing is in heat.
35. Speedball was the head of a wild horse band.
36. Running Thing won first prize for beauty two months ago.
37. Mr. Davis owns a female horse who is one of the five winners.
38. Runner is a year younger than his/her mother, a Jupiter horse, when she foaled.

Note 1: The stalls are in this order: 1, 2, 3, 4, and 5. You must find out which horse goes in which stall.

Note 2: Placing means coming in first, second, third, fourth, or fifth place.

Solution:

Place	Owner	Stall number	Name of horse	Color of ribbon	Age	Sex	Color of hair
First	Evans	5	Curly	green	2	M	polka dot
Second	Davis	3	Prepon	black	1	F	white
Third	Smith	2	Running Thing	red	1	F	purple
Fourth	Jones	4	Speed-ball	blue	3	M	violet
Fifth	Peters	1	Runner	brown	2	F	brown

A Coin Trick

Suppose you have nine coins identical in size and appearance but one is counterfeit and heavier than the other eight. How can you with a simple balance scale locate the counterfeit coin in just two balance tests? For your first step place any three on one side of the balance and any other three on the other side. Now complete the test and find the counterfeit coin.

Paper, pencil, and book

HENRY LULLI

Henry Clay Junior High School, Los Angeles, California

This is a mathematical game that has been used successfully in the classroom. It was tried with various levels of seventh and eighth grade classes at an inner-city school.

The origin of the game dates back to medieval times. The earliest reference is in the eighth century. Many versions of the game have been published and sold since 1900.

In the present version of the game, the code key has been alphabetized, which makes it easier for the students to memorize. The game uses a box with twenty-four counters—paper clips can be used as the counters—and three items: a sheet of paper, a pencil, and a book. (Any items could be used. Cards with the names or pictures of the items could also be used if desired.)

To play the game, choose three students, *A, B,* and *C*. Give *A* one counter, *B* two counters, and *C* three counters out of the box. A fourth student turns his back to the class while each of the first three students chooses one of the three selected articles. The student who chose the paper then takes from the box of counters as many counters as he was given initially; the one who chose the pencil takes twice as many counters as he was given; and the student who chose the book takes four times as many counters as he was given.

After the three students have each taken the appropriate number of counters, the fourth student turns around and counts the counters remaining in the box. With the aid of the code key in figure 1 he instantly tells which student took which article.

Number of counters	*Code key*	*Articles*
1	AERO	Paper, pencil, book
2	BEACON	Pencil, paper, book
3	CANOE	Paper, book, pencil
5	ENCOMPASS	Pencil, book, paper
6	FLOATER	Book, paper, pencil
7	GOLDLEAF	Book, pencil, paper

Fig. 1

Number of counters taken by			*Counters left over in box*
A	*B*	*C*	
1	4	12	1
1	6	8	3
2	2	12	2
2	6	4	6
3	4	4	7
3	2	8	5

Fig. 2

Suppose three counters remain in the box. The code key for 3 is CANOE. This means that the first student, *A*, took the paper; the second student, *B*, took the book; and the third student, *C*, took the pencil.

Each code key has three vowels. Each vowel corresponds to one of the three items—in this case, A corresponds to the paper, E to the pencil, and O to the book. The order of the vowels in the code key indicates the distribution of the three items.

Notice that there is no code word for 4. It is impossible for four counters to remain in the box. Why? A careful analysis of the situation provides the answer.

Take any of the six possible combinations of the numbers of counters taken by *A*, *B*, and *C*. (See fig. 2.) The total number of counters remaining in the box after the first 6 are distributed (1 to *A*, 2 to *B*, and 3 to *C*) is 18. If the sum of the numbers of counters taken by *A*, *B*, and *C* is subtracted from 18, the difference is from 1 to 7, but never 4.

References

Ball, W. W. R. and H. S. M. Coxeter. *Mathematical Recreations and Essays*. London: Macmillan, 1942.

Gardner, Martin. *Mathematics Magic and Mystery*. New York: Dover Publications, 1956.

Gibson, Walter. *Magic with Science*. New York: Grosset & Dunlap, 1968.

9

Multipurpose Games and Puzzles

By the time you have reached this point, you have probably said to yourself many times over, "You know, if we just changed this a little, we could use that game for . . ." To encourage this kind of thinking, we included in this last section a collection of games whose utility seems multifaceted. Articles appearing in this section were selected because—

1. they describe several games;
2. they show how one game has many uses;
3. they describe a game with a multipurpose potential.

By this time, you have also found that games wisely used in the mathematics classroom provide—

1. both diversion and motivation;
2. a technique for accomplishing the necessary repetition for sound drill and practice;
3. an intellectually honest means for developing new mathematical ideas.

As you read through the articles of this last section, keep in mind the more serious purposes of instructional games—initial teaching, drill, and evaluation—as well as the obvious and much-desired characteristic of fun.

To create multipurpose games for mathematics instruction, we borrow heavily from games and diversions in our ordinary lives. We use action games, especially for the early grades. We adapt all kinds of games that use playing cards. Athletic games—baseball, football, the Olympics—are a rich source of ideas. Television has more games than you can count on your fingers. Children are fond of creating their own games patterned after popular TV shows. Familiar games such as dominoes, checkers, and Monopoly are highly versatile. And the ever-popular crossword puzzles and crostics find their matches in cross-number puzzles and "mathematicalosterms." Where would we be if we couldn't find some use for dice (which you may prefer to call "number cubes"). Familiar teaching aids, such as the geoboard and geometric shape cutouts, provide the raw material for many activities. For those who enjoy brainteasers, there is a multitude of number oddities, codes, and riddles to ponder. Even the bulletin board can do a turn as a mathematical game board!

We hope that you have found some of these games useful and that you have been able to create some of your own. Remember—be sure to share your ideas with other teachers. After all, this collection was created from the ideas of classroom teachers.

Active games: an approach to teaching mathematical skills to the educable mentally retarded

GEORGE R. TAYLOR and
SUSAN T. WATKINS

As associate professor and chairman of the Department of Special Education at Coppin State College in Baltimore, George Taylor teaches methods courses in special education to both undergraduate and graduate students. He has taught both educable and trainable retarded children in the District of Columbia Public Schools. Susan Watkins is a teacher of the educable mentally retarded at Shady Side Elementary School in Prince Georges County, Maryland. She has used the active-game idea for a number of years in her classes.

Pragmatic experience has shown that curriculum adjustment must be made if schools are to meet the everyday needs of educable mentally retarded children. Many retarded individuals have been unable to cope successfully with the curriculum of the regular grades; they are too different from the average child to adjust to the usual academic demands. For this reason various forms of instructional procedures have evolved. "Active games" are an example.

The active-game approach to learning is concerned with how children can develop skills and concepts in various school subject areas while actively engaged in game situations. Although all children differ in one or more characteristics, the fact remains that they are more alike than they are different. One common likeness of all children is that they move and live in a mobile world. The active-game approach to learning is based essentially on the theory that children will learn better when learning takes place through pleasurable physical activity; that is, when the motor component of an individual operates at a maximal level in skill and concept development in school subject areas traditionally oriented to verbal learning (Humphrey 1970).

Research (Cruickshank 1946) indicates that there are significant differences in the ways by which mentally retarded children and normal children learn mathematics. Educable mentally retarded children tend to learn certain mathematical skills and concepts better through active games than through traditional approaches and media.

The following activities are illustrative examples of active games related to major areas in elementary school mathematics.

Number system

Objective

The student will demonstrate his ability to match numerals and sets of objects with eighty percent accuracy.

Activity: Hot spot

Pieces of paper with the numerals 1 through 10 are placed in various spots around the floor or play area. There should be several pieces of paper with the same

numeral. The teacher has a collection of large posters picturing various numbers of different objects. He shows a poster to the class. An overhead projector can also be used to show different quantities of objects. The children must identify the number of objects on the poster and then run to that numeral on the floor. A child gets a point every time he finds the right numeral. Any child who is left without a spot gets no points for that round. Any child who has five points at the end of the period of play is considered a winner.

Expected outcomes

Children are helped to count from one through ten objects and then to identify numerals that represent the number of objects. After a game, the posters can be put on display around the room along with the correct numerals. Procedures may be repeated for children who do not master the skill.

Addition

Objective

Given numbers from 1 through 9, the student will correctly add them during a specified period of time.

Activity: Addition tag

Each child is given a card bearing a numeral from 1 through 9, which he keeps throughout the game. One child is then chosen to be the tagger. He may go to any child and tag him. The tagger then adds his number to that of the child he tagged, and so on.

Expected outcomes

Such a game provides practice in adding for those who need it most. There are no penalties for children who make mistakes and have to drop out of the game. For children who score perfectly for several days, some other enrichment work or games should be planned while the other children are playing addition tag.

Subtraction

Objective

When presented with simple problems in subtraction, the child will demonstrate his understanding by giving the correct answer eighty percent of the time.

Activity: Musical chairs

Chairs, one less than there are children playing the game, are arranged in a circle, facing out. As the music plays, children walk, run, or hop around the chairs. When the music stops, all children try to find a seat. The child left without a chair drops out of the game and takes a chair out of the circle of chairs. The teacher may also take out two or more chairs at a time. The activity may be repeated as many times as seems appropriate to the teacher.

Expected outcomes

Children see concrete examples of subtraction; chairs are taken away. The teacher may have the children determine the number of chairs left each time the game is repeated. Results may be recorded on the chalkboard, and later on chart paper, for children to review at their seats.

Multiplication

Objective

Given a pair of one-digit numbers, the student will demonstrate his knowledge of simple multiplication facts.

Activity: Call ball

The children stand in a circle. The teacher stands in the center of the circle with a ball. The teacher calls out a combination (for example, 2×3) and bounces the ball to a child in the circle. The child must try to catch the ball and give the correct answer before the teacher counts to ten. One point is given for knowing the fact and another for catching the ball. The teacher may elect to have a child be the one in the center calling the problems and bouncing the ball. In such case, it should be emphasized that all children in the circle should be given a chance to catch the ball and answer problems.

Expected outcomes

This game is a little easier for children to play than the regular game of call ball, where they have to be very quick to remember the answer to a given problem. This activity helps to provide the repetition necessary to develop quick recall of the multiplication facts. Addition, subtraction, and division facts might also be used for this game.

Fractions

Objective

The student will demonstrate his knowledge of simple fractions by being able to verbalize values, to know that fractions are part of a whole, and to express wholes in simple fractional parts.

Activity: Hit or miss

Children are divided into teams. Each team is given an eraser, which is set on a table, a chalk tray, or the floor. The teams stand in rows a specified distance from the erasers. Each child is given three or four erasers. He tosses the erasers, trying to knock down the one set up on the table. As each child plays, he calls out his score, expressing it as a fraction. One hit in four tries equals one-fourth; three hits in four attempts would be three-fourths, and so on. The team with the highest score wins.

Expected outcomes

Children can learn to use the relationship between successful and unsuccessful attempts at making points to develop an understanding of fractions. The children can use this concept in those situations in which they are practicing skills; shooting baskets, for example.

Measurement

Objective

The student will demonstrate his knowledge of linear measurement (centimeters and meters) by being able to use a meter stick to measure distances from various points.

Activity: Ring toss

A regular ring toss game can be used for this activity. A meter stick is needed. The class is divided into two teams. Each team has a ring, and both teams try to ring the same post. The first child on each team takes a turn tossing his ring. Each child then takes the meter stick and measures the distance between the post and his ring. The ring closest to the post scores one point for the player's team. Children should use centimeters or meters, and tenths of hundredths of meters. The team with the highest score wins.

Expected outcomes

Each child has the opportunity to develop his measuring skills when he determines the closeness of his ring to the post. To get additional practice, a child could check his opponent's measurement.

Telling time

Objective

The student will demonstrate that he can correctly and with little hesitation tell time by the hour and that he understands the term *clockwise*.

Activity: Tick-tock

The class forms a circle that represents a clock. Two children, called Hour and Minute, are runners. The children in the circle chant "What time is it?" Minute then chooses the hour and calls it out (for example, six o'clock). Hour and Minute must stand still while the children in the circle call "one o'clock, two o'clock, three o'clock . . ." until they reach the time that Minute chose. When the count gets to the chosen hour, the chase begins. Hour chases Minute clockwise around the outside of the circle. If Hour can catch Minute before the children in the circle once again count up to the chosen number, he chooses another child to become Hour. The game can also be played by counting by half-hours.

Expected outcomes

Children are exposed to and given practice in telling time by the hour, as well as developing and understanding the term *clockwise*. Children may play this game independently to review or reinforce previously learned skills.

Monetary system

Objective

Given symbols that represent coins (pennies, nickels, dimes, quarters, and fifty-cent pieces) the student will demonstrate that he recognizes them and that he can correctly use their values in different combinations.

Activity: Banker and coins

The teacher has a set of signs denoting the basic coins or their values: 5¢, five cents, nickel, penny, 1¢, one cent, and so on. One child serves as the Banker. The Banker calls out a coin. The children then run and group themselves with other children until their group amounts to the value of the coin called. Every child who is a part of a correct group gets a point.

Expected outcomes

The children become aware of the terms for and value of different coins. The children get practice in combining different amounts of money to arrive at a specific amount. To check that a group is correct, the teacher can help the whole class count it out. In this manner children can be helped to see how it takes two nickels to make one dime, or five pennies and a nickel. And they learn that "five cents" is the same as "one nickel."

Implications

In planning for learning through active games, an attempt is made to arrange an active learning situation so that a fundamental skill or idea is being practiced in the course of participating in the active game situation. Activities are selected on the basis of their degree of involvement of (1) a skill or concept in a given school subject area, and (2) physical activity appropriate to the physical and social developmental level of a given group of children.

Carefully selected active games provide purposeful and pleasurable experiences. They involve the learner actively and give the child the opportunity to dramatize the idea. Active games reinforce a concept and develop skills by providing the repetitive drill that educable mentally retarded children need.

Since educable mentally retarded children are more severely handicapped in the study of mathematics than the majority of children, it is suggested that the use of purposeful active games for the development of mathematical concepts is a very effective instructional process.

References

Brueckner, Leo J. *How to Make Arithmetic Meaningful?* Philadelphia: John C. Winston Company, 1947.

Cruickshank, William M. "A Comparative Study of Psychological Factors Involved in the Responses of Mentally Retarded and Normal Boys to Problems in Arithmetic." Doctoral Dissertation, University of Michigan, 1946.

Hood, H. "An Experimental Study of Piaget's Theory of Development of Numbers in Children." *British Journal of Psychology* 53 (1962): 273–86.

Humphrey, James H. *Child Learning Through Elementary School Physical Education.* Dubuque, Iowa: William C. Brown Company, 1966.

——— and Dorothy D. Sullivan. *Teaching Slow Learners Through Active Games.* Springfield, Ill.: Charles C. Thomas Publisher, 1970.

More games for the early grades

EDWINA DEANS

Guessing and checking

Play a game of guessing and checking the weights of different children and of different objects in your room. Let several children lift an object in turn. Each will guess the weight and record his guess. Then the object is weighed to see which children came closest to the correct weight. Children are encouraged to use good judgment in making guesses. For example, if a child who has just been weighed stands beside the child who will be weighed next, there is some basis for comparison.

If children are to improve in the ability to make approximate estimates, they must discover and keep in mind that large objects are not necessarily heavy nor are small ones always light. They must learn to estimate weight by actually lifting objects or by some visual comparison of similar objects.

"Guessing and checking" can also be played with measuring. Mark off nine feet along the floor to serve as a model measure just as a ruler or yardstick. The door, a window, the chalkboard may be measured and marked as other model measures. Provide rulers, yardsticks, and tape for measuring stationary objects. Guess the width, length, and height of various objects, record the guesses, and measure to determine correct measurements.

Many of our decisions with regard to number experiences are reached, not by the pencil and paper method, but by rapid, rough calculation which gives a working estimate of the amount. Any number learning which leads the child to improve his ability to reason sensibly and arrive at a fairly accurate estimation will prove of infinite worth to him.

Button game

To play this game, assemble a paper cup, a paper plate, and six to ten buttons (two or three different colors will facilitate the recognition of groups). Glue the paper cup exactly in the middle of the paper plate and allow it to dry thoroughly.

Two to four children play. The players take turns throwing the buttons one at a time into the paper cup. The exact distance to stand from the plate is determined, measured accurately, and marked. Scoring for this game is done according to the desired scoring device. Each child playing will determine these facts: "How many did I throw into the cup?" "How many did I miss?" "Do the ones in the cup and the ones outside equal the total number I had to begin with?"

Hauling lumber

Individual children can benefit greatly from stimulating game-like experiences which are designed to help them overcome some difficulty in number understanding.

Anita Riess gives a report of a retarded child of eleven who was helped to build up an understanding of our number system and of money values by means of a game necessitating the use of money and of tens. A miniature lumber yard was set up and the lumber (wooden sticks) was bundled into tens. This enabled the lumber man to ship his lumber more easily. A toy truck was used for making deliveries. The boy, pretending that he was the truck driver, did the bundling, the loading and unloading, and kept account of his deliveries. At the beginning he drew heavy black lines to stand for each bundle of ten and thin lines to stand for the single sticks. Payment was made in pennies only, or in dimes for the bundles and pennies for the single sticks. Finally, figures were used to sum up the number of bundles and single sticks which were checked with the number of dimes and pennies.

I'm thinking of a number

One child says, "I am thinking of two numbers that make six. What are they?"

The leader answers "No" until the correct guess has been made. The child who guesses correctly then takes the place of the leader.

Some concrete means of working out or demonstrating wrong answers should be provided for the children. For example, if a child should say, "Are you thinking of four and three?" the leader will challenge him to prove or demonstrate by a request such as, "Show us four and three. See if it *could* be six."

The child will use sticks, blocks, or other available materials to demonstrate and to check his thinking.

This game can be played by any number of children. After it is well learned by the children, it may be played in small groups of six to ten to give all children an opportunity for active participation. It can also be adapted for use with larger addition combinations and with column addition and multiplication.

"I am thinking of two numbers that make thirteen."

"I am thinking of three numbers that make nine."

"I am thinking of two numbers that multiplied together will make twenty-four."

It is recommended that children be encouraged to use the language form "eight three's are twenty-four," or "six four's are twenty-four," which immediately suggests to the child a method of checking his answer by the use of drawings, a counting frame, or a similar device.

The guessing element comes into play as the children attempt to guess which of the combinations that make up a given number the leader has in mind. The teacher helps the child analyze the guesses that show lack of understanding so that the individual child who needs help will profit from the game.

Number puzzles

Make a set of disks the same size. Draw sets on the disks—four dots, five crosses, six squares, five cars, etc. Divide each circle into two parts. Be careful that no two circles are divided in exactly the same way (Fig. 1).

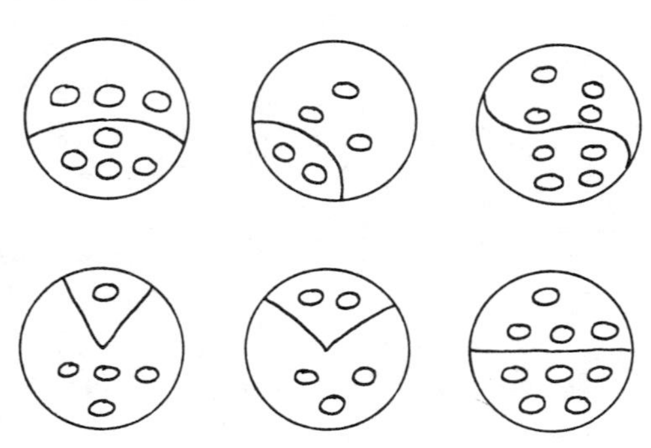

Figure 1

Two children work together as partners. They mix the pieces well and place them face down. Then they take turns drawing and matching the pieces. When the circles are all together, the children name the sets on each of the two pieces of the puzzle and the total number represented on the disks. If one player can catch the other one in a mistake, he can take the circle. When all the disks have been named, the person who has the most disks is the winner.

Similar puzzles can be made from other shapes, such as squares and oblongs.

Shooting rubber bands

Games invented by teacher and children to serve a specific purpose are often great fun for all. A precocious kindergarten youngster arrived at school one morning with a broken broom handle and a supply of rubber bands. He was anxious to demonstrate his skill in shooting the rubber bands off the end of the broom handle. The teacher, wary of his ability to keep within bounds, drew a target on tagboard. The outside area of the circle was painted red, the next circle blue, and the inside circle white (Fig. 2). The children and the teacher decided that the little circle should count the most because it was hardest to hit. Values decided upon for the white, blue, and red areas were five, four, and three points, respectively.

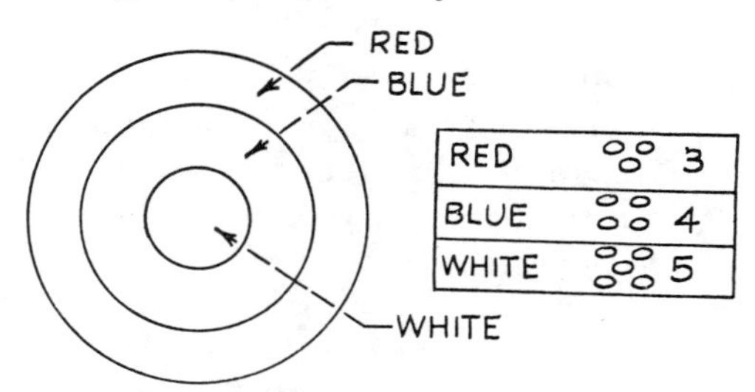

Figure 2

The target was placed on the floor. A chart on the blackboard served as a reminder of the number values. The children took turns shooting the rubber bands. The idea was to hit the bull's eye and get a score of five if possible. Interest in this game lasted for days.

An ingenious teacher had turned a potentially troublesome situation into a worthwhile learning activity.

A deck of cards, a bunch of kids, and thou

PETER K. GURAU
Springfield College, Springfield, Massachusetts

Peter Gurau is an assistant professor of education at Springfield College.

One of the more frustrating problems faced by the kindergarten and first-grade classroom teacher is the problem of providing for individual differences.

All too often the range of initial abilities of a kindergarten class is just large enough to invalidate the best-planned lesson. In arithmetic, for example, we may have one child who cannot even count when he enters the kindergarten or first-grade classroom, while another may actually be able to do simple addition and subtraction, and more. The problem is complicated by the fact that youngsters of this age group have not as yet learned how to sit still for any great length of time, and thus the more we can individualize their instruction, the better the chances that we will be able to hold their attention. Finally, if a lesson can be presented in pure game form, the lesson is much more likely to go down smoothly and without emotional or academic indigestion.

The following series of card games is offered almost as a curriculum in beginning mathematics. It presupposes only that the youngsters have all learned to recite the counting numbers up through 10—an event which is usually quite easy to achieve through a series of counting songs.

The stage that follows rote counting is usually considered to be symbol recognition. To implement the first stage, seat the youngsters around tables, with three players at each. Every table will have a deck of cards from which all the jacks, queens, and kings have been removed. In each instance, divide the deck into the four distinct suits and give each of three children a suit, keeping one suit aside. Now arrange the cards in the extra suit in order of size, from ace to 10, and place them face up on the table. Explain to the children that the object of the game is to match all the cards in the players' decks with the cards already on the table. Each child in turn places a card from his deck on top of the appropriate card on the table. The first few times, the children should receive their suits in order of size so that they will be able to associate their rote counting with the game. Each child will then in turn place his first card (the ace) on the ace on the board and call out, "One!"[1] In the same manner, they will next place their 2's on the 2 on the board, and call out, "Two!"

After having the children play this game once, twice, or three times, you can add a variation by having the cards preshuffled. Now each child will come up with a different card and should try to call out the right name for the card as he places it on the right card on the board. However, no great stress should be placed on calling out the correct name of the numeral, since

[1] It should be noted that little trouble is encountered in the ambiguous labeling of the ace as a "1." However, if you feel that it is preferable to have a "1" symbol, it should be separately constructed.

at this stage the primary objective will still be to have the youngsters recognize the visual symbol subverbally. You, as a teacher, can go around acting as referee. In the process, you will have an automatic coding device telling you which youngster is not placing the right numeral on the appropriate pile (because of the suit designation). Thus, if Johnny has the hearts at Table 3 and is placing a 6 on the 9's pile, you can make a mental note of that fact. The most common errors at this stage are reversals of 6 and 9, reversals of the ace ("A") and 4 (or of the ace and 7), and an occasional reversal of 3 and 8.

Within a very short time the children can not only recognize the symbols but also name them. Those children who, at this stage, still need more practice may be regrouped for more practice with this game, while the remainder of the class go on to the next game.

The next game can still be played in a group, but it requires each child to work with his own suit in front of him. Receiving the suit in preshuffled form, each youngster is instructed to lay the cards out before him in order of size (i.e., A, 2, 3, 4, 5, 6, 7, 8, 9, 10). If competition is deemed appropriate, then a speed contest can be instituted, with the winners from each of the tables competing with each other in an elimination contest.

Once the children have gained proficiency with this game, a game of "magic" is in order. This game starts off exactly like the previous one, with each child laying the cards out in front of himself in order of size. Next, all ten cards are turned face down on the table, still in order, however. Some youngster starts off the game by pointing to one of the cards in front of the child to his left. The child to his left must then "predict" what numeral is under the card, with points or prizes of some kind being awarded for correct responses, *if* the teacher desires it. In effect, each child is a magician who can tell what is under a given card by secretly counting. Youngsters enjoy this kind of game immensely and are likely to want to play it for long periods of time, especially for visiting "firemen" such as parents, the principal, or other teachers.

An alternate game—one which also stresses symbol recognition—is "Concentration." This game can be played with two to five youngsters working on one deck of cards. The deck is shuffled and laid out in a face-down, 5-by-8 array. One youngster starts the game off by turning over any two cards. If both cards are numerically equal, he takes them as his winnings and may go again. As long as he continues to pick up equals, he may continue to go on. However, once he misses, the pair of cards are again turned face down, and the next youngster goes. Memory plays a major part in this game, and you will find that children are better at it than adults. I know that I have been beaten repeatedly by my youngsters, despite my most sincere efforts to win.

By this time, youngsters have mastered symbol recognition and the appropriate name for each symbol. In addition, they are able to place the numerals in correct order of size. This stressing of the serial order of the first ten numbers leads to the concept of inequality and size.

Children who have mastered the previously mentioned games usually have no trouble whatsoever with the game called "War." While War can be played with more than two participants, for the sake of simplicity it will be explained for two players. Each child gets half the deck in a preshuffled form. Both players hold their decks face down and simultaneously reveal one card apiece. The player whose card has the higher numerical value wins both cards, which he places on the bottom of his deck. To prevent controversy over which of two numbers is the greater, each youngster could be provided with a picture of the number line. Ties are settled in the following manner. If both players display a three, for example, each player counts out four more cards, which he places face down in front of himself, and then places a fifth card face up. These fifth cards then determine the outcome of the tie. Of course, should another tie result, the procedure is repeated. Children tend to enjoy this game immensely and are likely to wish to play the game long after the apparent pedagogic values seem to have disappeared. Again, it is possible to make this sort of situation into a competition by having the winners from one group play the winners from another group until all but one individual are eliminated. A natural extension of this game, as well as of all the previous games, would be to create more cards by adding the numbers from 11 through 15, or even from 11 through 20.

A follow-up of the game "War" actually involves the child in a form of subtraction, or at least in thinking in terms of addition and subtraction. Actually, the rules of the game are identical to those of "War" except that the player with the card of higher value gets as many cards from the loser as are represented by the numerical difference between the two cards. Thus, if the players show respectively a 3 and a 7, the player holding the 7 gets four cards from the player holding the 3, plus the originally played cards. Chips can, of course, be substituted; and variations on the rules can lengthen or shorten the game.

The presentation of these materials can be so designed as to allow advanced learners and youngsters new to numerical concepts to work together in harmony, thus providing for productive peer learning.

Arithmetic card games

MARTIN H. HUNT

Boston Public Schools, Boston, Massachusetts

We all realize how important it is to provide youngsters with a variety of reinforcing opportunities that promote the acquisition of a knowledge of number facts. Playing arithmetic card games that have been adapted from games familiar to children and young adults not only is an enjoyable social experience for pupils, but serves as an interesting school or home activity that enhances facility with the number facts.

Because each card is imprinted with both numerals (the ace represents a one) and the equivalent number of elements (spades, hearts, clubs, and diamonds), the child who has not memorized a particular fact can compute it by counting. Any involvement in games, however, soon makes it clear that "knowing the number facts well" makes the game more enjoyable—and winning easier.

The five games described below were initially introduced at a recent summer program and should not be considered a complete list by any means. Both students and teachers should feel free to design new games or modify any of the given ones. While we encourage children to play the games, we would never compel any pupil to participate, as this would detract from the value of the activity as a game to be enjoyed. Generally, the children will be employing the aces through the 9's. However, if the goal is to achieve a combination equaling a number less than 10, the highest valued card used would be that which is one less than the desired goal. For example, if the purpose is to combine cards equaling 5, the 4 would be the highest card used.

War

The deck is divided equally among the players. Each player in succession turns over a card from his pack, which is held face down. The player with the highest valued card takes in the set, and either the player who ends up with all the cards or, if a time limit is set, the player with the most cards at the end of the time, is the winner. If two or more players turn over cards of equal value, "war" results. Two cards from the tops of the packs belonging to these latter players are discarded facedown, and the third card is played faceup. The player engaged in "war" whose third card is greatest in value takes in all the cards played. This game enables the children to learn to recognize numerals and their ordered relationship. It promotes an understanding of the concept of "equals," "greater than," and "less than."

Fish

Depending upon the needs of the children, the goal might be to obtain "runs" of three cards (A, 2, 3; 2, 3, 4; 8, 9, 10) to find two cards whose difference equals a specific number, or to obtain combinations of cards whose values equal a given number. Each player in turn asks another player for the needed card. If the card is not available, the requesting player selects the top card from the closed deck. The player with the most "books" at the end of the game is the winner.

Number Trick

Each player discards one card at a time, adding the value of his card to the one previously thrown by the player before him. The "trick" is made when a player, through the card he plays, succeeds in getting the cards to total a given number, such as 10 or 21. An alternate version involves having the cards totaled until the sum falls between 14 and 20, then subtracting the value of each successive discard until the value of 13 is reached. If one wishes to concentrate on subtraction skills, the players might begin with 25 and then subtract the value of each card until 3 is reached. Chips may be given each time a "trick" is completed.

Number Scrabble

This game is played in a manner similar to the commercial game of "Scrabble" in that a player attempts to complete combinations of numbers (instead of letters) equaling a designated sum such as 5 or 10. For more able students, players may attempt to make the combination of cards equal 85, using any combination of the arithmetical processes. Each time a player successfully combines the cards, he receives a chip and draws the appropriate number of cards from the deck to replace those he has just used. After all the cards have been removed from the deck, the student with the most chips is declared the winner.

SAMPLE COMBINATIONS

(*a*) $(9 \times 9) + 4 = 85$
(*b*) $(3 \times 4 \times 7) + 1 = 85$
(*c*) $(7 + 6) \times 7 - 6 = 85$

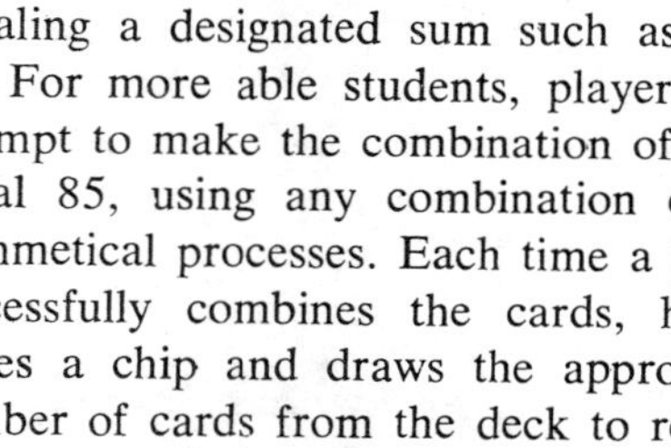

(*b*)
[3]
(*a*) [9] [9] [4]
(*c*) [7] [6] [7] [6]
[A]

Number Rummy

Each player is dealt seven cards. The first card from the remaining closed deck is turned over. The children in turn choose a card from the open pile of discards or from the closed deck in an attempt to obtain combinations of two or more cards that equal 5 (in the simplest version) or that calculate, when using any combination of the arithmetical processes, to 100 (a more complicated version). Again, the player with the most "tricks" at the end of the game is the winner.

EDITORIAL COMMENT.—Sporting events often provide a successful way to motivate students to do practice exercises. In the spring, baseball is appropriate. A playing field is needed—a square piece of tagboard with the bases and a pitcher's mound marked out will do. Place a spinner on the pitcher's mound.

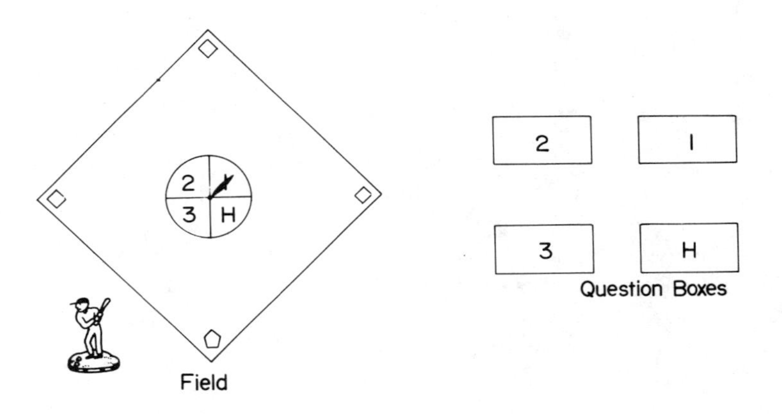

Place questions in boxes marked "1," "2," "3," and "H." The hardest problems, corresponding to homeruns, should be placed in the box marked "H." The easiest problems, singles, should be placed in the box marked "1."

To play the game, two teams are needed (the game is easier to manage if there are no more than four students on a team). Each player needs a marker. Players spin to see from which question box to choose. If the question is answered correctly, the student gets a single, double, and so on, according to the number marked on the box. Play changes hands when a team misses three questions.

Supply answers to an "umpire," give him a hat marked UMPIRE, and you're off!

Arithmetic Football

EARL A. KARAU
Saginaw, Michigan

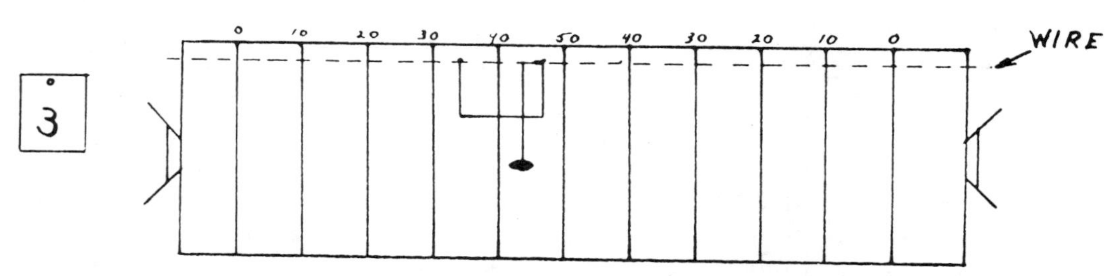

ARITHMETIC FOOTBALL is a game of mental arithmetic. It is played without the aid of pencil or paper. A football field made of roll paper or oilcloth is laid out as shown in the drawing. It is 120 inches long and 36 inches wide. The line markers are ten inches apart for easy layout of the field.

Over the top edge of the field a wire is strung with marks one inch apart. Put a small dent at every inch mark so it is easy to set the ball wire in place. One football on a moveable wire is used. The wire should be at least twelve inches long and the size of the football should be about two inches by three-quarters of an inch. This ball moves up and down the field as each team plays the part of the offensive team.

Four "down" cards are used in place of a "down" box. You hang up the number of the "down" so the quarterback will know what play to call. A wire ten inches long is used like the chain measure to mark off the ten yards to a first "down." This hangs from the top wire and must be moveable. It is placed and used as in an actual game. It shows how many yards are needed for a first "down."

You will need enough problem play cards to carry on the game. The cards are put in piles according to the type of play thereon. The following types of plays were found suitable:

		Gain
Off-guard	(Easiest problems)	1 to 3 yds.
Off-tackle	(Easy problems)	1 to 8 yds.
Reverse		1 to 25 yds.
End run		1 to 15 yds.
Pass	(Harder problems)	1 to 60 yds.
Punts		
Field goals		
Kick-off		
Runbacks		

Here are some examples of plays and their problems:

OFF-GUARD
50% of $24 = \underline{12}$
G. 3 yds. L. Fumble

END RUN
$\frac{7}{8} \div 4 = \frac{7}{32}$
G. 8 yds L. No Gain

PUNTS
$90 - 69 = 21$ Subtract
G. 45 yds L. 20 yds

PASS
$24 \times 31 = \underline{744}$
G. 51 yds. L. Inc.

RUNBACK
$\frac{1}{2} \times \frac{1}{16} = \frac{1}{32}$
G. 39 yds. L. No Gain

KICK-OFF
$4 + \frac{1}{2} + 7 = 11\frac{1}{2}$
G 35 yds. L 20 yds

Each card carries a gain or a loss. If the problem is answered correctly within the 30 second time limit you use the gain section. If the time runs out or the wrong answer is given you use the loss section. The loss may be: no yards gained, yards loss, fumble, or in case of a pass, it may be incomplete or intercepted. The cards are turned over, one at a time, so even the teacher doesn't know what problem is coming up next in any of the play stacks.

A stop watch is fine for timing but any sweep-second-hand watch will do. A fair knowledge of football is needed by the teacher as he acts as referee of the game.

Two teams are picked and a quarterback from each team is elected to run the team. The quarterback should know a little about football as he calls all the plays including the person to carry the ball. The teacher flips a coin to see which team shall kick-off. If team *A* kicks-off the moveable football is placed on the *A*'s forty-yard line. A kick-off card is picked up by the teacher and quarterback *A* picks a team member to answer the problem on the card. The problem is given, and if answered by the player, the ball is moved down the field the number of yards shown on the card. Now team *B* has the chance to run the kick-off back so the *B* quarterback picks a team member to answer the problem the teacher has picked up off the top of the run-back stack. The player *B* has 30 seconds to answer the problem. If he answers the problem correctly the ball is moved the number of yards shown on the card, but if he misses the problem the loss may be: ball down at that point or fumble, depending on the card.

Team *B* now has the ball and the moveable ten yard marker is placed so that one end is at the point of the ball and the other end is ten yards away. This is the same principle as the chain measure used at a real game. "Down" 1 is put up on a hook and team *B* is ready to try its first play of the game. Quarterback *B* picks a team player and the type of play he wants him to run. The quarterback can't call on one team member again till every one on his team has had a chance to carry the ball. The quarterback has five different types of plays he may call: off-guard, off-tackle, end-run, reverse, and pass. The easier problems give less yards on a play. The pass plays are the hardest but the greatest gains are made on this type of play. On each card is written the number of yards to gain if answered correctly and the card also carries the loss if the wrong answer is given. The teacher makes up his own problems and adds new problems as new work is learned in class. You can fit your cards to any grade level you wish. For seventh grade it is best to have more than a hundred play cards and use them over again. The students work outside of class to learn what types of problems are most difficult for them. They work on their weak points so as not to let their team down.

Field goal cards are used for the extra point when a touchdown is made. You may also let a team try a field goal if within the twenty-yard line for these cards the point is made if the problem is answered correctly and the point is missed if the problem is missed.

Penalties can also be applied, for example, if one team member helps another team member it is fifteen-yards and the "down" remains the same. If team *A* bothers team *B* by talking, team *A* is penalized five yards and the "down" remains the same. Not many penalties are marked off as each team really wants to win. The teacher may set any time limit he wishes for the duration of the game.

The use of this game in seventh and eighth grade classes has been highly successful in stimulating rapid and accurate computation without pencil and paper. Both the boys and girls enjoyed it and worked hard to improve their skill in mental arithmetic.

Television games adapted for use in junior high mathematics classes

DORIS HOMAN

Vallivue Junior and Senior High School, Caldwell, Idaho

Today, as in the past, the use of mathematical games is an effective way to provide meaningful, interesting drill for junior high students. Highly competitive by nature, students at the junior high level respond with enthusiasm to activities of this kind. When these games can be associated with familiar, pleasant experiences, there is an added motivational factor. Why not, then, relate mathematics to the most familiar activity of all—viewing television. Three games that have been adapted from television shows are described here.

For each of the games, the class is divided into two teams. Each team numbers off, and each member writes his number on a small card and puts the card into his team's box. Players take their turns at play as their numbers are drawn from the boxes.

Mathword

"Mathword," a game based on the television show "Password," provides practice with mathematical terms.

Materials:

A list of mathematical terms of either one or two words. (Examples: *equation, product, numeral, associative, commutative, ratio, inverse, prime, set, diagonal, polygon, integer, square, rectangle, prism, triangle, obtuse triangle, right triangle, re-*

ciprocal, terminating decimal, decimal, vertex, altitude, disjoint, volume, area, parallel, circle, perpendicular, quadrilateral, cylinder, circumference, diameter, radius, factor.)

Instructions for play:

1. Place a chair at the head of each team so that the occupant will face the class and his back will be turned to the overhead screen or the chalkboard.

2. Choose the first two contestants by drawing a number from each team box. The two contestants come forward and sit in the team chairs.

3. Write a mathematical term on the overhead projector (or chalkboard) so that the class can see the term, but the contestants cannot. Then turn off the overhead projector (or erase the term). Tell the contestants whether the term is one word or two words.

4. Draw another number from each team box. The two students whose numbers are drawn are the first to give a clue; the student with the lower number goes first. The first player can give the first clue, or he can pass and allow the other team to give the first clue.

5. The clues may be only one word or one number. The opportunity to give clues alternates between the teams. After one player gives a clue, another member of his team is selected, by drawing, to give the next clue for the team.

6. The contestants stay the same until the term has been identified, or until the five-point clue has been given. The game is over when the list of terms has been exhausted.

Scoring:

When a contestant gets the term on the first clue, his team receives ten points. The team gets nine points if its contestant gets the term on the second clue; eight points, on the third clue; and so on. When the five-point clue is given and missed, the game is over. New contestants are drawn, and another term is chosen.

Eye Guess Math

"Eye Guess Math" is a game based on the television shows "Eye Guess" and "Concentration" that provides drill with integers, fractions, decimals, whole numbers, and mathematical terms.

Materials:

Eye Guess Math board on an overhead projector transparency, as in figure 1 (a large piece of posterboard can be used), nine problems, overlay transparency for the Eye Guess Math board (on the overlay, as shown in figure 2, an answer to one of the selected problems is written in each of the eight outside squares. The ninth answer, the answer for the Eye Guess square, is not put on the overlay until later), and Nine cardboard squares to cover the answers on the Eye Guess Math board.

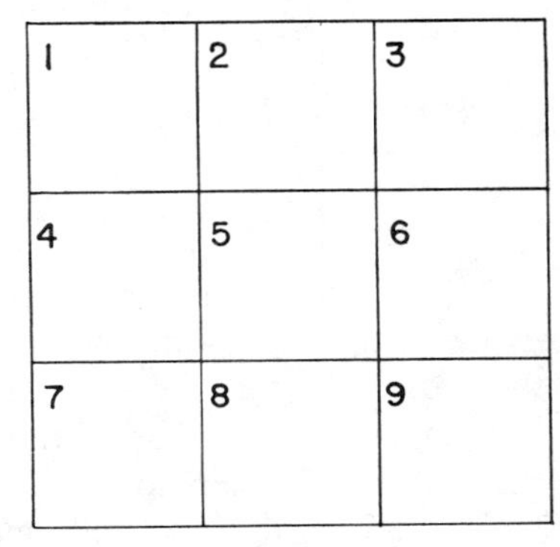

Fig. 1

Instructions for play:

1. Show the blank Eye Guess Math board to the students. (See fig. 1.) Explain to the students that answers to problems will be put in the squares on the board. The answers in the eight outside squares will then be revealed for fifteen seconds. The students are to try to remember the positions of as many answers as possible.

2. After the eight answers have been revealed, the ninth answer is put in the Eye

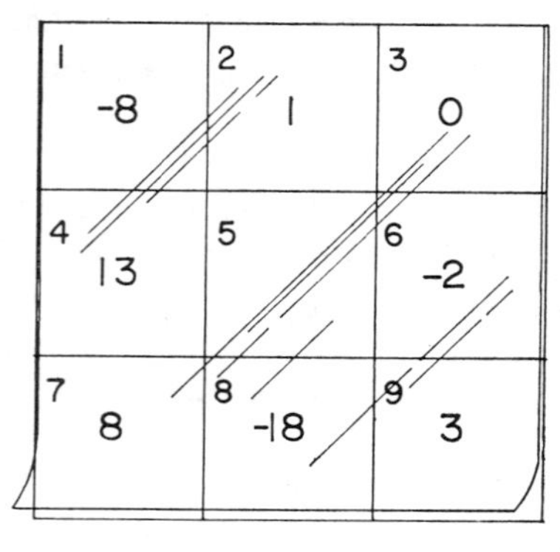

Fig. 2

Guess square, and all squares are covered with small cards as shown in figure 3.

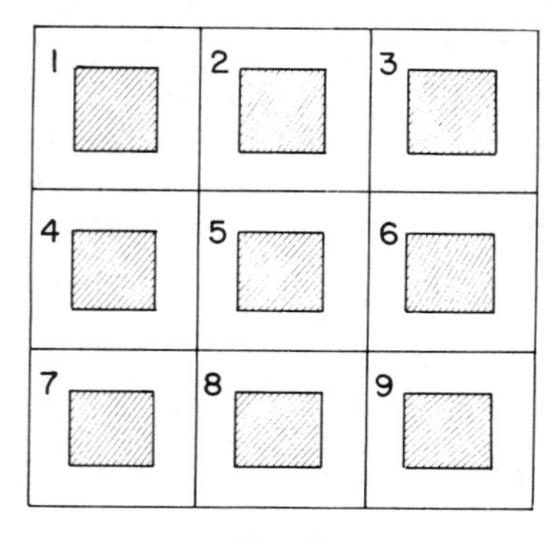

Fig. 3

3. The teacher then reads the first problem on his list. After sufficient time to work the problem has elapsed, the teacher draws a number from the box of the team chosen to go first. The team member whose number is drawn will then call out the number of the square in which he thinks the answer is located. The teacher lifts off the cover and reveals the number in that square. If the number in the uncovered square is the right answer, then the team receives one point. Right or wrong, the square is quickly covered again. The teacher then proceeds to give the next problem to the other team, following the same procedure as before.

4. When a problem is answered correctly, the teacher crosses off that problem on his list. If a problem is not answered correctly, the teacher leaves the problem unmarked on his list and goes on to the next problem. After going through the list of problems once, the teacher then goes back and uses the problems missed in the first round. This continues until all problems have been answered correctly. The team with the larger score is declared the winner.

Celebrity Math

"Celebrity Math," which is a game based on the television show "Celebrity Game," provides drill in fractions, decimals, or whole numbers. It also encourages mental calculation and approximations.

Materials:

A list of thirty problems that cannot be easily worked without pencil and paper. (Examples: $53 \times 79 =$ ____; $3798 - 59 =$ ____; $5800 \times 500 =$ ____; $19 + 78 + 53 + 98 =$ ____.)

Instructions for play:

1. Each team numbers off 1, 2, 1, 2, and so on. Each *one* is a "celebrity" for his team, and each *two* is a contestant.

2. Each student who is selected to be a celebrity must think of some celebrity that he would like to be. He writes the name of that person on the small card provided. The cards are collected, and the celebrities for each team are listed on the chalkboard.

3. The contestants (the *twos*) for each team number off 1, 2, 3, 4, and so on, and place their numbers in their team box.

4. One team is selected to go first. The game begins as the teacher draws a number from the starting team's box of contestant numbers. The contestant whose number is called selects one of the opposite team's celebrities to answer his problem.

5. The teacher then reads the problem or writes it on the overhead projector. The celebrities can use pencil and paper, but the contestants cannot. The selected celebrity works the problem as quickly as possible and gives his answer. The contestant

must immediately agree or disagree. If the contestant is correct, his team receives the point. If the contestant is not correct, the celebrity's team receives the point. The next problem is given to a contestant on the other team. A number is drawn from that team's box of numbers, and the contestant chooses a celebrity from the other team's list to answer the problem.

6. The game continues in this manner until all the problems from the list have been used, or until the time allotted for the game has elapsed.

7. Since the object of the game is to secure points for the team, a celebrity may deliberately give a wrong answer in an attempt to trick the contestant from the other team. The contestant cannot use pencil and paper; but he can approximate the answer or work the problem mentally, if he can do so in the time it takes the celebrity to give an answer.

EDITORIAL COMMENT.—In addition to the many television games of the contest type that may be adapted for use in mathematics classrooms, you should consider incorporating children's favorite TV characters into board games. Spending a few hours watching the Saturday morning cartoons will give you many ideas. If you feel that you cannot draw the cartoon figures, you can usually find a coloring book from which to trace the characters.

The concentration game

HOMER F. HAMPTON *Central Missouri State College, Warrensburg, Missouri*

The use of drill has fallen into disfavor in recent years. In spite of the shortcomings of drill in and of itself, it does seem to make a worthy contribution to a student's learning experience. Perhaps we can improve the setting in which drill experiences are conducted and benefit from drill without incurring its faults. The current image of drill is vague, but its theme could well be "reoccurring but varied contacts with a concept, skill, or procedure."

Extensive use of physical models and gamelike activities will assist a teacher in carrying out this theme. I should like to describe a game that provides for drill activities in a setting that can be both inviting and satisfying to pupils from grades two to six.

This game is patterned after a television program called "Concentration." The principal piece of equipment is a board that can be constructed from a piece of ⅜-inch plywood 18 inches wide and 24 inches long. Eyelets are used to attach small cards in a seven-by-seven array. Whole numbers are written on one side of these cards, and the cards are attached so that the numbers are not shown when the cards hang down but appear when they are turned up. The space arrangements are such that a turned-up card does not cover another card. The rows and columns are numbered, so that a pupil selects a card to be turned up by choosing an ordered pair of numbers. (See fig. 1.) With a little practice you can adjust the opening of the eyelet so that the cards can be easily and quickly replaced if a need should arise.

The concepts or skills that can be pursued with this game are limited only by your imagination. I shall describe only one in some detail to serve as a prototype.

Suppose you have a fourth-grade class and wish to improve their facility with multiplication facts of the sixes, sevens, and eights. Choose forty-nine whole numbers so that you can find a large number of multiplication facts utilizing three of them at a time. I suggest the inclusion of a symbol for a placeholder (variable), but it should be used as a factor only—not as a product.

Locate these cards, with the numbers selected as described above, at random on

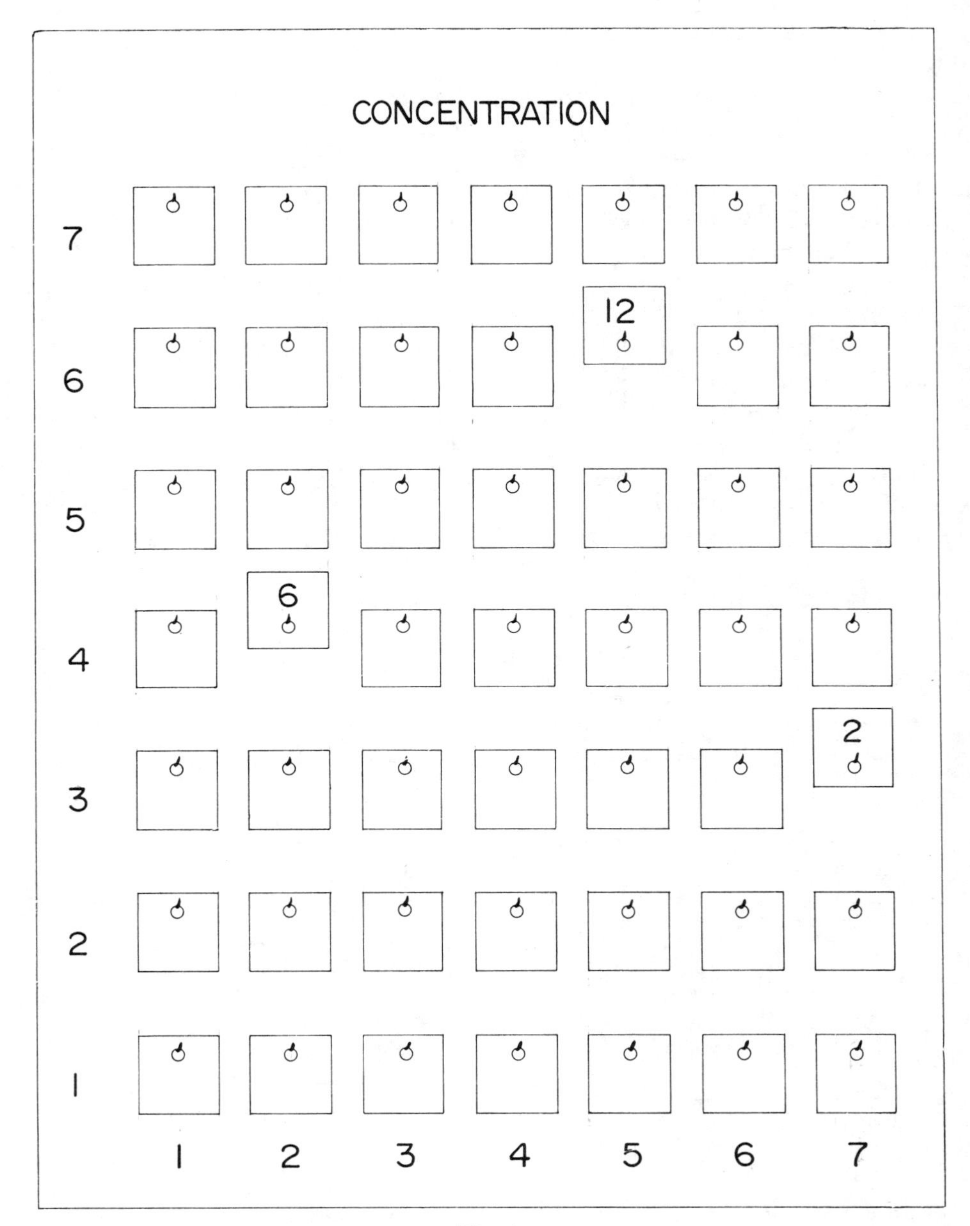

Fig. 1

the board. Divide your class into two teams and each team into subgroups of two or three players each. Alternate plays from team to team and rotate among the subgroups.

We are now ready for the procedure and rules. Each subgroup selects three cards to be turned up by choosing an ordered pair for each card. They are to "make a product" (state a multiplication fact) using the numbers on those three cards. If they are successful, their team scores a point and all cards are turned down. If it is not possible to make a product, the cards are turned down and the play moves to the other team.

If a product is made, then that particular product and its commutative counterpart are not to be used again. They should be written on the chalkboard for all to see.

But one or two of those numbers can be used to form a different product in another situation. For example, suppose a subgroup chose 6, 2, and 12. The product $6 \times 2 = 12$ (or $2 \times 6 = 12$) could not be used again later, but if a subgroup found a 3 and could recall the locations of the 6 and the 2, then they could score with $2 \times 3 = 6$.

The excitement really starts when the placeholder appears. It plays the role of a "wild card," and the pupil's ingenuity is now unleashed. If used as a product, however, it would permit scoring almost at will.

Of course, there are many possible variations, and you are encouraged to make use of them. A subgroup may wish to pass on their third choice of a card during their play if the two numbers already showing suggest that a successful third choice is very unlikely. You might permit a challenge from the opposing team to an attempt to make a product or a failure to recognize one, and award or take away a point depending on the outcome of such an effort.

If a class is not strong, a subarray, say a five-by-five array is suggested. You can build up several "decks" of cards and be prepared to "reload" the board for different situations.

This game can also serve as a readiness activity for a unit on graphs. The notions of ordered pairs of numbers representing points and perpendicular lines as references are the basic essentials of graphs. This alone will almost justify the use of this game in your classroom. Have fun.

EDITORIAL COMMENT.—Concentration games can be constructed using two boards, one for questions and the other for answers.

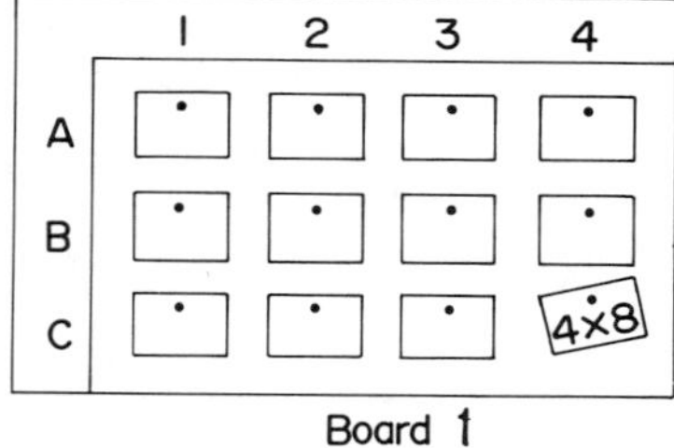

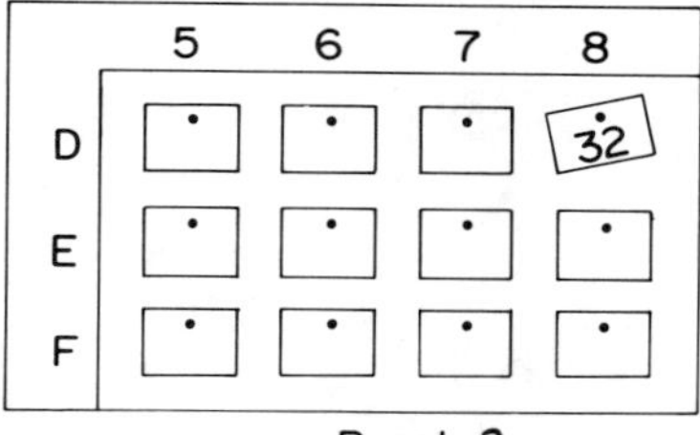

Students call for one card from board 1 (*example:* C,4) and one from board 2 (*example:* D,8). If the answer matches the problem, a point is scored.

The Match Game

LARRY HOLTKAMP

Mariemont Middle School
Cincinnati, Ohio

After the concept is understood, children need much drill in changing percents to decimals and fractions or changing fractions to decimals and percents. However, needless to say, drill can be very boring unless there is some means of motivation. To provide such motivation, I developed a game that I call the "Match Game."

Preparation

You will need fifty three-by-three-inch squares (I cut them from yellow construction paper) and between forty and fifty three-by-five-inch plain file cards. Take twenty-five of the squares and write a fraction or a decimal on each. On the remaining squares write an equivalent percent, one for each of the fractions and decimals, thus making a total of twenty-five matches. Mix the squares and, using double-stick Scotch tape, attach them face down on the blackboard in a five-by-ten arrangement. I call this the Match Board.

On the file cards write various problems that the children have had in the past and put them in some type of container. I call this container Potluck.

Explanation

Before the game begins, divide the class into two teams, A and B, and turn over any four squares on the Match Board.

To start the game, a member of team A picks a card from Potluck and hands it to the teacher. The player then goes to the blackboard and the teacher dictates the problem to him and the rest of the class. (I have all the children do the problems for practice.) If the player answers the problem correctly, his team scores one point and he advances to the Match Board. If the problem is incorrect, there is no score and the other team takes its turn.

At the Match Board the player turns over any *one* of the squares and says either "Match" and makes his claim or "No match."

Several possibilities now exist:

1. The player claims a match correctly and scores one additional point.
2. The player claims a match incorrectly and loses one point.
3. The player claims no match and scores no points.

Anytime a player says "No match," the teacher asks the other team for a challenge—there is a possibility that a match is overlooked. If the opposing team challenges and makes a correct match, it scores one point. If they challenge and make an incorrect match they lose one point.

The teams alternate turns for as many rounds as are needed to give each child a chance at the Match Board. At the completion of the play, the team with the most points is the winner.

It is possible that you may uncover a match before the game when you expose the four squares. This is fine. In fact, there is a slim possibility that in exposing the four squares two matches occur. This is also fine, but it is a rule that a player may make only one match a turn, no matter how many matches there may happen to be.

Remember, the only time a challenge may occur is when a player at the Match Board states, "No match."

A Cross Number Puzzle for St. Patrick's Day

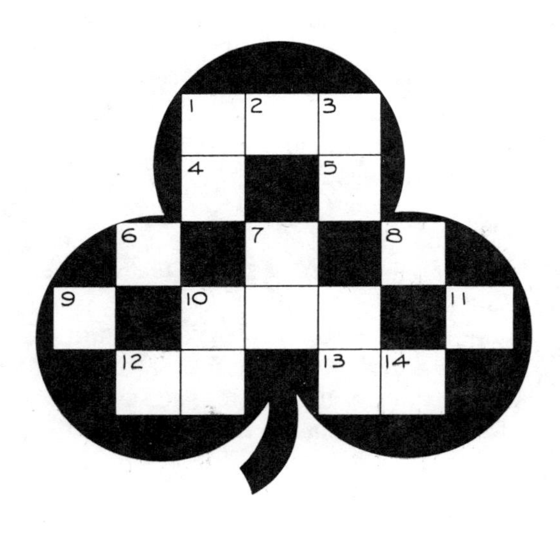

ACROSS

1. Product of $3 \times 9 \times 7$.
4. The unit's digit in the date of St. Patrick's Day.
5. The ten's digit in the date of St. Patrick's Day.
6. March is the ______ month of the year.
7. The number of prime numbers between 1 and 20.
8. The least common denominator of $\frac{1}{3}$, $\frac{1}{2}$, and $\frac{5}{6}$.
9. The altitude of a triangle whose base is 12 units and whose area is 42 square units.
10. The number of days from St. Patrick's Day to Christmas. (Do not count St. Patrick's Day or Christmas Day.)
11. The number of letters in the name of the color associated with St. Patrick's Day.
12. The square root of 1936.
13. The number of degrees in a right angle.

DOWN

1. The day of the month on which we celebrate St. Patrick's Day.
2. The number of sides an octagon has.
3. The average of 97, 98, 94, 80 and 86.
6. The number of sides of a triangle.
7. Two angles of a triangle are 42° and 50°. How many degrees are in the third angle of the triangle?
10. Two dozen.
11. A pentagon has ______ sides.

Contributed by MARGARET WILLERDING of San Diego State College, California

Dominoes in the mathematics classroom

TOM E. MASSEY
P. K. Laboratory School, University of Florida, Gainesville, Florida

During a summer mathematics program for ninth-grade students at the P. K. Yonge Laboratory School of the University of Florida, the mathematics staff found a traditional game to be a very popular and effective learning aid. The game of dominoes was enjoyed and played frequently by students of all levels of ability.

The game was an effective vehicle for practice in addition and recognition of multiples. The game as played traditionally (by two to four players) provides practice in addition and recognition of multiples of five. This practice results from the rule that a player scores by producing an addition total, on the exposed halves of the dominoes, which is a multiple of five. An example of play resulting in a score of ten is shown in figure 1.

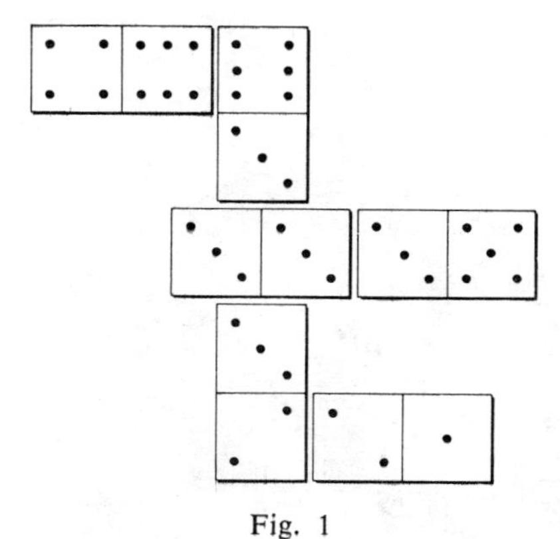

Fig. 1

Practice in recognition of multiples of numbers other than five may be obtained by requiring that the total be a multiple of a number other than five to score.

There seems to be no limit to the variations of rules of the game that will produce practice in a particular elementary mathematics skill. In one particularly effective variation the basic ideas of dominoes were transferred to playing tiles with fractions rather than spots. The fraction variation of dominoes, also, was well received by the students participating in the program. The play of the game is like dominoes in that tiles are played by matching the adjacent halves of the tiles. Scoring is accomplished by a player when he produces an additional total that is a natural number. Figure 2 shows play that resulted in a score of two.

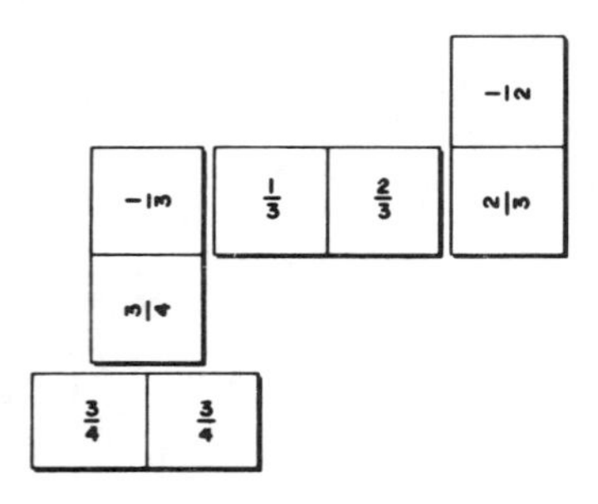

Fig. 2

The level of difficulty can be altered by the choice of fractions to be used in constructing the tiles and the number of different fractions used. A set consisting of twenty-eight tiles with the numbers 0, $\frac{1}{4}$, $\frac{1}{3}$, $\frac{1}{2}$, $\frac{2}{3}$, $\frac{3}{4}$, and 1 was used in initial experiments with the game. Another set of twenty-eight playing tiles consisted of the numbers 0, $\frac{1}{6}$, $\frac{1}{3}$, $\frac{1}{2}$, $\frac{2}{3}$, $\frac{5}{6}$, and 1. If the numbers of both of the above sets are used in constructing a set, it will contain forty-five playing tiles. Again, there seems to be no limit to the number of useful variations.

Any teacher can manufacture a crude set of plywood tiles for use in the classroom. The experimental tiles were constructed out of $\frac{3}{8}$-inch plywood measuring one inch by two inches. A light coat of varnish was applied to preserve the markings on the tiles. The students in the program were eager to assist in the determination of which fractions were utilized in the construction of the sets. Each set constructed is expected to provide many hours of enjoyable practice with fractions.

A Cross Number Puzzle for Valentine's Day

ACROSS

1. Date of Valentine's Day, February ______.
3. Valentine's Day is the ______th day of the year.
5. There are generally ______ days in the month of February.
6. There are ______ months in a year.
7. St. Valentine lived in the ______ century.
8. Seven months of the year have ______ days.
10. Number of letters in the month in which Valentine's Day is.
11. The number of days in a normal year multiplied by $3\frac{1}{5}$.
13. Four months of the year have ______ days.

DOWN

1. Two less than the product of $5 \times 5 \times 5$.
2. $8 \times 6 =$ ______.
3. 246 divided by 6.
4. The product of 22 and 24.
8. In a normal year there are ______ days between Valentine's Day and Christmas. (Do not count Valentine's Day or Christmas Day.)
9. The number of quarters in $40.
11. The number of days in the year we send Valentines.
12. The number of days in a week plus one.

Contributed by Margaret Willerding of San Diego State College, California

A CROSS-NUMBER PUZZLE FOR JUNIOR HIGH SCHOOL

MARGARET F. WILLERDING
San Diego State College

ACROSS

1. How many yards are in 4,890 feet?
4. 15^2
6. 8 dozen
7. Write five minutes past eight in figures
8. How many acres are in one square mile?
10. XLIV represents what Hindu-Arabic numeral?
13. What is the total cost of 4 and $\frac{3}{4}$ yards of cambric at $.40 a yard, 1 package of pins at $.10, and 3 spools of thread at $.05 each?
15. What is the average of 728, 964, 247, 425, and 316?
17. What is 90% of 100?
18. What is the perimeter of an equilateral triangle with a base of 104 feet?
21. What is the cost of 1 yard of bunting if 27 yards cost $11.61?
22. How many quarters are there in $33.75?
23. Write $4\frac{1}{2}$% as a decimal

Omit decimal points and percent signs in the puzzle. Just write the figures.

DOWN

2. What is the perimeter of a garden 180 feet by 168 feet?
3. Write the number of days in a regular year from January 1 through December 30
4. Express 1/5 as a per cent
5. Compute the interest for one year on $127.00 at 2%
9. Reverse the digits in $\frac{1}{4}$ miles (express in feet)
11. What is the area in square feet of a rectangle 1 yard 2 feet by 9 feet?
12. What is the total cost, including 2% state sales tax, for 1 dozen cream puffs at $.06 each, 2 loaves of bread at $.17 a loaf, and 1 coffee cake at $.40?
14. How many ounces are in one pound?
16. How many dozen cookies are needed to feed a troop of 33 Boy Scouts if each boy gets 4 cookies?
18. How many days remain in a normal year after January has passed?
19. What is the age in 1957 of a man born in 1930?
20. Write $\frac{3}{4}$ as a decimal
21. Write 2/5 as a decimal

A CROSS-NUMBER PUZZLE FOR INTERMEDIATE GRADES

MARGARET F. WILLERDING
San Diego State College

ACROSS

1. 100 − 15
3. 3,240 ÷ 36
4. 7 + 8 + 9 + 6 + 5 + 3
5. 2,650 ÷ 25
7. 11,820 ÷ 60
8. 8 × 7
9. 315 ÷ 9
11. 52 × 92
12. 12 − 4
13. 8 × 8
14. 14 × 7
16. 7 × 7
17. 121 ÷ 11
19. ½ of 164
21. From 10,272 subtract the year Columbus discovered America

DOWN

1. 9 × 9
2. 884 − 379
4. 2,765 ÷ 7
6. 3,356 + 3,288
7. 2,408 − 1,059
10. 63 ÷ 9
13. 780 − 711
15. 616 ÷ 7
16. 6 + 7 + 8 + 9 + 10
17. 2 × 4 + 9
18. Reverse the digits in the product of 9 × 9
20. 100 − 80

A CROSS-NUMBER PUZZLE FOR PRIMARY GRADES

MARGARET F. WILLERDING
San Diego State College

ACROSS

1. What number comes after 110?
3. What time does this clock say?

4. How many tens are there in 40?
5. What time does this clock say?

7. When counting by 5's what number comes after 20?
9.

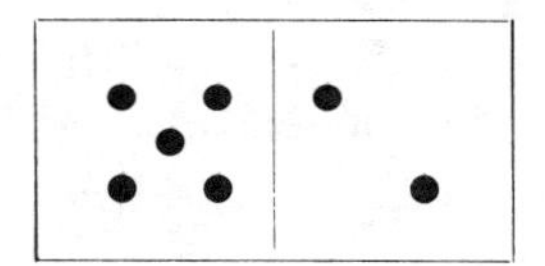

This is a number picture of what number?

10. When counting by 2's what number comes after 150?

DOWN

2. How many eggs in a dozen?
3. What number comes after 117?
4. What three numbers come after 3?
7. $1+1=?$
9. What number comes after 74?
10. What is 6 take away 5?
11. What is 7 take away 5?

EDITORIAL COMMENT.—Here is another word activity.

Unscramble these mathematical words:

NOTEQUAI	(equation)
EST	(set)
DADNOITI	(addition)
CRAFTINO	(fraction)
MADELIC	(decimal)
MEETGORY	(geometry)
RASQUE	(square)
RIMPE	(prime)

Mathematicalosterms

SALLY MATHISON

North Junior High School, Portage, Michigan

If you had trouble deciphering the title of this article, perhaps you should read no further. Robert E. Reys' "Mathematical Word Search" that appeared in the April 1967 issue of THE MATHEMATICS TEACHER gave me the idea for this puzzle. I was sure that my junior high students would enjoy a puzzle of this kind, but Mr. Reys

E	S	U	B	T	R	A	C	T	A	P	E	R	I	M	E	T	E	R	M
S	R	N	I	N	E	V	B	C	D	H	U	N	D	R	E	D	E	T	E
T	A	R	E	A	S	E	F	U	P	G	F	H	D	R	A	T	E	R	A
I	C	I	O	S	J	R	F	A	N	E	G	A	T	I	V	E	E	I	S
M	C	P	E	R	K	A	R	R	L	I	R	M	C	Z	G	G	N	A	U
A	U	L	D	I	A	G	O	N	A	L	T	C	O	T	E	I	P	N	R
T	R	A	I	Q	R	E	S	A	T	C	U	H	E	T	O	R	T	G	E
E	A	C	S	P	O	W	E	R	O	F	T	E	N	N	V	R	O	L	W
X	C	E	T	A	D	I	V	I	D	E	F	I	X	Y	T	Z	C	E	A
P	Y	V	A	R	B	C	T	T	D	E	O	G	O	W	P	R	I	M	E
O	F	A	N	A	G	A	H	H	I	J	U	H	B	N	I	K	L	D	M
N	N	L	C	L	R	N	U	M	B	E	R	T	A	C	O	D	D	P	Q
E	D	U	E	L	R	S	T	E	M	P	T	Y	S	E	T	O	T	A	L
N	I	E	L	E	M	E	N	T	U	O	U	V	E	Q	U	A	L	H	E
T	U	N	C	L	V	W	X	I	L	S	Y	R	V	U	U	Z	D	E	N
W	W	X	V	I	Y	Z	A	C	T	I	H	A	E	A	C	A	M	D	G
B	C	O	F	E	M	A	T	H	I	T	N	Y	N	T	X	U	R	D	T
E	N	U	M	E	R	A	L	F	P	I	G	E	H	I	L	I	B	E	H
S	P	H	E	R	E	S	L	J	L	V	K	L	S	O	M	N	O	E	P
G	R	E	A	T	E	R	E	Q	Y	E	R	S	V	N	P	O	I	N	T

made use of words such as ellipse, congruence, calculus, polynomial, contrapositive, etc. Junior high students would find these words as nonsensical as the whole array of letters. So I decided to create my own puzzle using mathematically related words that a junior high student would recognize. It is a challenge to hide as many of these words as possible in a 20-by-20 square. Try it and see.

As I anticipated, the students loved working with this (and so did some of the teachers!) I gave both my seventh- and eighth-grade classes half of one class period (about twenty-five minutes) to work on the puzzle, and the better students found as many as fifty math-related words and several others that are not mathematical and therefore did not count. At the end of the class period the students were so captivated that they asked me to postpone giving them the complete solution until they had more time to search on their own. The next day I posted a solution. The students were delighted with themselves over the many words they had found, and surprised to have revealed to them some words that were hidden so well that they missed them. I'm quite sure you will find this an interesting class activity if you try it.

The words are located horizontally, vertically, and diagonally. All of the words are spelled forwards. How many words can you find? That depends on how hard you look! But if you look very hard you can find seventy-five words that I've placed there, and maybe even more that I haven't discovered myself! See if you can find the math word that appears three different times in the puzzle—once horizontally, once vertically, and once diagonally. (Incidentally, this only counts as one word!) Can you find all of the numbers from zero through ten? Watch for words within words, for example, the word "exponent" contains the word "one." Both of these are mathematical, so you can count this as two words. Watch for words that overlap, such as those in the title of this article. It might be fun to challenge your students to make a puzzle of this kind that contains all of the numbers from one to twenty, using as few letters as possible. In the meantime, check your own ability to locate mathematical words in this puzzle. If you get stuck, look for help in my list of hidden terms.

This is a list of the words that can be found in the word puzzle—in no particular order. The word that appears three different times in the puzzle is the word "odd."

zero	rate	divide
one	ratio	add
two	meter	term
three	total	arc
four	fraction	length
five	width	prime
six	area	integer
seven	multiply	inverse
eight	set	parallel
nine	empty set	diagonal
ten	average	distance
triangle	cube	element
angle	place value	factor
unit	base	graph
digit	negative	numeral
height	positive	sphere
measure	ray	math
estimate	decimal	error
number	line	greater
hundred	even	less
arithmetic	odd	perimeter
volume	pi	power of ten
subtract	point	circle
equation	percent	square
equal	exponent	accuracy

Editorial feedback

VERNE G. JEFFERS
Mansfield State College, Mansfield Pennsylvania

I noted with interest the article entitled "Mathematicalosterms" by Sally Mathison in the January 1969 issue of THE ARITHMETIC TEACHER. I have been using a version of this puzzle for a number of years but had never attached a name to it. However, seeing the idea in print did give rise to another idea which, like "Mathematicalosterms," may not be original, but has not had popular usage. It involves using nu-

9	×(3 +	6))	= 81	60	7	12	19	57	76	23
	×	=								
45	2	3	12	48	6 +	8 =	14	25	33	58
	×	=								
3	35	18	4	72	42	30	56	3	44	66
	+									
15 −	5	=10	9	12	5	28	11	17	35	16
		=								=
72	38	8	36 ÷	6 =	6	4	27	31	5	1
										+
48	8	80	4	20	11	7	4	9	36	2
			=							+
64	4	24	8	3	5	15	45	40	3	10
		+								+
7 +	13+	12	=32	35	19	8	3	6	18	3
32	33	28	4	7	6	9	4	19	21	5
				×	=					
58 +	18 −	40 =	36	3	7	21	28	36	4	9
				÷	=					
24	54	37	26	63	13	49	57	20	25	45
3	72	17	4 ×	(7 +	6)=	52	2	26	50	72

merals rather than words in a matrix. My graduate students in "Mathematics for the Elementary School Teacher" thought a puzzle of the following type had possibilities and encouraged me to develop it. I am sure that none of us realized its potential at that time.

Number combinations appear in the grid vertically, horizontally, and diagonally. If you examine the matrix closely you will find many of the basic facts for addition ($6 + 8 = 14$), subtraction ($8 - 3 = 5$), multiplication ($9 \times 2 = 18$), and division ($27 \div 9 = 3$). Also, inverse operations ($28 = 4 \times 7$ and $28 \div 4 = 7$) and commutative, associative, and distributive properties can be shown. Decade addition ($19 + 9 = 28$), column addition ($7 + 13 + 12 = 32$), simple subtraction ($28 - 11 = 17$), and compound subtraction ($44 - 17 = 27$) may also be found.

Variations may be used in marking combinations, possibly depending on grade level and purpose. These may range from merely encircling the combinations for identification purposes to inserting the appropriate signs to complete a number sentence. A few combinations are encircled here for illustration.

I first tried this on my own children, who are third- and fifth-grade students. Both were intrigued and found much enjoyment in seeing how many familiar combinations they could find. My wife's fourth-grade class found the puzzle equally interesting and challenging.

EDITORIAL COMMENT.—Try a mathematicalosterm using the names of famous mathematicians.

S S A U G A U S G O
P Y T H A G O R A S
D I L C U E E U U N
C A T A S U U S S O
A S W N S A C S T T
N I E T S N I E O W
T O N O W E N L R E
N T O R P Y T L I N

Can you find these names:

Gauss
Newton
Einstein
Pythagoras
Euclid
Russell
Cantor

Two mathematical games with dice

RONALD G. GOOD

Florida State University, Tallahassee, Florida

Each of the games described here uses a pair of dice. The two dice must be distinguishable by color, one die of one color and the other of a different color. The variations in color can be in the dice or in the spots on the dice. Such dice can be purchased commercially or they can be produced by making wooden cubes and then using magic markers to make the required spots on the faces of the cube. Only two colors are needed and it might be a valuable activity to allow students to make their own dice before learning how to play the games.

In the descriptions of the games that follow, one die is red and the other is black.

Game 1

A number line (see fig. 1), a marker for each player, and a pair of colored dice are needed to play the game. The winner's spot, marked by $\times$ in figure 1, could be either to the left or to the right of 0. The number that turns up on the black die means a move to the right. The number that turns up on the red die means a move to the left. Each player begins at 0 and on each turn rolls a red die and a black die. A few sample plays will illustrate the game.

Suppose player *A* rolls first and rolls a red 2 and a black 5. He ends up on black 3. Player *B* then rolls and follows the same directional moves. Suppose *B* rolls a red 6 and a black 2. *B* then ends up on the red 4. On his next turn *A* rolls a red 4 and a black 2, which puts him at the black 1. *B* then rolls a black 5 and a red 1, which takes him back to 0. The play continues until a player gets a roll that places him on or beyond the win slot. That player is then the winner and the game is over.

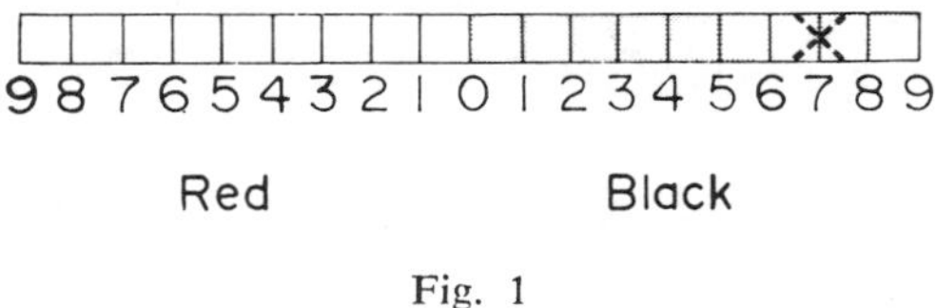

Fig. 1

The main intent of the game is to introduce subtraction in a concrete way. Although negative numbers as such do not appear on the game board, the fact that there are numbers to the left of 0 is one way of looking at negative numbers on a number line.

The game can be varied in several ways. Number lines of different lengths can be used. A "lose" spot could be substituted for the "win" spot—children might want to see if this changed their luck."

Because the "laws of chance" are operating when the dice are being rolled, the children will likely gain an intuitive idea of such laws because of the tendency for the play to remain near 0. In fact, if number lines of different lengths are used, children will probably notice that it is more difficult to win with a longer line than with a shorter line.

Game 2

Coordinate paper with axes drawn; pins, map tacks, or other small markers; and a pair of colored dice are needed to play the game. The coordinate paper (see fig. 2) should be prepared by the teacher in advance so that children can take the dice and begin the game after minimal initial instructions.

The object of this game is to move the playing pieces from 0 to the point on the paper marked by an × (see fig. 2). A player rolls one red die and one black die. The black die means move either right or left and the red die means move either up or down. A player cannot move both up and down in the same move, nor can he move both left and right in the same move. The following series of moves illustrates the game.

Player *A* rolls a red 4 and a black 3. He proceeds from 0 to the right 3 spaces and then up 4 spaces. He puts his marker at (3, 4). Player *B* rolls a red 6 and a black 5 and puts his marker at (5, 6). On his next turn, *A* rolls a red 4 and a black 7, which puts him at (10, 8). Then *B* rolls a red 6 and a black 3, which puts him at (8, 12). *A* then rolls a red 3 and a black 2. Since *A* is already beyond the ×, he

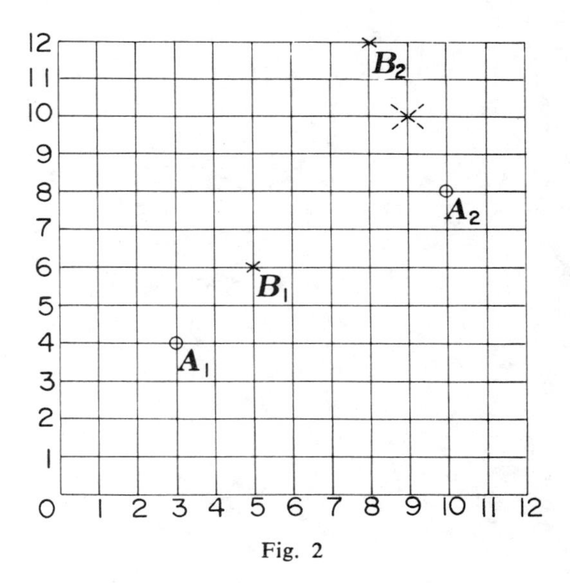

Fig. 2

chooses to move to the left 2 spaces and up 3. This puts him a (8, 11). (The direction of a player's move may be dictated by a boundary. Thus, if *A* had rolled a black 3 he would have been forced to move to the left. A move of 3 to the right would have put him "out of bounds.") On his next turn *B* rolls a red 2 and a black 1. This enables *B* to move 1 space to the right and 2 spaces down, which puts him on ×. *B* thereby wins the game.

Number pairs are obviously involved in

this game, as well as rudimentary ideas of the whole process of graphing. Adding and subtracting, as useful processes, are emphasized. Perhaps most importantly, the game requires some strategy, planning ahead and considering the consequences of different moves.

As with any game with predetermined rules, the children should be allowed, and perhaps even encouraged, to "see what happens" if they change the rules in some way. The "game board" can be extended to include all four quadrants of the graph thereby involving negative numbers. The × could be put in any of the quadrants.

Minimal computational skills are required to participate with understanding in both dice games and yet the outcomes in terms of other kinds of mathematical thinking are potentially great. If left to their own devices, it is likely that children will devise other variations of the two games which might prove to be even more valuable to their mathematical thinking abilities.

EDITORIAL COMMENT.—Ordinary dice do not have enough faces to practice all the basic facts with one pair of dice. Dodecahedron dice will do the trick. Make models for two dodecahedra and write the numerals 1 through 12, one to a face, on each die. Roll the dice and add, subtract, multiply, or divide with the two numbers named on the top faces.

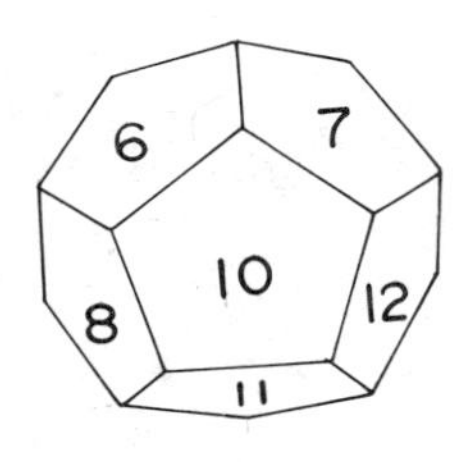

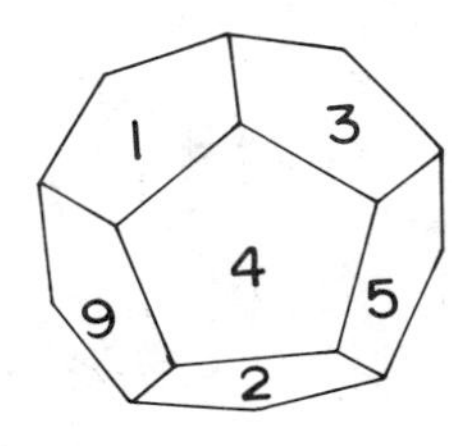

Three games

BRUCE F. GODSAVE

Bruce Godsave is associate professor of mathematics at Gallaudet College, Washington, D.C. All his undergraduate students are deaf. The ideas for these games were developed while he was teaching a graduate course in mathematics education to future teachers of the deaf.

When René Descartes sat up in bed one day and created coordinate geometry, he did a fantastic thing for mathematics. Many students and teachers believe that much of mathematics came to us in a manner similar to that of the Ten Commandments. In fact, something like the coordinate system was made up by a mortal man, in much the same way that man created Monopoly. It would be difficult, if not impossible, to devise an intuitive way of learning the rules of Monopoly. We are told the rules, and after we learn them, we use our skill and intelligence to send our friends to the "poorhouse." The same is true with coordinate geometry and most other mathematical concepts: we need to learn the rules.

The idea of a point, how to name a point, where to begin numbering, and the direction the numbers go are all part of the rules of a coordinate system. Assume now that we have a class that has been given the rules; the children recognize the axis and have been given a method for finding a point. We need to add "experience."

The first two games described below are geoboard activities that provide such experience in naming points, checking points, writing the names of points, and organizing data. This will help children become familiar with coordinate systems before we complicate the idea with lines, circles, graphs, and things like that. The activities will also be fun for the class.

The third game described disguises general review as a game of tag.

Treasure Hunt

This is like the game "Battleship," except it is nonviolent! Use 4×4 or 5×5 geoboards, balls of clay, and two teams. Each team will put five balls of "gold" (clay) on the pins or nails of the geoboard without letting the other team know where the gold is (see fig. 1). The teams take turns naming points. When one team names a point that has gold on it, this team gets the gold. The first team to lose its five pieces of clay loses the game.

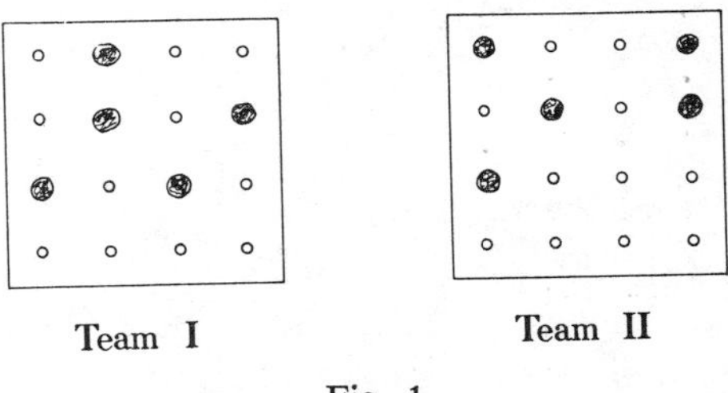

Fig. 1

The rules of the game are simple, but the skills required to win efficiently involve practice with using ordered pairs to name points and organizing data so that students know whether or not a point has already been named. While one team gets practice by naming points, the other team gets practice by checking each point to see if it has any gold on it.

For the first game or so, after naming ten or more points the students will forget whether they have already called a particular point or not. This should lead to writing down the ordered pairs as they are named so that the players don't have to waste time checking the same point over and over. If children don't do this on their own, suggest it. Here again is the practice they need. Later, their list will be rather long and they may spend a long time checking the list. This should lead to organizing the list as shown in figure 2. Again, if children don't do this on their own, suggest it.

First List

(0,0)	(2,2)
(3,4)	(4,4)
(2,1)	(0,1)
(1,3)	(1,1)
(4,2)	
(1,2)	
(0,3)	
(4,1)	
(2,3)	

Organized List

(0,0)	(1,3)	(2,1)	(3,4)	(4,2)
(0,3)	(1,2)	(2,3)		(4,1)
(0,1)	(1,1)	(2,2)		(4,4)

Fig. 2

Through the Maze

This is also a team game involving group versus group, although it may be played by one student against another. Each team constructs a maze, using rubber bands and a geoboard. By calling off ordered pairs, one team is to get through the other team's maze without seeing the maze. Figure 3 shows two mazes with their solutions.

In this game, no rubber bands are to be crossed, nor may anyone land on a peg with a rubber band on it. The beginning is at (0,0), and the goal is to get to (n,n).

To play, someone on team 1 names a point on the maze. If there is a straight-line path between where he is and where he wants to go, he is told he can move. For example, on Maze II he could make one of two possible first moves, (2,1) or (3,1). All other choices are blocked.

This game requires a different recording method from that used in Treasure Hunt. Points that are blocked while a player is in one position will not be blocked if he is in another position. Here are some suggestions for this game:

1. Use a marker to show the position of the team in the maze.

2. The first time the game is played, make up a maze and play while the class looks at the maze. Explain why various moves can or can't be made. Be sure there is a path through the maze.

3. For the next game, make up a maze and show it only briefly to the class. Then have the class work together to choose points. Once they understand the method of playing, they can divide into groups or pairs to play.

Children find this a very interesting game, although it can't be played fast. It can be played during the time before or after school, or during lunchtime. It's a quiet game that provides needed practice in finding points on a plane.

Another game that could be used to provide experience in naming points is "Four-in-a-Row," a game used in the Madison Project. It is similar to tic-tac-toe, except that there is an extra column and row.

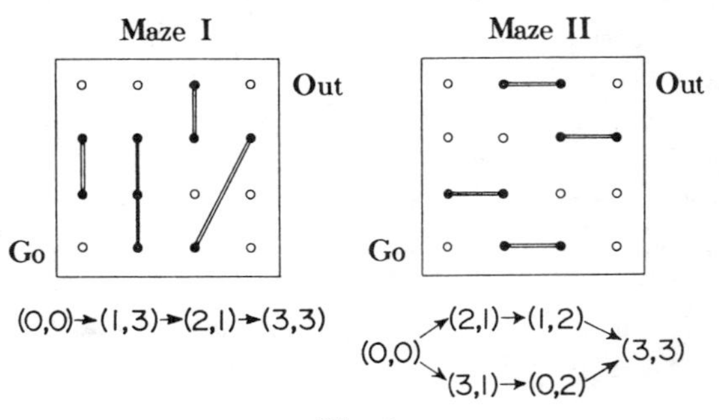

Fig. 3

The teams take turns naming points where an × or an ○ is to go.

Mathematical Tag

Sometimes in the teaching of mathematics we feel compelled to do a little review. I'm sure most of you have tried to find different ways to review that don't seem like review. I was faced with a review and also needed to keep a promise that we could have mathematics outside someday. My solution to both problems was a variation on the game of tag. I called it "Mathematical Tag," since that described very well what was going on.

Mathematical Tag is based on the rules of the tag you played (or still play). Here is how it works:

Everyone in class chooses a number from 1 through the number of students in the class. No two pupils can have the same number. Each child will wear his number so it can be seen from the front or the back. Wide tape does a good job. You will have already prepared a pack of 3 × 5 cards on which are written expressions like the following:

□ is not equal to 5.
3 is greater than □.
□ + 3 is an odd number.
□ is a member of {5, 10, 15, 20}.
□ + 4 = □ + 4.

You will need to keep a tally of your solution sets to check that each number is used about the same number of times. If after checking you find you need another 1, 16, and 19, you can make up a card with the expression

□ is a member of {1, 16, 19}

—and you're all set, so to speak. Try to avoid cards whose solution set will have only one or two members. It's hard on whoever is "it" to go after only two people.

Now comes the hard part—deciding who will be "it." Let me suggest that you be "it." This shows that you're a real sport, and it will show the children how the game is played.

Now "it" draws a card, reads it to himself, and figures out the solution set. Without telling anyone what the solution set is, he must tag a person whose number is a member of the solution set. The person tagged is then "it," and the new "it" draws a card. This can continue until the desire to hold class outside is eliminated from everyone, or until the kids are tired enough to do seatwork.

What do you do when "it" tags someone whose number is not a member of the solution set? You could impose a penalty, but that is not necessary. The truth (set) will out.

Just a few more comments. The level of difficulty of the cards depends on the abilities of the students. For example, compare (□ is not equal to 5) with ($x \neq 5$) and (□ is between -3 and 3) with ($|y| < 3$).

After the game is played once, the class can be asked to make up a pack of cards for the next time. They will see how hard it is to get the number of occurrences of each number the same.

Try to work out a way of using negative numbers.

Try to keep the playing area restricted to a small space.

This game can be played indoors; but it is a noisy game, so choose a place accordingly.

I have played this with my classes, and it is a lot of fun. If your class size is about twenty-five, you could use the following sentences for your deck of cards. Copy each sentence on a card by itself. If I figured right, each number will appear in seven different solution sets.

□ is even.
□ is odd.
□ is not even.
□ is not odd.
□ is a multiple of 2.
□ is a multiple of 3.
□ + 1 is a multiple of 2.
□ is a prime.
□ is not a prime.
□ is less than 11.
□ is greater than 17.
□ is between 10 and 18.
□ is a multiple of 5.
□ is a member of {1, 2, 4, 7, 8, 11, 13}.
□ is a member of {14, 16, 17, 19, 22, 23, 25}.
□ is not 15.

Have fun with mathematics!

Take a mathematical holiday

ELAINE McCALLA
Christian Academy in Japan, Tokyo, Japan

The day before vacation usually brings an air of carefree excitement into the classroom as the children come, bubbling over with thoughts of "No school tomorrow!" To the teacher, educationally speaking, I'm afraid it often means a wasted day. I have found a way to put some of this restlessness to work, and so I invite you to join me as we "take a mathematical holiday"—the day before Christmas vacation.

The ideas presented here can be made as simple or complex as desired, so they can be adapted to most grade levels. They can also be altered to fit the particular mathematical skill or concept you wish to incorporate; i.e., from basic operations of whole numbers, fractions and decimals, to geometry and probability. One holiday idea can be used for almost any other holiday, simply by adapting the figure.

Good old St. Nick (Fig. 1) can be used to construct and to name different geometric figures. Halloween could use a few witches made from squares, circles and triangles; Thanksgiving vacation could easily fatten up a geo-turkey; and spring vacation could bring geometric flowers into bloom.

Santa Claus (or any other holiday figure) can easily be drawn by "joining the dots" (Fig. 2), employing mathematical problems of varying difficulty.

I hope that these ideas will give birth to many more mathematical holidays as you use your imagination to create some vacation plans of your own. Happy holiday!

FIGURE 1

GEO-SANTA

1. Draw the following geometric figures;* cut them out; put them together to make a Santa.
2. For Santa's hat, make an EQUILATERAL TRIANGLE measuring 1 inch on each side. The brim is made from a RECTANGLE that measures ¼ by 1 inch.
3. Santa's face is made from a SQUARE 1 inch on each side. Give him two little TRIANGLES

* Younger children can more easily draw these shapes on graph paper.

for eyes, a little SQUARE for a nose, and a little CIRCLE for a mouth.

4. Make Santa's beard with an ISOSCELES TRIANGLE, measuring 1 inch on one side, 1½ inches on each of the other two sides.

5. His body is made from an ISOSCELES TRAPEZOID measuring 3 inches on the two equal but nonparallel sides, 3 inches on one parallel base and 1 inch on the other base.

6. Santa's arms are outstretched with two RECTANGLES that are 1 inch long and ½ inch wide.

7. His hands are made from two CIRCLES that are ½ inch in diameter.

8. Each of Santa's legs is made from a RHOMBUS, 1 inch on a side.

9. His feet are made from two RECTANGLES, 1½ by ½ inches.

10. Give him three CIRCLE buttons, and you've finished your task!

SANTA CLAUS DOTS

Join the dots from 1 to 29 in the correct order.

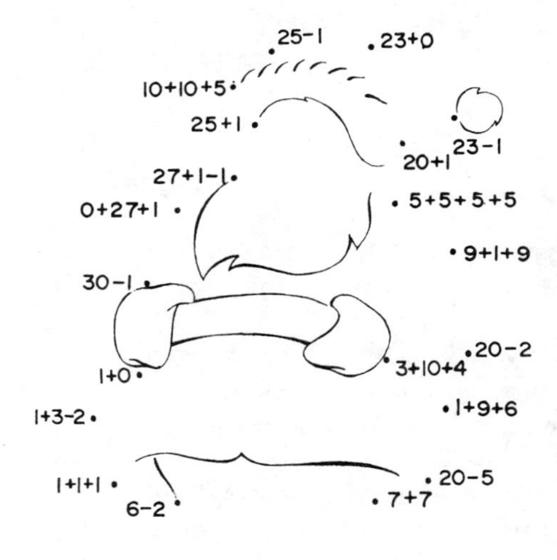

FIGURE 2

EDITORIAL COMMENT.—Worksheets can be turned into personalized puzzles. Do the problems at the bottom of this sheet. Then color in the letters corresponding to the answers. Can you make your children one of these using each child's name?

Classroom Experiences with Recreational Arithmetic

RUTH H. NIES*
Sixth Grade, Wright School, St. Louis County, Mo.

MUCH OF THE SO-CALLED DISLIKE for arithmetic in the intermediate grades is only pretense. Attitudes of apathy and hatred can be changed to contagious enthusiasm when methods of recreational arithmetic are employed. When pupils participate in the healthy enjoyment of number games, they are not ashamed to admit their enjoyment to the world in general. Original puzzles, tricks, and recreational units devised by pupils prove that working with numbers is not really distasteful.

Recreational mathematics can serve as a "pepper-upper" to start a school day or to begin an arithmetic period. A short impromptu game will sharpen wits on a dull day and will relieve tensions and boredom. Sometimes between periods, just before lunch, or before dismissal time there are five or ten minutes which seem to "dangle" and which are opportune for a game which can build a worthwhile interest in working with numbers. An occasional interruption of the usual routine with a puzzle, joke, story, riddle, or game not only provides a pleasurable break but may also be stimulating for interest in learning. There are many stories, games, and puzzles with a mathematical content.

In our area some children must come to school early and have free time before school and must stay indoors when weather does not permit outdoor play. A "number oddity" or a number progression or some pattern of numbers written on the chalk board will fill the pupil's time pleasantly and to good advantage.

HERE ARE A FEW EXAMPLES:

1. Select a number less than 10
   Multiply it by 9
   Use the result as a multiplier of 123456789
   (The final product will be a succession of the digit originally chosen)
   (Expect for a zero in the 10's place.)

```
  7      123456789
 ×9            ×63
 63      370370367
        740740734
        7777777707
```

2. Select a three-digit number
   reverse the order of digits
   subtract the smaller number from the larger
   divide the result by
   (The answer can be read backward or forward)

```
 453
−354
  99÷9=11
```

3. Add:

```
123456789
987654321
123456789
987654321
        2
222222222
```

   (Don't show the answer, let pupils find it.)

Boys and girls like to see for themselves how these exercises "work out." They will test one repeatedly, try it with other

* Mrs. Nies developed an interest in and a basis for recreational arithmetic in a course with Dr. Margaret F. Willerding at Harris Teachers College, St. Louis, Mo.

numbers, and in the meantime they are developing speed and accuracy without sensing any drill. They think it is "kinda fun." Some days I will put a number oddity or riddle on the chalk board or bulletin board and make no comment about it. Many children will "surprise" me by solving the oddity in their spare time. Like many teachers, I have a stock of seasonal items. Occasionally codes are used and both the boys and girls have a great fondness for them.

1. Code Problem:

Add: In the sum each letter represents a distinct digit. Find the numbers.

	S	E	N	D
	M	O	R	E
M	O	N	E	Y

2. Hocus Pocus Halloween:

Add. In the sum each letter represents a distinct digit. Find the numbers.

	H	O	C	U	S
	P	O	C	U	S
P	R	E	S	T	O

3. April Fool's Puzzle:

Write a number of three digits on your paper. Be sure that the difference between the first and last digit exceeds one. Reverse the digits. Find the difference between the two numbers. Reverse the digits of this difference. Add these two numbers. Multiply by a million. Subtract 966,685,433. Substitute these letters for figures: under every figure 1, write the letter L, under every figure 2, write the letter O, under every figure 3, write the letter F, under every figure 4, write the letter I, under every figure 5, write the letter R, under every figure 6, write the letter P, under every figure 7, write the letter A. Read the result backwards.

I have observed that frequently pupils will take a greater interest in numbers and measures if they know something about how these originated. The history of mathematics in a simplified version fascinates the elementary school child. My pupils particularly like such things as the "stick method" of counting, ancient ways of telling time, early calendars, etc. These topics also help them to appreciate modern arithmetic. The formula for finding area seems more useful to a child when he sees how troublesome area problems were to the Egyptian rope-stretchers. The story of our modern standardized measures and how these were developed can be both dramatic and worthwhile. The humor of impractical methods of measuring yards and feet in the days of King Henry II helps emphasize the importance of accuracy.

The interest of almost all children can be captured by stories, but some will prefer action. When pupils experiment with an abacus, geometric paper folding, or a set of Napier's "bones" which they themselves construct, I find their concepts of numbers and processes strengthened. Certain games and puzzles afford action, but not every child will enjoy arithmetic games immediately. He may not "catch on" to the tricks and short computations or he may not "see" a pattern or relationship in number progressions, or he may not grasp the essential point of a riddle. On the other hand, there are times when recreational materials will help a child to sense number relationships which he cannot understand through basic techniques. In certain remedial cases, when all other methods have failed, "light has dawned" during use of recreational devices. But it would be a mistake to force a child to participate in number games. Let him observe passively, he may soon take an active interest along with his classmates.

Likewise, it is a mistake to "explain" some short computations which may only confuse some of the children. A trick or a puzzle which calls for processes beyond the current learning level of the class is not useful. Conversely, one which makes application of a child's current stock of number facts provides him with a thrill when he sees it work. The fourth grader who sees for the first time the interesting patterns made by multiples of 9 is actually gleeful when he reaches grades five and six and more advanced patterns he can work out with "Magic 9," "Tricky 3," and "Lucky 7."

TRICKY 3

37× 3=111
37× 6=222
37× 9=333
37×12=444
37×15=555
37×18=666
37×21=777
37×24=888
37×27=999

LUCKY 7

15,873× 7=111,111
15,873×14=222,222
15,873×21=333,333
15,873×28=444,444
15,873×35=555,555
15,873×42=666,666
15,873×48=777,777
15,873×56=888,888
15,873×63=999,999

MAGIC 9

123456789× 9=111111111
123456789×18=222222222
123456789×27=333333333
123456789×36=444444444
123456789×45=555555555
123456789×54=666666666
123456789×63=777777777
123456789×72=888888888
123456789×81=999999999
222222222×9=1999999998
333333333×9=2999999997
444444444×9=3999999996
555555555×9=4999999995
666666666×9=5999999994
777777777×9=6999999993
888888888×9=7999999992
999999999×9=8999999991

I have no wish to over-emphasize arithmetic in our course of study. Any sixth grade teacher knows that she must cover much ground in teaching many other subjects. Occasionally correlation with arithmetic proves valuable. In science the study of the solar system has provided information and entertainment for pupils who have devised recreational type units. They prepare these in committees and small study groups beyond the minimum assignment. Our language arts activities include dramatizations. We have written and presented several dramatizations based upon arithmetic. In the social studies our units the Latin American countries, the British Isles, Canada, and Australia. Possibilities for recreational units involving arithmetic are obvious. Comparisons and contrasts relative to size in area, population, etc. always strengthen knowledge in both fields. I have had many such units worked out by pupils. Subject matter of the sixth, seventh, and eighth grades is especially good for incidental correlation of other subjects with arithmetic.

Drills and number games for fun can be used in lower grades, but recreational arithmetic begins to come into its own in grades four through eight, after basic fundamentals have been mastered. Of course, these same fundamentals need constant review.

Because sixth-graders love guessing games, I may start one out by saying, "Think of a number between 1 and 10, and I'll bet I can guess your number! Write it down, but don't let anyone see it. Now multiply your number by 3. Add 1. Multiply the result by 3. Add the number which you selected in the beginning and be ready to tell me your final answer. I'll promise to guess your original number if you have made no careless mistake!" By striking out the units digit in the final answer, the child's number will remain, but of course I don't tell that "secret" to the class. Until they discover it, they are amazed at my ability, and they all clamor to have me guess their chosen numbers. I may add a few more difficult guessing tricks with two or three digits. When the secrets are exposed, the pupils all try the tricks on Mom, Dad, Big Brother, or Sister when they go home. It gives my sixth-graders a chance to "show-off" at home, perhaps to gain added prestige. Often, as Dr. Willerding has pointed out, children can establish, through recreational arithmetic, a new type of comradeship with the elder members of their families. This leads to an exchange of tricks and puzzles, which stimulates interest in mathematics. Eyes and ears are alerted for new tricks to try. I am always delighted when my pupils bring a trick worked out at home or spotted in a newspaper or magazine.

Guessing ages, birth dates, small change, pages in a book, etc. is sufficient fun to prove to an eleven year old that working with numbers is not all dreary business.

On the basis of personal experience, I feel impelled to reiterate that, going beyond the point of having fun, working with magic squares and cross number puzzles can actually strengthen mathematical ability. When I present a magic square to my class, I explain what one is and how it is constructed. I demonstrate with one simple 9-celled square and tell the class that there are seven other possible combinations. Within a few minutes many pupils come up with the other squares filled in; they beg for more patterns, larger and more difficult squares.

Cross-number puzzles serve as good review of basic processes. I teach area, perimeter, percents and decimals as applied to practical, every-day life, of course. But the necessary processes and formulae can be reviewed in cross-number puzzles. These are not usually included in text-books, and, while the text-book is important in an arithmetic program, materials not found between its pages seem to have more magnetism for the minds of sixth-graders. If the pupils like the puzzles I give them for review, they make puzzles of their own to exchange with classmates. None of these has been required, but I have been surprised at the number of puzzles turned in on a voluntary basis. My files are full of really good recreational materials which are the original creations of pupils.

MAGIC SQUARE

4	9	2
3	5	7
8	1	6

When youngsters choose to use leisure time to "play" with numbers and when they beg to remain indoors at recess to work on number puzzles (which plea I do not grant in good weather!), I am sure that sixth-graders do like arithmetic. But of course the true test comes in evaluating the results of a program where recreational methods are used. Standard tests show the pupils' performance to be well above the norm. Each year the improvement in arithmetic achievement is gratifying. During this year the progressive growth in skills has been especially noteworthy. I must in no way imply that recreational arithmetic is solely responsible for a good achievement record. There are various contributing factors.

I recognize that many teaching procedures are necessary for a thoroughly rounded arithmetic program. Recreational devices are only among the many, but their value should not be overlooked. Some teachers will say that they have no time for fun in arithmetic. I think that they should allocate some time for it. I believe improvement in pupils' skills and attitudes will result.

→

EDITORIAL COMMENT.—Can you solve this code problem?

```
  FIVE
 -FOUR
 -----
  ONE
```

"Arithmecode" puzzle

DAVID F. WINICK *Minneapolis, Minnesota*

The dashes below, labeled with Roman numerals, represent the words of a quotation or phrase. Each dash stands for a single letter. For example, five dashes represent a five-letter word. To determine each word, the reader must examine the clues by the corresponding Roman numerals and identify the number described. Once the number is identified it must be decoded by selecting the appropriate code letter for each digit of the number. Each digit corresponds to exactly one letter of the word in the same order, but there may be two or three code letters for any given digit.

```
_ _ _ _ _ _ _ _ _   _ _ _ _ _   _ _ _   _ _ _ _   _ _ _ _   _ _
        I               II       III      IV        V       VI

_ _ _   _ _ _ _ _ _   _ _ _   _ _ _ _   _ _   _ _ _   _ _ _ _ .
 VII       VIII        IX        X      XI    XII      XIII
```

—*FRANCOIS, DUC DE LA ROCHEFOUCAULD*

CODE

A 0
D 1
E 2
F 3

CLUES

I. 04,011,272
II. $(59 \times 10^3) + (47 \times 10^1) + 10^0$
III. 5 plus the number of feet in 148 fathoms
IV. $(14046)2^{-1}$

G	4
H	5
I	6
L	7
N	8
O	9
Q	0
R	1
S	2
T	3
U	4
W	5
Y	6

V. 64 plus the total number of degrees in all the interior angles of a 46-sided polygon

VI. $\sqrt{3,969}$

VII. $(26 + 25 + 24 + \ldots + 3 + 2 + 1) + 1$

VIII. $(5! \times 254) - 7$

IX. If 6 is subtracted from this number the result will be what Euclid called a "perfect number."

X. 79 plus the product of the greatest two-digit prime number and the least three-digit prime number

XI. 343_{five} expressed as a base-ten numeral

XII. If this number is X, then $\log_{\text{four}}(X + 42) = 5$

XIII. MMDCXII

SOLUTION

I. *04,011,272* .... QUARRELS

II. $(59 \times 10^3) + (47 \times 10^1) + 10^0 = 59{,}471$ .... WOULD

III. 5 plus the number of feet in 148 fathoms *893* .... NOT

IV. $(14{,}046)2^{-1}$ *7023* .... LAST

V. 64 plus the total number of degrees of all the interior angles of a 46-sided polygon *7,984* .... LONG

VI. $\sqrt{3,969}$ *63* .... IF

VII. $(26 + 25 + 24 + \ldots + 3 + 2 + 1) + 1$ *352* .... THE

VIII. $(5! \times 254) - 7$ *30,473* .... FAULT

IX. If 6 is subtracted from this number the result will be what Euclid called a "perfect number." *502* .... WAS

X. 79 plus the product of the greatest two-digit prime number and the least three-digit prime number *9,876* .... ONLY

XI. 343_{five} 98_{ten} .... ON

XII. If this number is X, then $\log_{\text{four}}(X + 42) = 5$ *982* ONE

XIII. MMDCXII *2,612* .... SIDE

EDITORIAL COMMENT.—Here is an easier one:

___ ___ ___ ___   ___ ___   ___ ___ ___

I II III IV   V VI   VII VIII IX

(Everyman)

Clues:

I. 2×3	IV. 3×3	VII. 5×1
II. 2×2	V. 3×1	VIII. 1×1
III. 2×1	VI. 4×2	IX. 7×1

Code:

A	4	I	3	S	8
F	5	M	6	T	2
H	9	N	7	U	1

Arithmetical brain-teasers for the young

HENRY WINTHROP, *University of South Florida*

Dr. Winthrop is professor and chairman of the Department of Interdisciplinary Social Sciences at the University of South Florida, Tampa, Florida.

An interest in numbers is sometimes stimulated in the elementary grades by arithmetical brain-teasers. Provided below are a number of arithmetical brain-teasers that range in difficulty from items that may be employed in the lower grades to items that can be used in those schools than have introduced a little elementary algebra in the seventh and eighth years. Without more ado let me present these in an order that reflects, in my judgment, their increasing difficulty.

1. *Q*. Perform any operations you wish with the digit 9, which is to be used *four times,* so that the result of all the operations employed will give the number 100.

A. $99 + 9/9$.

2. *Q*. Name two coins that add up to 11¢, but *one of them mustn't be a penny;* and give the reason for your answer.

A. One dime and one penny.

Many children and adults fail to distinguish between the condition which demands that *neither of them must be a penny* and the milder condition which demands only that *one of them mustn't be a penny.*

3. *Q*. If a bottle and cork together cost \$1.05 but the bottle costs \$1.00 more than the cork, how much does the cork cost?

A. 2½¢.

Many individuals develop a mental block to this problem, seeking for a solution that shall be an integer.

Where youngsters have been taught how to obtain the square and cube of the digits from 1 through 9, the following brain-teaser will be in order.

4. *Q*. Notice that $43 = 4^2 + 3^3$, that is to say, the digits that make up 43 are so related that the square of the first digit plus the cube of the second digit will yield the number 43 itself. How many other numbers can you find exhibiting the following property? A two- or three-digit number is to be found such that if its digits from left to right are raised to the first, second, and third powers, respectively—and in that order—the sum of these digits raised to the first, second, and third powers will give the number itself.

A. Sample answers: $1^1 + 3^2 + 5^3 = 135$; $5^1 + 1^2 + 8^3 = 518$; $6^2 + 3^3 = 63$; $1^1 + 7^2 + 5^3 = 175$; and $5^1 + 9^2 + 8^3 = 598$.

5. *a) Q*. A number has three digits. One of these digits multiplied by the square root of the other two digits is equal to the square root of the number itself. What is the number?

A. $\sqrt{135} = 3\sqrt{15}$.

b) Q. A number has three digits. The square root of this number is equal to the product of one of these three digits multiplied by the square root of the other two digits. What is the number?

A. $\sqrt{783} = 3\sqrt{87}$.

If the notion of *factorial* is explained to youngsters, then the following brain-teaser can be given to them.

6. *Q*. Notice that some factorials are the product of two or more other factorials. Thus $6!\ 7! = 10!$ Using *nonconsecutive*

numbers, how many more examples of this type of relationship can you supply?

A. Sample answers: 4! 23! = 24!; 2! 4! 47! = 48! 2! 3! 4! 287! = 288!; and 3! 5! 7! = 10!

7. "Take a number."

Ask a person to take a number but not to let you know what it is. Then ask him to perform the following operations in sequence.

a) Double the number thought of.

b) Square the result of doubling.

c) Multiply the result obtained in Step 2 by the number 3.

d) Ask the person involved to divide the result of Step 3 by the number he or she originally thought of.

e) Now ask this person to subtract, from the result of Step 4, an amount equal to four times the product of the original number.

f) Now ask the person in question to give you his last result from Step 5. As soon as he does so you can tell him the number he originally thought of. As an example, let us assume the person originally thought of 6. Then the results of the first five operations described above would be the following: (1) 12; (2) 144; (3) 432; (4) 72; (5) 48. If you now divide 48 by 8, you will obtain the number originally thought of, 6.

What is the rationale of this procedure?

For a person whose task is to guess the number that was originally thought of by the other, all that is required is a slight acquaintance with the most elementary aspects of beginning algebra. Suppose you are the person whose task is to guess the number. In your own mind, think of the number selected by the other party as x. Then the five results that you would get from performing the steps outlined can be represented as follows: (1) $2x$; (2) $4x^2$; (3) $12x^2$; (4) $12x$; (5) $8x$. When the person who selected a number gives you the answer corresponding to $8x$, you quickly divide mentally by 8. This immediately will give you the number that this person thought of. Thus, at the end of Step 5, the selector will give you his result, 48. Dividing 48 by 8 will immediately give you the number he originally selected, 6.

Obviously this procedure is such that the number of operations to which the selector can be submitted is unlimited, while the nature of the operations can be more varied than the few operations shown in our example.

The whole procedure tends to mystify those who select a number. Once shown what has been done, however, the effort to duplicate this type of performance with others tends to sharpen the individual's sense of the meaning of arithmetical operations. This is all to the good. This last type of brain-teaser is, of course, most appropriate for individuals in the seventh and eighth years, provided they have been exposed to a modicum of elementary algebra.

The toughest brain-teaser I can think of giving youngsters, one that will really keep them going on unfinished business for quite a while, is to interest them in *automorphic* numbers and then see if any of them can find automorphic numbers for themselves. *An automorphic number is one whose square ends with the number itself*. Two large examples of automorphic numbers are 43,740,081,787,109,376 and 56,259,918,212,890,625. In asking youngsters to find other automorphic numbers, many individuals will be surprised to discover that a final few or more digits of the two automorphic numbers already given will, themselves, be automorphic numbers. Thus 76^2 ends with 76; 90625^2 ends with 90625; and $87{,}109{,}376^2$ ends with 87,109,376.

It is an interesting exercise to present the first two large automorphic numbers given above to youngsters; then to tell them that there are sequences of successive digits in these two numbers that are themselves automorphic numbers, and to ask them to find these new automorphic numbers.

"Parallelograms": a simple answer to drill motivation and individualized instruction

BENNY F. TUCKER

Benny Tucker is mathematics consultant for grades K–12 for the Parkway School District, Chesterfield, Missouri. He is the author of a handbook of activities for elementary school mathematics for use in the Parkway schools.

To drill or not to drill—that is the question that "bugs" many teachers of elementary mathematics. Some who would have us teach "modern" mathematics have said that drill and memorization have no place in contemporary mathematics programs. Experience and common sense tell us that these spokesmen of "modern math" are either wrong or misunderstood. Indeed, few have actually said that memorization and drill do not have their place in mathematics education. Most experts agree that both must be an integral part of our programs but that they must be kept in proper perspective.

Drill, and the memorization and skills development that can result, should come only after the student adequately understands underlying concepts. Drill should never be used as a substitute for teaching and should not be a club to be used on children who have trouble understanding mathematics teachers.

Teachers sometimes forget that if a child is to maintain a receptive attitude toward drill over any significant period of time, every effort must be made to keep the drill procedures varied and interesting. There are many drill games on the market which can be of valuable help to the teacher who wants to maintain interest.

A game that has been particularly successful in the Parkway elementary schools is a game similar to tic-tac-toe called "Parallelograms."

Nearly every child knows how to play tic-tac-toe. By requiring the plays to answer questions correctly at the risk of losing turns, a teacher can effectively use tic-tac-toe for drill or review. Out of a search for a game that is similar to tic-tac-toe but that would seem quite different to the student, the game Parallelograms evolved. The object of the game is to be the first player to mark the four vertices of one of the many parallelograms contained in figure 1. Of course one's opponent will be trying to do the same thing, so each player must try to anticipate his opponent's strategy and block him. Although it will be obvious to the teacher that squares are also parallelograms, it may be necessary to point out to the students that the vertices of any of the four squares in the figure could also be a winning combination.

In the game at the left in figure 1, "X" needs one more move to win. In the second game, "O" needs one more move to win.

In addition to the obvious benefits from the drill, the use of this figure gives students practice in the visualization of geometric shapes and in the use of game strategies. Parallelograms is simple enough for third-grade students but challenging enough for the most talented sixth-grade students.

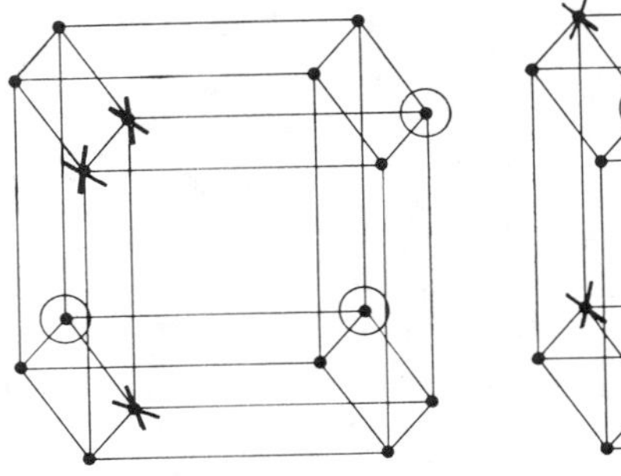
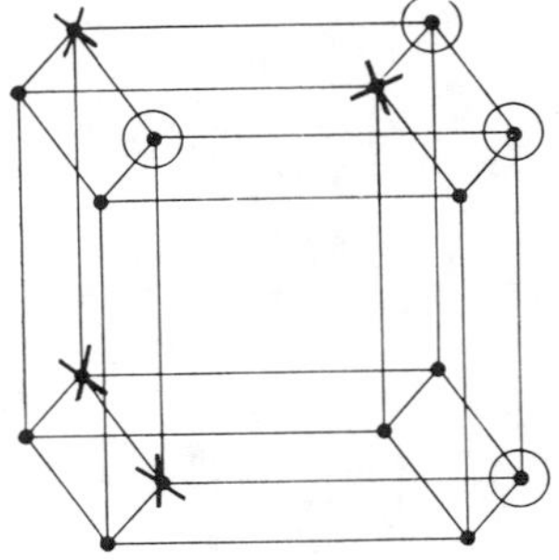

Fig. 1

Since the figure for Parallelograms is rather complicated and not easy to draw, it is suggested that dittoed copies be provided for the students. Or an overhead transparency can be used as a game sheet and marked with a crayon or wax pencil. When the game is finished the acetate can be wiped clean with a tissue to make ready for the next game.

Perhaps the greatest value of the game is its adaptability to many situations. A sequence of questions can be used, with the players answering them alternately. If the player answers correctly, he may mark a vertex. If he misses, he loses his turn. The questions used in the game could just as well come from an academic area other than mathematics (for example, spelling, science, or geography). Questions could be used from several academic areas all in the same game. The teacher can, by using carefully selected questions, teach new material in the context of the game. The best application of the game may be in the individualization of mathematics. If a teacher who is assigned thirty students is to be effective in dealing with individual needs, he must have at his disposal a wealth of materials that are worthwhile but that require only indirect teacher supervision. Parallelograms fits this description. A set of multiplication flash cards can be the source of questions for a pair of students who need drill on their multiplication facts. If the teacher sees a need for work in an area for which there are no commercially prepared cards, a deck of cards with questions on one side and answers on the other—prepared by the teacher to meet the specific need—can be used.

A question that elementary teachers perpetually face is what to do with the student who is having trouble with topics normally learned in an earlier grade. Certainly these students should be taught what they need to learn, but many teachers do not find the time that they feel is necessary to satisfy the needs of such students. Another common problem is the student who always seems to finish his assignment ahead of the rest of the class. The pat answer to this problem is that such a student should be allowed to continue at his own rate—he should not wait for the other students but should go on with the next "lesson."

The game Parallelograms can supply one solution for these problem students. Working in pairs, both students needing remedial tasks and accelerated students can enjoy playing the game, and a deck of cards containing a programmed sequence of questions on a needed or interesting topic will allow those students to effectively teach themselves while requiring only indirect supervision from the teacher. Work with programmed materials is far more interesting to the student when it is placed in the context of a game like Parallelograms.

In the following game between Bill and Jerry, the questions are drill questions on decimals. Bill is marking with an X, and Jerry is marking with an O.

QUESTION: Where would you place the decimal point in the numeral 1 2 3 4 5 so that the 3 would be in the hundreds place?

BILL: After the 5. (Correct; Bill marks a vertex as shown in fig. 2.)

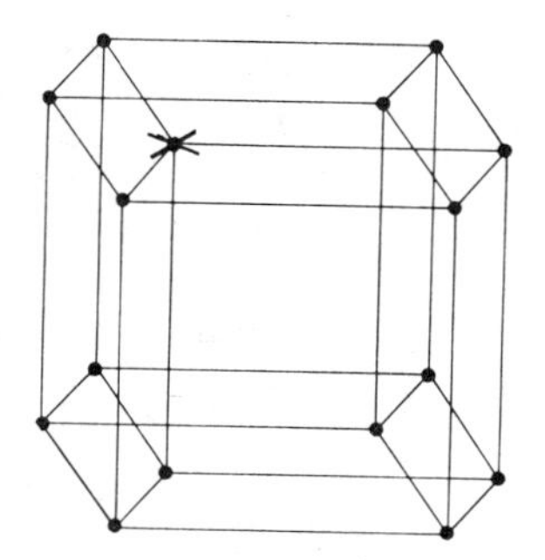

Fig. 2

QUESTION: What is .023 – .009?

JERRY: .0194. (Wrong; Jerry loses his turn.)

QUESTION: Which is greater, .0098 or .012?

BILL: .012. (Correct; see fig. 3.)

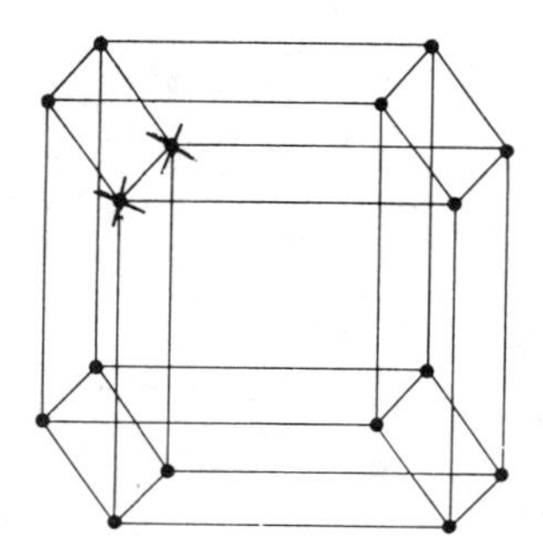

Fig. 3

QUESTION: Where would you place the decimal point in the numeral 1 2 3 4 5 so that the 4 would be in the hundredths place?

JERRY: Between the 2 and the 3. (Correct; see fig. 4.)

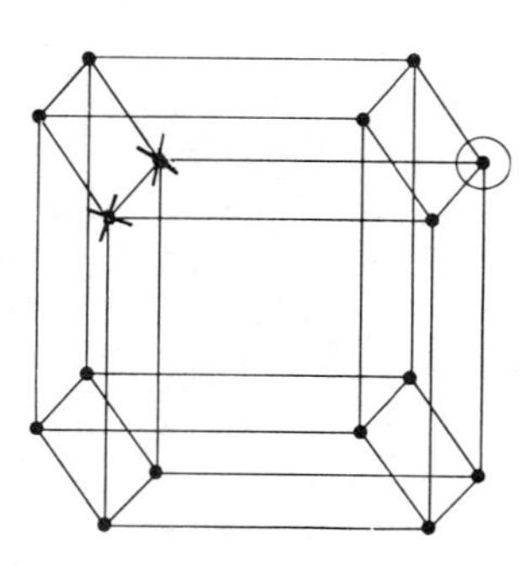

Fig. 4

QUESTION: What is 5.62 – .13?

BILL: 5.49. (Correct; see fig. 5.)

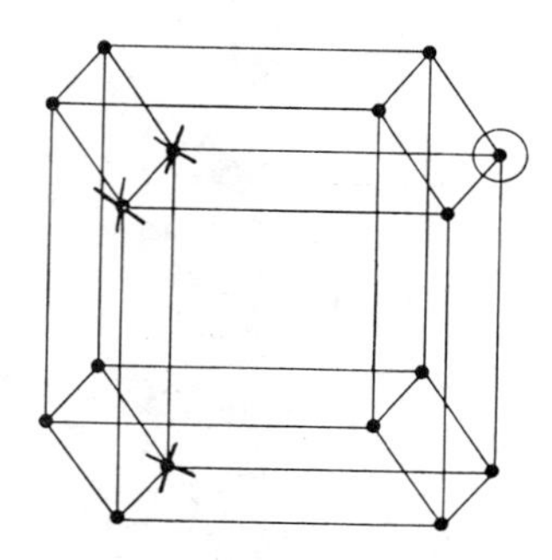

Fig. 5

QUESTION: What is .984 ÷ 6?

JERRY: .164. (Correct; see fig 6.)

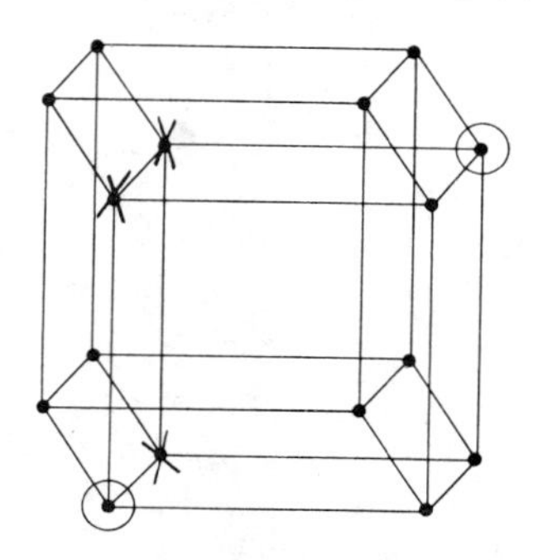

Fig. 6

QUESTION: Answer the following: 7.96 + □ = 24.3.

BILL: 32.26. (Wrong; Bill loses his turn.)

QUESTION: What is .32 + 1.5?

JERRY: 1.82. (Correct; see fig. 7.)

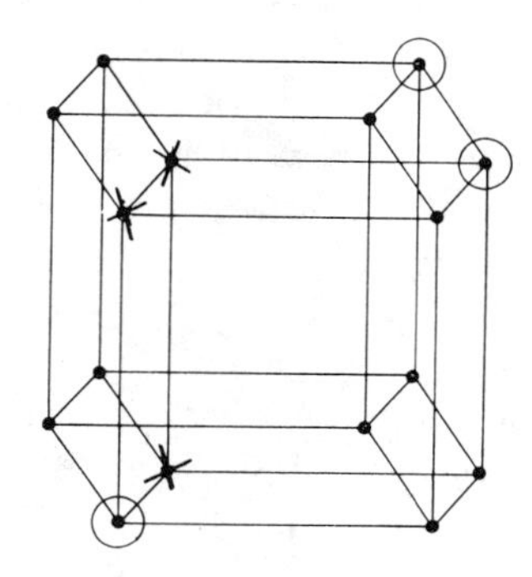

Fig. 7

QUESTION: What is 5 + .2?

BILL: 5.2. (Correct; see fig. 8.)

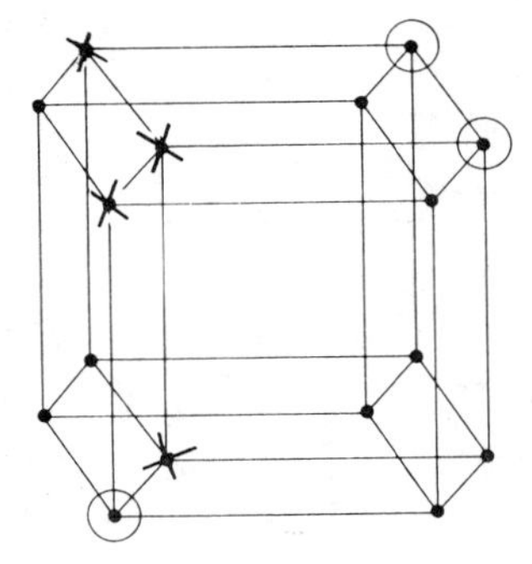

Fig. 8

QUESTION: What is $.32 \times 1.5$?
JERRY: .48. (Correct; see fig. 9.)

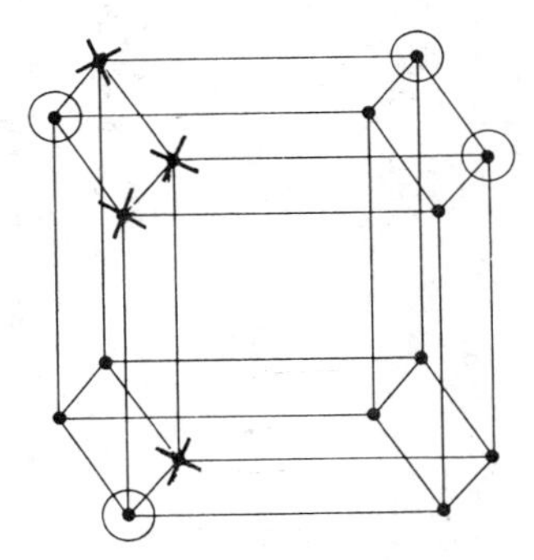

Fig. 9

QUESTION: What is 6.5×12.4?
BILL: 80.60. (Correct; see fig. 10.)

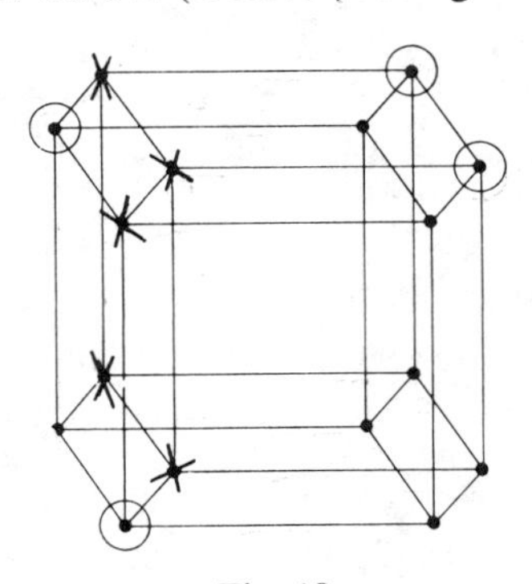

Fig. 10

Bill wins the game.

The following is an example of a programmed sequence of questions, for third-grade students or for fourth-grade students needing remedial work, developing the distributive property as a tool to be used in multiplying one-digit numbers by two-digit numbers.

1.

0 0 0 0	0 0 0 0
0 0 0 0	0 0 0 0
0 0 0 0	0 0 0 0
0 0 0 0	+ + + +
0 0 0 0	+ + + +

5 fours {is the same as} 3 fours + $\square$ fours

2.

0 0 0 0 0 0
0 0 0 0 0 0
0 0 0 0 0 0
0 0 0 0 0 0
+ + + + + +
+ + + + + +
+ + + + + +

7 sixes

is the same as

$\square$ sixes + 3 sixes

3.

+ + + + +
+ + + + +
0 0 0 0 0
0 0 0 0 0
0 0 0 0 0
0 0 0 0 0

$6 \times 5 = (2 \times 5) + (4 \times \square)$

4.

0 0 0 0
0 0 0 0
0 0 0 0
0 0 0 0
0 0 0 0
0 0 0 0
0 0 0 0

$\square \times 4 = (4 \times 4) + (3 \times 4)$

5. 9 threes = 5 threes + $\square$ threes

6. $9 \times 5 = (4 \times 5) + (\square \times 5)$

7. $6 \times 3 = (1 + 5) \times 3$
$= (1 \times 3) + (\square \times 3)$

8. $6 \times 9 = (3 + 3) \times 9$
$= (3 \times 9) + (3 \times 9)$
$= 27 + \square$

9. $7 \times 8 = (5 + 2) \times 8$
$= (5 \times 8) + (2 \times 8)$
$= \square \times \triangle$

10. $12 \times 7 = (10 + 2) \times 7$
$= 70 + \square$

11. $14 \times 6 = (10 + 4) \times 6$
$= \square + 24$

12. $13 \times 8 = (10 + 3) \times 8$
$= 80 + 24$
$= \square$

13. $15 \times 5 = (10 + 5) \times 5$
$= (10 \times 5) + (5 \times 5)$
$= \square$

14. $32 \times 6 = (30 + 2) \times 6$
$= (30 \times 6) + (2 \times 6)$
$= \square + \triangle$

15. $27 \times 4 = (20 + 7) \times 4$
$= (20 \times 4) + (7 \times 4)$
$= \square$

Like most games, Parallelograms will only be as effective as the ingenuity and creativity of the classroom teacher allows. It is hoped that the reader will adapt the game to his own needs and feel free to use it.

NOTE. The author would appreciate hearing from readers about any success or failure experienced with the game.

Take a chance with the wheel of fortune

BARBARA ROSSER
Sparta Elementary School, Sparta, New Jersey

Through the years I have noticed that any of my friends who have chosen to make mathematics their life's work have thought of numbers in terms of a game. Working out a mathematical problem is as exciting and challenging to them as deciding on a move in a chess tournament.

Why not, then, inject a little of the intrigue of the game into the math class? Undoubtedly there are distasteful aspects of any subject, and obviously we can't make it *all* fun and games.

Take multiplication facts, for instance. I don't know of a single child, gifted or otherwise, who has enjoyed learning them. What's more, in the day of "concept teaching" the teacher is hesitant to take class time for old-fashioned drill. But in the few years that I have used the "game of chance," the chore has been incorporated into a suspenseful game that takes no more than five minutes out of the period and turns drudgery into a sport. I use a "wheel of fortune" for fact-study motivation. Yes, we're gambling in the classroom —and the children love it!

Here's how it works. A spinner from any game is simply converted into a number wheel by cutting a mask to fit around the outer edge of the board so as not to interfere with the movement of the spinner (mine comes from the family's "Twister" game).

Enough divisions are made on the mask to accommodate a number for every child in the class. Each child is assigned the number that corresponds to his name in my grade book.

When the spinner lands on a number, the child who has that number gets the "oral test of the day," consisting of five randomly-chosen multiplication facts, with about three or four seconds allowed for an answer. I usually give three of these tests a day. Whichever facts are missed are practiced ten times for homework in addition to the regular assignment.

Going one step further, it's an easy matter to make up index cards with facts on them along the left edge, with space at the top for pupils' numbers. An × or √ in the appropriate box on your ruled grid provides an at-a-glance record on Johnny or Mary for quick checks on their progress.

I usually limit my tests to three per marking period so that everyone has at least two or three grades to average in with the others.

At first the idea of being tested "out of the blue" produces groans of horror, but it's surprising how after a week or two the children beg for more: "Let's do four today!" "Oh, please! I really studied last night—get me today." "We should do six today instead of three; remember we skipped a day last week."

Sure—it's an easy 100 if they know their facts, and the oral quiz (five facts to a customer) carries the weight of a regular quiz. One thing to keep in mind is that everyone has a bad day now and then. To cut down on "test tension" I usually tell my pupils that their lowest mark will be discarded. They all want to "try for high," so nightly review becomes a must.

A gimmick? Perhaps, but no more so than many forms of motivation. It's a game. It's exciting, and everyone wants to win. It's suspenseful; it adds excitement to what could easily be another dull math routine—the learning of multiplication facts.

It's a challenge. No one knows when it will be his turn to be tested, so everyone studies every day. Sometimes a few minutes' study time is given in class for those few children who let the home study slip. Seeing almost everyone else cramming for the "big maybe" is usually enough to prod the one or two laggards into action.

Immoral? Not at all. Life's a gamble. The math teacher could call on anyone for an answer. She could also be accused of "picking on" someone. The wheel of fortune is impartial.

One final advantage is that conscientious youngsters learn the lesson of cause and effect by watching the upward trend of their graphs of weekly speed tests, as well as being able to reap the immediate rewards of winning at a game of chance.

Using functional bulletin boards in elementary mathematics

WILLIAM E. SCHALL
State University College, Fredonia, New York

Visual aids—films, still pictures, models, bulletin boards, and so on—are among the most useful tools in education, but they do not teach without intelligent planning and use (Glenn O. Blough and Albert J. Hugget, *Elementary School Science and How to Teach It* [New York: Dryden Press, 1957], pp. 33–34). Bulletin boards can play an important role in today's mathematics program. However, a bulletin board, if it is to be successful in achieving its purpose, must gain and be *worthy* of the class's attention.

The writer wishes to extend his appreciation to Mary Jane Koepfle and Deborah Lewis, students in his mathematics methods course at the University of Cincinnati, for their help with this article.

A good bulletin board should also be supportive of and adaptable to classroom activities in a particular subject area, elementary mathematics in this case. Since children like activities or games in which they participate or are actively involved, a game approach is suggested for use here.

The rest of this paper describes several bulletin boards in various areas of elementary school mathematics. Each suggested bulletin board includes a short discussion of the bulletin board's purpose, the appropriate grade level, concepts and objectives that the bulletin board is intended to develop or reinforce, suggested questions that the teacher might use in connection with the bulletin board to stimulate additional thought and discovery, and a short description of a class activity that could involve pupils with the bulletin board.

The first bulletin board is shown in figure 1.

A. PURPOSE: To review and reinforce the basic mathematical operations as well as the concept of renaming numbers. The code or the message can be changed frequently, depending on the class, activities, season, and so on.

B. GRADE LEVEL: Fourth or higher depending on the code used. For higher grades a rational-number code can be used.

C. BEHAVIORAL OBJECTIVES:
1. The children will be able to work the problems and read the message.
2. The children will be able to recognize that a number can be renamed in many ways.
3. The children will be able to rename numbers in different ways.

D. DISCUSSION QUESTIONS:
1. What is a code?
2. Why do people sometimes write in codes?
3. How do you read a code?

E. DESCRIPTION OF THE ACTIVITY:
1. The children work the problems to read the message.
2. For additional practice in renaming, each child can write his own message in code and rename the letters as different mathematical problems. Besides reviewing renaming, this will also review the basic mathematical operations.

Another bulletin board is shown in figure 2. It is used as follows:

A. PURPOSE:
1. To use the written numerals as a daily practice in counting
2. To serve as a daily practice to match a set with the appropriate cardinal number
3. To use the displayed objects as a basis of comparing sets

B. GRADE LEVEL: This bulletin board could be used in the kindergarten and first grade. Counting, the introduction of sets, and the comparison of sets are usually introduced in the kindergarten and could be used for review purposes in the first grade. (However, the specific time it is used depends on the children's progress.)

C. BEHAVIORAL OBJECTIVES:
1. The child will be able to count to 10.
2. The child will be able to identify each numeral; this means knowing a "5" or a "7" when he sees it.
3. The child will be able to name the geometric figures that are the elements of the sets.
4. The child will be able to match the

CAN YOU READ THIS MESSAGE??

$3 + (4 + 3)$ $(5 + 5) - 3$ $(8 + 2) + 0$ $(5 + 8) + 13$

$3 + 20$ $(5 + 5 + 5) - 10$ $(7 + 7) - 13$ $(25 + 4) - 26$

$(1 + 1) + 0$ $2 + 2 + 2 + 2$ $(6 + 6) - 7$ $(4 + 4) + 0$

$(17 + 10) - 6$ $(24 - 8) + 5$ $(1 + 1) - 1$ $(5 + 6 + 6) - 12$

$(1 + 2) + 0$ $3 + (28 - 27)$ $79 - 53$ $(18 - 7) - 2$

$(3 + 1) + (1 + 3)$ $(8 - 8) + 1$

a = 8	f = 25	k = 20	p = 21	u = 2
b = 14	g = 12	l = 4	q = 6	v = 24
c = 22	h = 5	m = 19	r = 15	w = 10
d = 9	i = 26	n = 17	s = 23	x = 16
e = 7	j = 13	o = 3	t = 11	y = 1
		z = 18		

Fig. 1

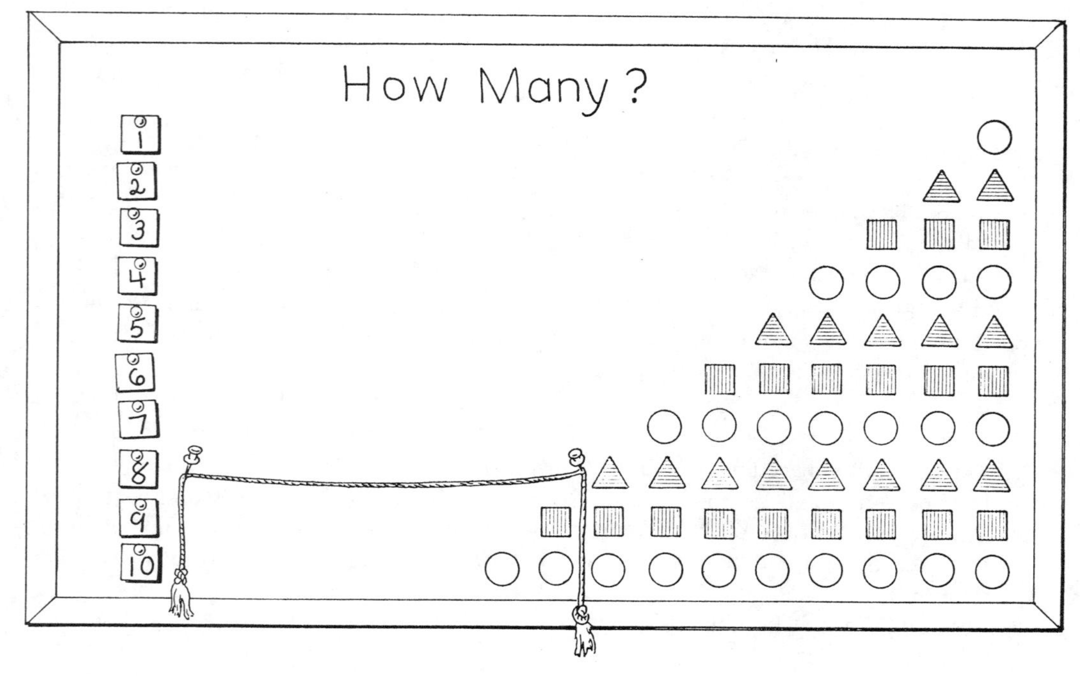

Fig. 2

sets with the correct cardinal number. Rope mounted on the bulletin board can be used to indicate the correspondence.

5. He will be able to compare the sets, thus using the ideas of more or less.

D. QUESTIONS TO STIMULATE THOUGHT:

1. Can you name the numerals written on the bulletin board?
2. Can you show me where the "5" is (similarly for other numerals up to ten)?
3. Can you show me the set of four objects? Or, which group has the four circles?
4. Which set of objects has the largest cardinal number?
5. Which set of objects has the smallest cardinal number?
6. Are there more circles (○) than squares (□)? How can you tell?

E. DESCRIPTION OF THE GAME:

1. A child can select a number and point to the numeral that represents the number.
2. Another child can match the correct set with the cardinal number.
3. If the second child gets the correct answer, he gets to select a number.
4. The activity can be varied. The teacher can point to a numeral and ask children to identify the correct set.

The next bulletin board is shown in figure 3, and its use is outlined below.

A. PURPOSE:

This bulletin board is designed to stimulate thinking about geometric shapes—how they are made and what they are called. There is to be transfer of learning from the geometric shapes illustrated on the bulletin board to geometric shapes in the everyday environment.

B. GRADE LEVEL:

This particular bulletin board is designed for the primary grades; however, the basic idea of the bulletin board (the geo-board) can be used for all grade levels in the elementary school.

Fig. 3

C. Behavioral objectives:

1. The child will be able to identify the six geometric shapes pictured on the bulletin board.
2. The child will be able to copy the indicated shapes on the geo-board.
3. The child will be able to recognize the number of points connected in each shape.
4. The child will be able to distinguish between the geometric shape and its region.
5. The child will be able to recognize a wide variety of geometric shapes in the classroom.

D. Materials:

1. Individual geo-boards
2. Rubber bands
3. Yarn (for use on the bulletin board)

E. Questions to stimulate thinking:

1. The same geometric shapes that you made on the geo-board can be found within the classroom. Can you find some examples?
2. Can you make a given shape on the geo-board?
3. If given the number of points or sides in a geometric form, can you create the corresponding form?
4. How many different geometric shapes did you see on the way to school this morning?

The last bulletin board to be discussed here is shown in figure 4; its use is outlined below.

A. Purpose:

To motivate children as they work with the fundamental operations of arithmetic; also, to increase the learners' skill in computation in the fundamental operations

B. Grade level:

Most intermediate grade levels—depending on the difficulty of the computation and skills involved

C. Behavioral objectives:

1. The learners will work examples using the four fundamental operations of arithmetic with increased accuracy and speed.
2. The learners will demonstrate increased interest in arithmetic through participation in the "Grand Prix" activity and other mathematical activities.
3. The learners will demonstrate cooperative learning and working skills

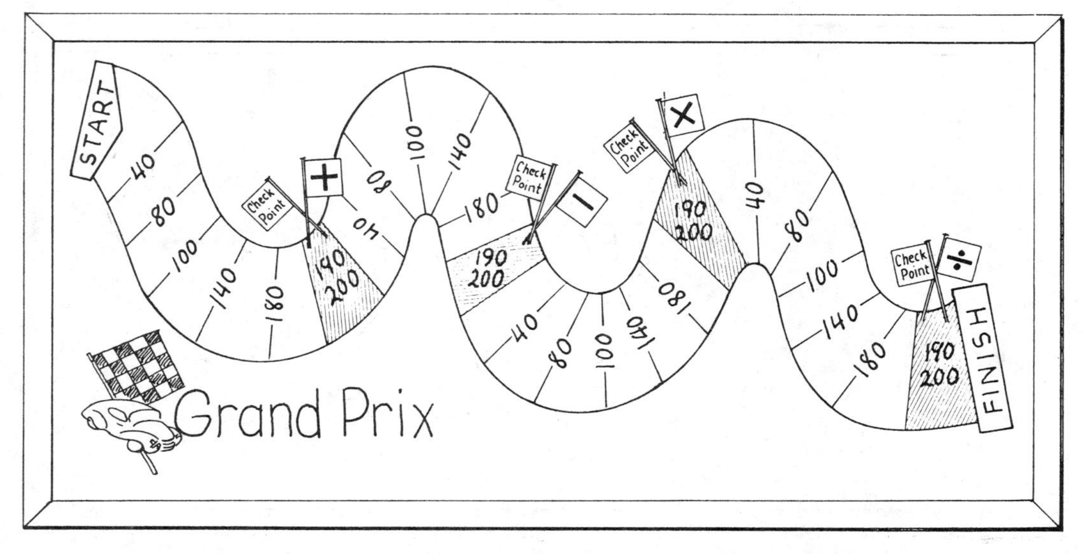

Developed by Mrs. Beth McCracken, Westwood Elementary School, Cincinnati, Ohio.

Fig. 4

through participation in the Grand Prix activities.

D. Description of the activity:

Grand Prix is designed to motivate children to attain a higher level of proficiency with the four fundamental operations of arithmetic—addition, subtraction, multiplication, and division. Children within each class are divided into teams of two. These teams

1. Are composed of one student from more skilled or more able groups and one from the less skilled or slower-moving groups to maintain a balance;
2. Are in twos so they can help each other with flash cards and other team activities;
3. Provide good "working together" experience.

These activities are done, of course, after good, meaningful experiences have been provided for the learners with the operation; the emphasis is on refinement of the skill, that is, accuracy, speed, retention, and operational ease. To begin the activities, the addition facts are given for the qualification day. The learners are then given a chance to have a "trial run." They have a week to practice with their partners before "race day," when their combined scores (these scores can be predetermined values) determine where their racer moves on the track. The same procedure is repeated for substraction, multiplication, and division. Trophies, certificates, and so on, are presented to teams that reach the "checkpoint." The checkpoints are team scores of 200 or whatever categories are chosen.

One corner of the room has letters suspended over it spelling "THE PIT," where children may go when other work is finished to get "tuned up." There is a box labeled "Mechanics' Tools" that contains flash cards and various other devices for practice with the basic facts.